Regular Extensions of Hermitian Operators

Regular Extensions of Hermitian Operators

A.V. Kuzhel and S.A. Kuzhel

///VSP///

Utrecht, The Netherlands, 1998

VSP BV
P.O. Box 346
3700 AH Zeist
The Netherlands

Tel: +31 30 692 5790
Fax: +31 30 693 2081
E-mail: vsppub@compuserve.com
Home Page: http://www.vsppub.com

First published in 1998

ISBN 90-6764-294-0

Printed in The Netherlands by Ridderprint bv, Ridderkerk.

Contents

Preface xi

Chapter 1. REGULAR EXTENSIONS 1

1. Linear Operators. General Concepts 1
1.1. Linear Operators 1
1.2. Extensions, Restrictions, and Closure of an Operator 2

2. Spectrum of a Linear Operator 4
2.1. Resolvent and Resolvent Set 4
2.2. Spectrum 5
2.3. Point Spectrum 6
2.4. Continuous Spectrum 7
2.5. Residual Spectrum 7

3. Hermitian Operators 9
3.1. Definition and Properties 10
3.2. Defect Subspaces 11
3.3. Defect Numbers 14
3.4. Defect Numbers of a Real Operator 17

4. Symmetric Operators 18
4.1. General Properties 18
4.2. Lineal $\mathcal{D}(A^*)$ 19
4.3. Symmetric (Self-Adjoint) Extensions of Symmetric Operators 20
4.4. Extensions to a Broader Space 24

5. Regular Extensions of Hermitian Operators 26
5.1. Hermitian Domain of a Linear Operator 26
5.2. Regular Extensions 28
5.3. Quasi-Hermitian Operator 31
5.4. Operator $T_{\mu\lambda}$ 32
5.5. Analog of the Von Neumann Formulas 34
5.6. Generalization of the Krein Resolvent Formula 37
5.7. Relatively Simple Extensions 44

6. Dissipative Extensions of Hermitian Operators 45
6.1. Dissipative Operator 45
6.2. Condition of Dissipativity of Linear Operators in Terms of the von Neumann Formulas 47
6.3. Maximal Dissipative Operators 49
6.4. Operator Adjoint to a Dissipative Operator 51
6.5. Other Results 52

7. Accretive Operators 53
7.1. Definition and Properties 53
7.2. Maximal Accretive Operators 56
7.3. Sectorial (θ-Sectorial) Operators 56
7.4. Accretive Operators and Contracting Semigroups 58

Chapter 2. REGULAR EXTENSIONS WITH RESTRICTIONS 61

1. Self-Adjoint Bound-Preserving Extensions of Semibounded Symmetric Operators. Theorem and Hypothesis of Von Neumann 61
1.1. Semibounded Operators 61
1.2. Theorem and Hypothesis of Von Neumann 62

2. Proof of the Von Neumann Hypothesis (Friedrichs Method) 64
2.1. Auxiliary Hilbert Space $\mathcal{D}_0$ 64
2.2. Friedrichs' Proof of the Von Neumann Hypothesis 65

3. Self-Adjoint Norm-Preserving Extensions of Hermitian Contractions 67
3.1. Preliminary Comments 67
3.2. Counterexample 67
3.3. Self-Adjoint Norm-Preserving Extensions of Hermitian Contractions 68
3.4. Description of Self-Adjoint Norm-Preserving Extensions of Hermitian Contractions 75
3.5. Condition under Which the Operator $\tilde{S}_0$ Coincides with S_- 79

4. Normal Norm-Preserving Extensions of Hermitian Contractions 81
4.1. Regular Normal Extensions 81
4.2. Condition under Which the Operator $R_1 = \operatorname{Re} T_1$ Coincides with R_- 84

5. Krein's Proof of the Von Neumann Hypothesis 85
5.1. Extensions of Nonnegative Symmetric Operators 85
5.2. Conditions under Which the Hard Extension A_μ Coincides with the Friedrichs Extension 88
5.3. Additional Properties of the Operators A_μ and A_M 90

6. Squared Symmetric Operators **92**
6.1. Operators A^*A and AA^* 92
6.2. The Hard and Soft Extensions of a Symmetric Operator B^2 94

7. Self-Adjoint Bound-Preserving Extensions of Semibounded Hermitian Operators **96**
7.1. Preliminary Comments 96
7.2. Sufficient Conditions for the Existence of Bound-Preserving Extensions 98
7.3. Necessary and Sufficient Conditions 99

8. Regular $\mathfrak{U}$-Invariant Extensions of Hermitian Operators **106**
8.1. Preliminary Concepts and Results 106
8.2. Some General Properties of $\mathfrak{U}$-invariant Hermitian Operators 107
8.3. Semibounded Operators. Phillips Theorem 108
8.4. Condition for the Existence of Regular $\mathfrak{U}$-Invariant Extensions Formulated in Terms of the Von Neumann Formulas 109
8.5. Condition for the Existence of Self-Adjoint $\mathfrak{U}$-Invariant Extensions Formulated in Terms of the Shtraus Characteristic Function 110
8.6. Symmetric $\mathfrak{U}$-Invariant Operator without $\mathfrak{U}$-Invariant Self-Adjoint Extensions 112
8.7. Self-Adjoint $\mathfrak{U}$-Invariant Norm-Preserving Extensions of Hermitian Contractions 117

9. Canonical Dissipative Extensions of Hermitian Operators **118**
9.1. Auxiliary Statements 119
9.2. Canonical Dissipative Extensions 122

Chapter 3. BOUNDARY-VALUE SPACES OF HERMITIAN OPERATORS **129**

1. Definition and General Properties of Boundary-Value Spaces **129**
1.1. Symmetric Operators 129
1.2. Hermitian Operators with Equal Defect Numbers 133
1.3. Hermitian Operators with Arbitrary Defect Numbers 137
1.4. Examples of BVS 141

2. Description of Regular Extensions of Hermitian Operators in Terms of Boundary-Value Spaces **146**
2.1. General Case 146
2.2. Special Cases 150

3. Characteristic Functions of Hermitian Operators **159**
3.1. Definition of Characteristic Function 159
3.2. Condition of Unitary Equivalence of Simple Hermitian Operators 163

4. Spectral Properties of Regular Extensions **166**
4.1. Auxiliary Statements 166
4.2. Description of Nonreal Spectra of Regular Extensions 168

Chapter 4. EXAMPLES AND APPLICATIONS **177**

1. Quasidifferential Operators **177**
1.1. Quasiderivatives 177
1.2. Quasidifferential Systems 179
1.3. Quasidifferential Operator A 179
1.4. Boundary-Value Spaces of the Operator A 184

2. Boundary-Value Spaces for Model of Zero-Range Potentials with Internal Structure **189**
2.1. Model of Zero-Range Potentials with Internal Structure 189
2.2. Boundary-Value Space of the Operator A 190
2.3. Description of Dissipative Extensions of the Operator A 193
2.4. Characteristic Function of the Operator A 196

3. Some Properties of Nonperturbed Operators **201**
3.1. Properties of Simple Maximal Symmetric Operators 201
3.2. Nonperturbed Operators 204
3.3. Examples of Nonperturbed Operators 210

4. Abstract Wave Equation **215**
4.1. Definitions and Preliminary Remarks 215
4.2. Outgoing and Incoming Subspaces 218
4.3. Translation Representation 221
4.4. Spaces of Test and Generalized Functions 223
4.5. Eventually Outgoing and Initially Incoming Elements 227
4.6. Wave Equation in $\mathbb{R}^n$ 231

5. Elements of the Lax–Phillips Scattering Theory for ρ-Perturbed Abstract Wave Equations **234**
5.1. Preliminary Remarks 234
5.2. ρ-Perturbed Abstract Wave Equation 235
5.3. Definitions of Scattering Operator and Scattering Matrix 240
5.4. Description of Singularities of the Scattering Matrix 241
5.5. Case of 0-Perturbations 244
5.6. Examples 254

REFERENCES **261**

SUBJECT INDEX **271**

NOTATION **273**

Preface

In this monograph, we investigate various properties of regular extensions of Hermitian (nondensely defined) operators and describe some applications of this theory.

The concept of regular extensions of an Hermitian operator was introduced by A. Kuzhel in 1980. This concept is a natural generalization of proper extensions of symmetric (densely defined) operators. The use of regular extensions enables one to study various classes of extensions of Hermitian operators without using the method of linear relations, which is rather cumbersome.

It should be noted that, since any linear operator is a regular extension of its Hermitian part, in fact, in this book, we consider the following problem: "To what extent does the Hermitian part of a linear operator determine its properties ?"

In Chapter 1, we present some results from operator theory and the theory of extensions, which are used in subsequent chapters. We also introduce the notion of regular extensions of Hermitian operators as follows:

Let a linear operator A be an extension of an Hermitian operator H in a Hilbert space $\mathcal{H}$. Operator A is called a *regular extension* of the operator H if

$$(Hx, y) = (x, Ay)$$

for any $x \in \mathcal{D}(H)$ and $y \in \mathcal{D}(A)$.

Clearly, if H is a symmetric operator, the notions of "regular extensions" and "proper extensions" (i.e., $H \subset A \subset H^*$) are equivalent. Finally, in this chapter, we establish an analog of the von Neumann formulas and the Krein resolvent formula for regular extensions. We show that dissipative extensions of an Hermitian operator are regular extensions and study this important subclass of regular extensions.

Chapter 2 is devoted to the investigation of regular extensions of Hermitian (symmetric) operators with certain restrictions. In particular, for a semibounded symmetric operator, by using the Friedrichs approach and Krein approach, we give two different proofs of the existence of self-adjoint bound-preserving extensions. For a semibounded Hermitian operator, we discuss necessary and sufficient conditions for the existence of self-adjoint bound-preserving extensions. We give an improved version of the proof of the Krein theorem on self-adjoint norm-preserving extensions of Hermitian contractions and investigate regular $\mathcal{U}$-invariant extensions of Hermitian operators in detail.

In Chapter 3, we investigate regular extensions of Hermitian operators with the use of boundary-value spaces (BVS). The structure of BVS enables one to define regular extensions in terms of abstract boundary values, which is convenient for applications to the theory of boundary-value problems for differential equations.

We also analyze various definitions of BVS of an Hermitian (symmetric) operator and study their general properties. In terms of BVS, we define the characteristic function of an Hermitian operator with arbitrary defect numbers, determine its regular extensions, and describe the nonreal spectra of these extensions.

In Chapter 4, we apply the results presented in Chapters 1–3 to the investigation of quasidifferential operators and models of zero-range potential with internal structure.

We also define the special class $\mathfrak{M}_B$ of regular extensions of a symmetric operator B^2, where B is a simple maximal symmetric operator, and study the properties of these extensions. We show that, for a proper choice of the operator B, the classical Laplacian belongs to the set $\mathfrak{M}_B$ and, moreover, any operator from $\mathfrak{M}_B$ possesses many common properties with the classical Laplacian. Operators from the class $\mathfrak{M}_B$ are called nonperturbed. We consider an abstract wave equation

$$u_{tt} = -Lu,$$

where L is a nonperturbed operator, and establish that, from the viewpoint of the Lax–Phillips approach in scattering theory, the abstract and classical wave equations possess similar properties. In particular, we describe the structure of incoming and outgoing subspaces of the group of solutions of the abstract wave equation, construct a translation representation, and investigate the structure of test and generalized functions associated with this representation.

Further, we present a brief sketch of the Lax–Phillips scattering scheme for an abstract wave equation. For this purpose, we introduce the concept of ρ-perturbed abstract wave equation. These perturbations can be regarded as an abstract analog of perturbations of the Laplace operator in bounded domains.

For the most part, Chapters 1 and 2 were written by A. Kuzhel, and Chapters 3 and 4 were written by S. Kuzhel.

A. Kuzhel thanks the ISSEP for support. We also express our deep gratitude to Dr. P. Malyshev and Dr. D. Malyshev for their careful work in the translation of this book.

A. Kuzhel and S. Kuzhel
Simferopol–Boyarka
July 1998

1. Regular extensions

In Chapter 1, we consider the general properties of linear (Sections 1, 2), Hermitian (Section 3), symmetric (Section 4), dissipative (Section 6), and accretive (Section 7) operators. We introduce the concept of regular extensions, establish an analog of the von Neumann formulas, and generalize the Krein resolvent formula.

1. Linear Operators. General Concepts

In this section, we recall standard notation and terminology related to linear operators acting in normed spaces (in particular, in Hilbert spaces). Only the concept of N-operators is nonstandard.

1.1. Linear Operators. We consider only linear operators (mappings) acting in a normed space X (or from X into another normed space Y). The domain and the range of an operator T are denoted by $\mathcal{D}(T)$ and $\mathcal{R}(T)$, respectively. We write $T\colon X \to Y$ if $\mathcal{D}(T) = X$ and $\mathcal{R}(T) \subset Y$. In this case, we say that T maps X into Y. The set of all bounded operators mapping the space X into Y is denote by $\mathcal{B}[X, Y]$. In particular, $\mathcal{B}[X] := \mathcal{B}[X, X]$ is the set of all bounded operators acting in X and defined on all elements of a space X.

The set

$$\operatorname{Ker} T = \{x \in \mathcal{D}(T) \mid Tx = 0\}$$

is called the *kernel* of an operator T.

An operator T is called *closed* if its graph is closed or, equivalently, if for any sequence $\{x_n\}$ of elements from $\mathcal{D}(T)$ such that[1] $x_n \to x$ and $Tx_n \to y$, we have $x \in \mathcal{D}(T)$ and $Tx = y$.

[1] We write $x_n \to x$ if $\|x_n - x\| \to 0$ as $n \to \infty$.

An operator T is *densely defined* in X if $\overline{\mathcal{D}(T)} = X$, where $\overline{\mathcal{D}(T)}$ is the closure of the lineal $\mathcal{D}(T)$. A closed densely defined operator T is sometimes called an N-operator. We denote the set of all N-operators acting in X by $N(X)$.

If $T \in N(\mathcal{H})$, where $\mathcal{H}$ is a Hilbert space, then, it is known that $T^* \in N(\mathcal{H})$ and $T^{**} = T$, where $T^{**} := (T^*)^*$. Moreover, we have

$$\mathcal{H} = \overline{\mathcal{R}(T)} \oplus \operatorname{Ker} T^* \tag{1.1}$$

Let $\mathcal{M}_1, \mathcal{M}_2, \dots$ be a collection of subsets (not only lineals) of a space X. There exists at least one lineal containing these subsets, e.g., the lineal X. The intersection M of all these lineals is also a lineal. This lineal is called the *linear span* of the subsets $\mathcal{M}_1, \mathcal{M}_2, \dots$ and denoted by

$$\mathcal{M} = \langle \mathcal{M}_1, \mathcal{M}_2, \dots \rangle,$$

In particular, if $x \neq 0$ is a fixed vector from X, then $\langle x \rangle$ is a one-dimensional subspace of the space X.

In the case where the sets $\mathcal{L}_1, \mathcal{L}_2, \dots, \mathcal{L}_n$ are lineals, their linear span $\mathcal{L}$ is called the *sum (algebraic sum)* of these lineals and denoted by

$$\mathcal{L} = \mathcal{L}_1 + \mathcal{L}_2 + \dots + \mathcal{L}_n.$$

In particular, if the lineals $\mathcal{L}_1, \mathcal{L}_2, \dots, \mathcal{L}_n$ are linearly independent, then we say that the set $\mathcal{L}$ is their *direct sum* and write

$$\mathcal{L} = \mathcal{L}_1 \dotplus \mathcal{L}_2 \dotplus \dots \dotplus \mathcal{L}_n.$$

The following statement is true:

The sum $\mathcal{L}$ of lineals $\mathcal{L}_1, \mathcal{L}_2, \dots, \mathcal{L}_n$ is the direct sum of these lineals if and only if, an arbitrary vector x from $\mathcal{L}$ admits a unique representation in the form

$$x = x_1 + x_2 + \dots + x_n, \quad x_k \in \mathcal{L}_k, \quad k = 1, \dots, n.$$

The closure of the linear span of lineals $\mathcal{L}$ and $\mathcal{M}$ is denoted by $\mathcal{L} \vee \mathcal{M}$. Thus,

$$\mathcal{L} \vee \mathcal{M} = \overline{\langle \mathcal{L}, \mathcal{M} \rangle}.$$

1.2. Extensions, Restrictions, and Closure of an Operator. If $\tilde{T}$ and T are operators acting from X into Y such that $\mathcal{D}(T) \subset \mathcal{D}(\tilde{T})$ and $Tx = \tilde{T}x$ for all $x \in \mathcal{D}(T)$, then the operator $\tilde{T}$ is called an *extension* of the operator T and T is called a *restriction* of the operator $\tilde{T}$. For the sake of brevity, we write $T \subset \tilde{T}$ or $T = \tilde{T}|_{\mathcal{D}(T)}$.

We say that an operator T *is closable* if it has a closed extension. The minimal closed extension $\overline{T}$ of an operator T is called its *closure.* (The extension $\overline{T}$ is minimal if any closed extension $\tilde{T}$ of an operator T is also an extension of the operator $\overline{T}$.) In this connection, it is natural to ask whether any linear operator is necessarily closable. In what follows, we show that the answer is negative.

If an operator T is closable, then, for any sequence $\{x_n\}$ of elements from $\mathcal{D}(T)$ such that $x_n \to 0$ and $Tx_n \to y$, we have $y = 0$. This property enables us to construct the examples of nonclosable operators.

Example 1.1. Let $\mathcal{H} = l_2$ be a space of sequences $x = (x_1, x_2, \ldots, x_n, \ldots)$ such that

$$\sum_{k=1}^{\infty} |x_k|^2 < \infty.$$

A scalar product and a norm in l_2 are defined by the equalities

$$(x, y) = \sum_{k=1}^{\infty} x_k \bar{y}_k, \quad y = (y_1, y_2, \ldots, y_n, \ldots),$$

$$\|x\| = \sqrt{(x, x)} = \left(\sum_{k=1}^{\infty} |x_k|^2 \right)^{1/2}.$$

We denote the set of vectors x from l_2 such that

$$\sum_{k=1}^{\infty} |x_k| < \infty$$

by $\mathcal{D}(T)$ and consider the operator T

$$Tx = \left(\sum_{k=1}^{\infty} |x_k|, 0, 0, \ldots \right), \quad x \in \mathcal{D}(T).$$

This operator is defined, in particular, for all elements with finitely many nonzero components. Thus, the operator T is densely defined.

We consider a sequence $\{x_n\}$ such that

$$x_n = \left(\frac{1}{n}, \frac{1}{n}, \ldots, \frac{1}{n}, 0, 0, \ldots\right).$$

For any $n \in \mathbb{N}$, we have $x_n \in \mathcal{D}(T)$. Moreover,

$$\|x_n\| = \frac{1}{\sqrt{n}} \to 0 \quad \text{as} \quad n \to \infty$$

and, therefore, $x_n \to 0$. At the same time, for all $n \in \mathbb{N}$, we can write

$$Tx_n = (1, 0, 0, \ldots).$$

Thus, the sequence $\{\|Tx_n\|\}$ does not converge to zero. Therefore, the operator T is not closed and not closable.

2. Spectrum of a Linear Operator

We now recall the definitions of the resolvent set $\rho(T)$, the resolvent R_λ, the regularity set $\pi(T)$, and the spectrum $\sigma(T)$ of an operator T. We also consider the classification of the spectrum which consists of the point spectrum $\sigma_p(T)$, continuous spectrum $\sigma_c(T)$, and residual spectrum $\sigma_r(T)$. Note that the decomposition of the point spectrum into the stable and unstable parts is not generally accepted.

2.1. Resolvent and Resolvent Set. A complex number λ is called a *regular point* of an operator T if there exists the operator $(T - \lambda I)^{-1}$ bounded and defined in a set dense in $\mathcal{H}$. The set of all regular points of an operator T is called the *resolvent set* of this operator. We denote the resolvent set by $\rho(T)$. The set $\rho(T)$ is open and, for any $\lambda \in \rho(T)$, there exists a number $k = k(\lambda) > 0$ such that

$$\|(T - \lambda I)x\| \geq k\|x\|, \quad x \in \mathcal{D}(T). \tag{2.1}$$

A complex number λ is called a *point of regular type* of an operator T if inequality (2.1) holds for some $k = k(\lambda) > 0$. The set $\pi(T)$ of all points of regular type of an operator T is called the *regularity set* of this operator.

Thus, $\rho(T) \subset \pi(T)$. Moreover, $\pi(T)$ is also an open set.

Theorem 2.1. *Let T be a closed operator acting in a Hilbert space $\mathcal{H}$ and let*

$\lambda \in \pi(T)$. *Then the lineal* $\mathcal{R}(T-\lambda I)$ *is closed and, therefore,* $\mathcal{R}(T-\lambda I)$ *is a subspace of the space* $\mathcal{H}$.

Proof. Let $\{y\}$ be a fundamental sequence of elements from $\mathcal{R}(T-\lambda I)$. Then $y_n = (T-\lambda I)x_n$, where $x_n \in \mathcal{D}(T)$. By virtue of inequality (2.1), we have

$$\|x_n - x_m\| \le \frac{1}{k}\|(T-\lambda I)(x_n - x_m)\| \le \frac{1}{k}\|y_n - y_m\|$$

and, consequently, $\{x_n\}$ is also a fundamental sequence. Let $x_n \to x$ and $y_n \to y$. Then

$$Tx_n = y_n + \lambda x_n \to y + \lambda x.$$

Thus, $x \in \mathcal{D}(T)$ and since the operator T is closed, we have $Tx = y + \lambda x$. Consequently, $y = (T-\lambda I)x \in \mathcal{R}(T-\lambda I)$ and this proves that the lineal $\mathcal{R}(T-\lambda I)$ is closed.

■

Let $\lambda \in \rho(T)$. The operator $(T-\lambda I)^{-1}$ is called the *resolvent* of an operator T and denoted by R_λ. Note that the Hilbert identity

$$R_\lambda - R_\mu = (\lambda - \mu)R_\lambda R_\mu \qquad (\{\lambda, \mu\} \subset \rho(T))$$

is true for the resolvents.

2.2. Spectrum. The set

$$\sigma(T) = \mathbb{C} \setminus \rho(T)$$

is called the *spectrum* of an operator T.

Since the resolvent set $\rho(T)$ is open, its supplement $\sigma(T)$ is a closed set. Furthermore, the following assertions are true.

Theorem 2.2. *If* $T \in \mathcal{B}[\mathcal{H}]$, *then* $\sigma(T)$ *is contained in a circle of radius* $r(T)$ *centered at the origin of coordinates, where*

$$r(T) = \lim_{n\to\infty} \left\|T^n\right\|^{1/n}. \tag{2.2}$$

The limit in equality (2.2) exists and is called the spectral radius of the operator T. *Furthermore,* $r(T) \le \|T\|$.

It is well known that the following theorem holds:

Theorem 2.3. *If* $\mathcal{H}$ *is a complex Hilbert space and* $T \in \mathcal{B}[\mathcal{H}]$, *then* $\sigma(T) \neq 0$.

The spectrum of an operator T consists of three disjoint sets $\sigma_p(T)$, $\sigma_c(T)$ and $\sigma_r(T)$. They are called point, continuous, and residual spectra, respectively.

Let T be an N-operator and let $\lambda \in \rho(T)$. Then

$$[(T-\lambda I)^{-1}]^* = \left(T^* - \bar{\lambda} I\right)^{-1} \tag{2.3}$$

and, hence, $\bar{\lambda} \in \rho(T^*)$ or, in other words,[2] $\lambda \in \rho^*(T^*)$. Similarly, if $\lambda \in \rho^*(T^*)$, then $\lambda \in \rho(T)$. Thus, for an N-operator, we get $\rho(T) = \rho^*(T^*)$ or, which is the same, $\rho(T^*) = \rho^*(T)$.

Therefore,

$$\sigma(T^*) = \sigma^*(T). \tag{2.4}$$

*2.3. **Point Spectrum.*** The set $\sigma_p(T)$ of eigenvalues of an operator T is called the *point spectrum* of this operator. Thus,

$$\sigma(T) = \{\lambda \in \mathbb{C} \mid \operatorname{Ker}(T-\lambda I) \neq \{0\}\}.$$

If the operator T^* exists, then it is reasonable to split the point spectrum of the operator T into two sets

$$\sigma_s(T) = \sigma_p(T) \cap \sigma_p^*(T^*), \tag{2.5}$$

$$\sigma_u(T) = \sigma_p(T) \setminus \sigma_p^*(T^*) \tag{2.6}$$

which are called the *stable* and *unstable* point spectra of the operator T, respectively.

Theorem 2.4. *If* $T \in N(\mathcal{H})$, *then* $\sigma_s(T^*) = \sigma_s^*(T)$.

Proof. Let $\lambda \in \sigma_s(T^*)$. Then, by virtue of (2.5), we obtain $\lambda \in \sigma_p(T^*)$ and $\lambda \in \sigma_p^*(T^{**}) = \sigma_p^*(T)$. Therefore,

$$\bar{\lambda} \in \sigma_p^*(T^*) \cap \sigma_p(T) = \sigma_s(T),$$

---

[2] Here and in what follows, if M is a set of complex numbers, then $M^*\{\bar{\lambda} \mid \lambda \in M\}$.

i.e., $\lambda \in \sigma_s^*(T)$. By analogy, one can easily show that if $\lambda \in \sigma_s^*(T)$, then $\lambda \in \sigma_s(T^*)$.

■

*2.4. **Continuous Spectrum.*** The set

$$\sigma_c(T) = \left\{\lambda \in \sigma(T) \setminus \sigma_p(T) \,\middle|\, \overline{\mathcal{R}(T-\lambda I)} = \mathcal{H}\right\}$$

is called the *continuous* spectrum of an operator T. Thus, $\lambda \in \sigma_c(T)$ if and only if the operator $(T-\lambda I)^{-1}$ exists and is densely defined and unbounded.

Theorem 2.5. *If $T \in N(\mathcal{H})$, then $\sigma_c(T^*) = \sigma_c^*(T)$.*

Proof. Let $\lambda \in \sigma_c(T^*)$. Then $\lambda \in \sigma(T^*)$ and, therefore, by using equality (2.4), we obtain $\bar{\lambda} \in \sigma(T)$. In addition, since $\overline{\mathcal{R}(T^* - \lambda I)} = \mathcal{H}$, then, in view of (1.1), we have $\operatorname{Ker}(T - \bar{\lambda} I) = \{0\}$ and, consequently, $\bar{\lambda} \notin \sigma_p(T)$.

Similarly, since $\lambda \notin \sigma_p(T^*)$, we have $\operatorname{Ker}(T^* - \lambda I) = \{0\}$ and, hence, by virtue of relation (1.1), $\overline{\mathcal{R}(T - \bar{\lambda} I)} = \mathcal{H}$. This enables us to show that $\bar{\lambda} \in \sigma_c(T)$, i.e., $\sigma_c(T^*) \subset \sigma_c^*(T)$. By using the same reasoning as above, we establish the inverse inclusion.

■

*2.5. **Residual Spectrum.*** The set

$$\sigma_r(T) = \left\{\lambda \in \sigma(T) \setminus \sigma_p(T) \,\middle|\, \overline{\mathcal{R}(T-\lambda I)} \neq \mathcal{H}\right\}$$

is called the *residual* spectrum of an operator T.

Thus, $\lambda \in \sigma_r(T)$ if and only if the operator $(T-\lambda I)^{-1}$ exists but its domain of definition is not dense in $\mathcal{H}$.

Clearly, the sets $\sigma_s(T)$, $\sigma_u(T)$, $\sigma_c(T)$ and $\sigma_r(T)$ are mutually disjoint and their union coincides with $\sigma(T)$.

Theorem 2.6. *If T is an N-operator, then*

(a) $\sigma_u(T^*) = \sigma_r^*(T)$;

(b) $\sigma_r(T^*) = \sigma_u^*(T)$.

Proof. Let $\lambda \in \sigma_u(T^*)$. Then $\lambda \in \sigma_p(T^*)$ and $\lambda \notin \sigma_p^*(T^{**})$, i.e., $\lambda \notin \sigma_p^*(T)$. Therefore, the operator $T - \bar{\lambda} I$ is invertible. Since $\operatorname{Ker}(T^* - \lambda I) \neq \{0\}$, we have $\overline{\mathcal{R}(T - \bar{\lambda} I)} \neq \mathcal{H}$ and, thus, $\bar{\lambda} \in \sigma(T)$, i.e., $\lambda \in \sigma_r^*(T)$. The inclusion $\sigma_u(T^*) \subset \subset \sigma_r^*(T)$ is proved.

Conversely, let $\lambda \in \sigma_r^*(T)$. Hence, $\bar{\lambda} \in \sigma_r(T)$. Then $\bar{\lambda} \notin \sigma_p(T)$ and, therefore, $\lambda \notin \sigma_p^*(T^{**})$. Since $\overline{\mathcal{R}(T - \bar{\lambda} I)} \neq \mathcal{H}$, by virtue of (1.1), we have $\operatorname{Ker}(T^* - \lambda I) \neq \{0\}$, i.e., $\lambda \in \sigma_p(T^*)$. Hence, $\lambda \in \sigma_u(T^*)$. Equality (a) is thus proved.

Equality (b) follows from equality (a) with T^* substituted for T.

■

Corollary 2.7. *A self-adjoint operator has neither residual spectrum nor the unstable point spectrum.*

Proof. Indeed, if $T^* = T$, then we can rewrite equality (a) from Theorem 2.6 in the form

$$\sigma_u(T) = \sigma_r(T). \tag{2.7}$$

Since the sets $\sigma_u(T)$ and $\sigma_r(T)$ have no common elements, equality (2.7) is possible only under the condition that $\sigma_u(T) = \sigma_r(T) = \varnothing$.

■

Theorem 2.8. *Let T be a bounded operator acting in $\mathcal{H}$ and defined on a subspace $\mathcal{H}_1 \neq \mathcal{H}$. Then all points λ located outside the circle of radius $r = \|T\|$ centered at the origin of coordinates belong to the residual spectrum of the operator T.*

Proof. Let $|\lambda| > \|T\|$. Then $\lambda \notin \sigma_p(T)$. Assume that $\lambda \in \sigma_c(T)$. Then the space $\mathcal{H}$ contains a sequence $\{x_n\}$ such that $\|x_n\| \leq 1$ and the unbounded sequence $\{y_n = (T - \lambda I)^{-1} x_n\}$. Furthermore, $\lambda y_n = T y_n - x_n$ and, hence,

$$|\lambda|\,\|y_n\| \leq \|T\|\,\|y_n\| + 1$$

It follows from the last inequality that

$$\|y_n\| \leq \frac{1}{|\lambda| - \|T\|} \qquad (\forall\, n \in \mathbb{N})$$

and we arrive at a contradiction. Therefore, $\lambda \notin \sigma_c(T)$. Thus, either $\lambda \in \rho(T)$ or $\lambda \in \sigma_r(T)$.

Assume that $\lambda \in \rho(T)$ and $y \in \mathcal{H} \ominus \mathcal{H}_1$ ($y \neq 0$). Then there exists a vector $x \in \mathcal{H}_1$ such that $y = (T - \lambda I)x$. Since $(y, x) = 0$, we have $(Tx, x) = \lambda \|x\|^2$ and, hence,

$$|\lambda| \|x\|^2 = |(Tx, x)| \leq \|T\| \|x\|^2,$$

i.e., $|\lambda| \leq \|T\|$. Contradiction. Therefore, $\lambda \notin \rho(T)$ which means that $\lambda \in \sigma_r(T)$.

∎

In conclusion, we note that the properties of spectra established above are not always preserved for nonclosed operators. In fact, if we consider the operator T from Example 1.1, then $\sigma_p(T) = \{0, 1\}$. For $\lambda \notin \{0, 1\}$, in view of the evident equality $T^2 x = Tx$ ($\forall x \in \mathcal{D}(T)$), we obtain

$$(T - \lambda I)^{-1} = \frac{1}{\lambda}\left[\frac{1}{1-\lambda} T - I\right]. \tag{2.8}$$

Since T is an unbounded operator, equality (2.8) implies that $\sigma_c(T) = \mathbb{C} \setminus \{0, 1\}$. Hence, $\rho(T) = \sigma_r(T) = \emptyset$. Moreover, it is clear that $\mathcal{D}(T^*)$ consists of those and only those elements y from l_2 whose first component is equal to zero. Obviously, $T^* y = 0$ ($y \in \mathcal{D}(T^*)$). Thus, T^* is a bounded (zero) operator whose domain of definition is nondense in l_2. Therefore, $\sigma_p(T^*) = \{0\}$ and $\sigma_r(T^*) = \mathbb{C} \setminus \{0\}$ (in view of Theorem 2.8). Then $\sigma_c(T^*) = \emptyset$, $\sigma_s(T) = \{0\}$, and $\sigma_u(T) = \{1\}$. We cannot consider the sets $\sigma_s(T^*)$ and $\sigma_u(T^*)$ because the operator adjoint to the operator T^* does not exist.

Thus, in the considered case, the properties of spectra established in Theorems 2.4–2.6 are not preserved.

Remarks. As a basis of our classification of spectra, we took the so-called "rough classification" (see, e.g., Dunford and Schwartz [1] or Richtmyer [1]). This classification was generalized to the case of associative algebras by A. Kuzhel in [5], where one can also find the definitions of stable and unstable point spectra and a relationship between different parts of the spectrum.

3. Hermitian Operators

In the present section., we consider some general properties of Hermitian (not only densely defined) operators, namely, defect subspaces, defect numbers, etc. The criterion of linear independence of defect subspaces (Theorem 3.8) is important for our further discussion.

3.1. Definition and Properties. We use the following terminology. An operator A acting in a Hilbert space $\mathcal{H}$ is called *Hermitian* if

$$(Ax, y) = (x, Ay) \qquad \forall \{x, y\} \subset \mathcal{D}(A). \tag{3.1}$$

We do not assume that the operator A is densely defined. (The equality $\overline{\mathcal{D}(A)} = \mathcal{H}$ is not necessarily true.) If $\overline{\mathcal{D}(A)} = \mathcal{H}$, then the Hermitian operator A is called *symmetric*.

A symmetric operator A is called *self-adjoint* if $\mathcal{D}(A) = \mathcal{D}(A^*)$.

Theorem 3.1. *All eigenvalues of an Hermitian operator A are real. The eigenvectors corresponding to different eigenvalues of the operator A are necessarily orthogonal.*

Proof. Let $\lambda \in \sigma_p(A)$ and let $x \neq 0$ be the corresponding eigenvector, i.e., $Ax = \lambda x$. Then

$$\lambda(x, x) = (Ax, x) = (x, Ax) = \bar{\lambda}(x, x)$$

and, hence, $\lambda = \bar{\lambda}$ because $(x, x) \neq 0$.

Assume that λ and μ are different eigenvalues of an operator A and x and y are the corresponding eigenvectors. Then

$$\lambda(x, y) = (Ax, y) = (x, Ay) = \mu(x, y)$$

but it is possible only for $(x, y) = 0$ since $\lambda \neq \mu$.

■

Thus, it follows from Theorem 3.1 that the point spectrum of an Hermitian operator A is real, i.e., $\sigma_p(A) \subset \mathbb{R}$.

Note that the case where $\sigma_p(A) = \varnothing$ is not excluded (see Example 3.1).

Theorem 3.2. *If A is an Hermitian operator, then $\sigma_c(A) \subset \mathbb{R}$.*

Proof. To prove the theorem, it suffices to show that if $\operatorname{Im}\lambda \neq 0$, then $\lambda \notin \sigma_c(A)$. Let $\lambda = a + ib$, where $b = \operatorname{Im}\lambda \neq 0$. It is easy to show that, for any $x \in \mathcal{D}(A)$,

$$\|(A - \lambda I)x\|^2 = \|(A - aI)x\|^2 + b^2\|x\|^2 + ib\,[((A - aI)x, x) - (x, (A - aI)x)].$$

The first term on the right-hand side of this equality is nonnegative and the last term is equal to zero by virtue of equality (3.1). Therefore,

$$\|(A-\lambda I)x\| \geq |b|\,\|x\|. \tag{3.2}$$

Since $\lambda \notin \sigma_p(A)$, the operator $A-\lambda I$ is invertible. Let $x=(A-\lambda I)^{-1}y$. Then, by using inequality (3.2), we obtain

$$\|y\| = \|(A-\lambda I)x\| \geq |b|\,\|x\|$$

and, thus,

$$\|(A-\lambda I)^{-1}y\| \leq \frac{1}{|\operatorname{Im}\lambda|}\|y\| \qquad (y \in \mathcal{R}(A-\lambda I)).$$

This means that

$$\|(A-\lambda I)^{-1}\| \leq \frac{1}{|\operatorname{Im}\lambda|} \qquad (\operatorname{Im}\lambda \neq 0). \tag{3.3}$$

In other words, the operator $(A-\lambda I)^{-1}$ is bounded if $\operatorname{Im}\lambda \neq 0$. Hence, $\lambda \notin \sigma_c(A)$. ■

3.2. ***Defect Subspaces.*** As earlier, let A be an Hermitian operator acting in $\mathcal{H}$. Consider the lineal $\mathcal{R}_A(\lambda) = \mathcal{R}(A-\lambda I)$ and the subspaces

$$\mathfrak{N}_\lambda = \mathcal{H} \ominus \mathcal{R}_A(\lambda), \qquad \mathcal{D} = \mathcal{H} \ominus \mathcal{D}(A). \tag{3.4}$$

The subspace $\mathfrak{N}_\lambda$ is called the *defect subspace* of the operator A.

Proposition 3.3. *Let Q be the orthogonal projector onto the subspace $\overline{\mathcal{D}(A)}$ in $\mathcal{H}$ and let $\lambda \notin \sigma_p(QA)$. Then the lineals $\mathcal{D}(A)$ and $\mathfrak{N}_\lambda$ are linearly independent.*

Proof. Note that the operator QA is Hermitian. Indeed, note that, for $\{x, y\} \subset \mathcal{D}(A)$, we have $Qx = x$ and $Qy = y$, therefore,

$$(QAx, y) = (Ax, Qy) = (Ax, y) = (Qx, Ay) = (x, QAy).$$

This proves that the operator QA is Hermitian.

Assume that $x \in \mathcal{D}(A) \cap \mathfrak{N}_\lambda$. Then, for any $y \in \mathcal{D}(A)$, we have

$$((QA - \bar{\lambda}I)x, y) = (Q(A - \bar{\lambda}I)x, y) = (x, (A - \lambda I)y) = 0.$$

The last equality holds because $(A - \lambda I)y \in \mathcal{R}_A(\lambda)$. Therefore, $(QA - \bar{\lambda}I)x \in \overline{\mathcal{D}(A)}$ and $(QA - \bar{\lambda}I)x \perp \overline{\mathcal{D}(A)}$. Hence, we obtain $(QA - \bar{\lambda}I)x = 0$. Moreover, if $x \neq 0$, then $\bar{\lambda} \in \sigma_p(QA)$. Thus, $\lambda = \bar{\lambda}$ (since the operator QA is Hermitian). As a result, we conclude that $\lambda \in \sigma_p(QA)$ but this contradicts the assumption. Therefore, $x = 0$, i.e., $\mathcal{D}(A) \cap \mathfrak{N}_\lambda = \{0\}$ and the linear independence of the lineals $\mathcal{D}(A)$ and $\mathfrak{N}_\lambda$ is proved.

■

Corollary 3.4. *Let A be a closed Hermitian operator acting in a Hilbert space $\mathcal{H}$ and let $\operatorname{Im}\lambda \neq 0$. Then the lineals $\mathcal{D}(A)$ and $\mathfrak{N}_\lambda$ are linearly independent and the closure of the lineal $\mathcal{D}(A) \dotplus \mathfrak{N}_\lambda$ coincides with $\mathcal{H}$.*

Proof. Since the operator QA is Hermitian, we have $\lambda \notin \sigma_p(QA)$ whenever $\operatorname{Im}\lambda \neq 0$. By virtue of Proposition 3.3, we obtain $\mathcal{D}(A) \cap \mathfrak{N}_\lambda = \{0\}$.

Let $h \in \mathcal{H}$ and $h \perp \mathcal{D}(A) \dotplus \mathfrak{N}_\lambda$. Then $h \in \mathcal{R}_A(\lambda)$ and, hence, $h = (A - \lambda I)x$, where $x \in \mathcal{D}(A)$. Since $(h, x) = 0$, we have $(Ax, x) = \lambda \|x\|^2$. The last equality holds only for $x = 0$. Therefore, $\overline{\mathcal{D}(A) \dotplus \mathfrak{N}_\lambda} = \mathcal{H}$.

■

Proposition 3.5. *If $\operatorname{Im}\lambda \neq 0$, then the lineals $\mathcal{R}_A(\lambda)$ and $\mathcal{D}$ are linearly independent.*

Proof. Let $y \in \mathcal{R}_A(\lambda) \cap \mathcal{D}$. Then $y = (A - \lambda I)x$, where $x \in \mathcal{D}(A)$. Since $y \in \mathcal{D}$, we can write

$$0 = (y, x) = (Ax, x) - \lambda(x, x),$$

whence it follows that $x = 0$ (as $\bar{\lambda} \neq \lambda$). Therefore, $y = 0$ and $\mathcal{R}_A(\lambda) \cap \mathcal{D} = \{0\}$.

■

Proposition 3.6. *If $\lambda \neq \mu$, then*

$$\mathfrak{N}_\lambda \cap \mathfrak{N}_\lambda = \mathfrak{N}_0 \cap \mathcal{D}. \tag{3.5}$$

Proof. Let $y \in \mathfrak{N}_\lambda \cap \mathfrak{N}_\mu$. Then, for all $x \in \mathcal{D}(A)$,

$$((A - \lambda I)x, y) = 0 \quad \text{and} \quad ((A - \mu I)x, y) = 0, \tag{3.6}$$

or

$$(Ax, y) = \lambda (x, y) \quad \text{and} \quad (Ax, y) = \mu (x, y). \tag{3.7}$$

Since $\lambda \neq \mu$, equalities (3.7) are true only under the condition that

$$(Ax, y) = (x, y) = 0 \quad (\forall \in \mathcal{D}(A)). \tag{3.8}$$

Thus, $y \in \mathcal{H} \ominus \mathcal{R}(A) = \mathfrak{N}_0$ and $y \in \mathcal{D}$, i.e., $y \in \mathfrak{N}_0 \cap \mathcal{D}$. Conversely, if $y \in \mathfrak{N}_0 \cap \mathcal{D}$, then equalities (3.8) and, hence, equalities (3.6) hold. Therefore, $y \in \mathfrak{N}_\lambda \cap \mathfrak{N}_\mu$.

■

Corollary 3.7. *If* $\lambda \neq \mu$, *then the subspace* $\mathfrak{N}_\lambda \cap \mathfrak{N}_\mu$ *does not depend on* λ *and* μ.

The proof of Corollary 3.7 follows from equality (3.5).

■

Theorem 3.8. *The defect subspaces* $\mathfrak{N}_\lambda$ *and* $\mathfrak{N}_\mu$ $(\lambda \neq \mu)$ *of an Hermitian operator* A *are linearly independent if and only if the closure of the linear span of the lineals* $\mathcal{D}(A)$ *and* $\mathcal{R}(A)$ *coincides with* $\mathcal{H}$:

$$\mathcal{D}(A) \vee \mathcal{R}(A) = \mathcal{H}. \tag{3.9}$$

Proof. Let $\mathcal{H}_1 = \mathcal{D}(A) \vee \mathcal{R}(A)$. Then it is easy to see that

$$\mathcal{H} \ominus \mathcal{H}_1 = \mathfrak{N}_0 \cap \mathcal{D}. \tag{3.10}$$

By using equalities (3.5) and (3.10), we conclude that the equality $\mathfrak{N}_\lambda \cap \mathfrak{N}_\mu = \{0\}$ holds if and only if $\mathcal{H}_1 = \mathcal{H}$.

■

Proposition 3.9. *Under condition (3.9), the subspaces* $\mathfrak{N}_\lambda$ *and* $\mathcal{D}$ *are linearly independent.*

Proof. Let $y \in \mathfrak{N}_\lambda \cap \mathcal{D}$. Then, for all x from $\mathcal{D}(A)$, we obtain

$$(Ax, y) = ((A - \lambda I)x, y) = 0$$

and, hence, $y \in \mathfrak{N}_0 \cap \mathcal{D}$. It follows from (3.9) and (3.10) that $\mathfrak{N}_0 \cap \mathcal{D} = \varnothing$ and, therefore, $y = 0$.

■

Example 3.1. Let $\mathcal{H}=\langle l_1, l_2\rangle$ be a two-dimensional Euclidian space and let l_1, l_2 be an orthonormal basis in $\mathcal{H}$. Consider an Hermitian operator A defined as

$$A l_1 = l_2 \qquad (\mathcal{D}(A) = \langle l_1\rangle).$$

Then $\mathcal{R}(A) = \mathcal{D} = \langle l_2\rangle$ and it is easy to see that

$$\mathcal{R}_A(\lambda) = \langle \lambda l_1 - l_2\rangle \quad \text{and} \quad \mathfrak{N}_\lambda = \langle l_1 + \bar{\lambda} l_2\rangle.$$

In particular, $\mathfrak{N}_0 = \mathcal{D}(A)$. If $\lambda \neq 0$, then the subspaces $\mathcal{D}(A)$ and $\mathfrak{N}_\lambda$ are linearly independent.

The other propositions in this subsection can be illustrated in a similar way. Note that, in the considered case,

$$\sigma_p(A) = \sigma_c(A) = \varnothing \quad \text{and} \quad \sigma_r(A) = \mathbb{C}.$$

*3.3. **Defect Numbers.*** The following statements are true:

Lemma 3.10. *If $\mathcal{H}_1$ and $\mathcal{H}_2$ are subspaces of a space $\mathcal{H}$ and* $\dim \mathcal{H}_2 > \dim \mathcal{H}_1$, *then* $\mathcal{H}_2 \ominus \mathcal{H}_1 \neq \{0\}$.

Proof. Let P_1 be the orthoprojector onto $\mathcal{H}_1$ in $\mathcal{H}$ and let $T = P_1|_{\mathcal{H}_2}$ be the restriction of an operator P_1 to the subspace $\mathcal{H}_2$. Thus, $\mathcal{D}(T) = \mathcal{H}_2$ and $\mathcal{R}(T) = T\mathcal{H}_2 \subset \mathcal{H}_1$.

Assume that $\mathcal{H}_2 \ominus \mathcal{H}_1 = \{0\}$. Let us show that, in this case, $\operatorname{Ker} T = \{0\}$. Let $x \in \operatorname{Ker} T$, i.e., $x \in \mathcal{H}_2$ and $Tx = P_1 x = 0$. The last equality means that $x \perp \mathcal{H}_1$ and, hence, $x \in \mathcal{H}_2 \ominus \mathcal{H}_1$ $(= \{0\})$. Thus, $x = 0$ and the equality $\operatorname{Ker} T = \{0\}$ is proved. Since the operator T is invertible, we get

$$\dim \mathcal{H}_2 = \dim T\mathcal{H}_2 \leq \dim \mathcal{H}_1,$$

but this contradicts the assumption. Thus, $\mathcal{H}_2 \ominus \mathcal{H}_1 \neq \{0\}$.

■

Proposition 3.11. *If P_1 and P_2 are, respectively, the orthoprojectors onto subspaces $\mathcal{H}_1$ and $\mathcal{H}_2$ in the space $\mathcal{H}$ and $\|P_1 - P_2\| < 1$, then* $\dim \mathcal{H}_1 = \dim \mathcal{H}_2$.

Proof. Assume that $\dim \mathcal{H}_2 > \dim \mathcal{H}_1$. Then it follows from the previous lemma that there exists a nonzero element $x \in \mathcal{H}_2 \ominus \mathcal{H}_1$. Thus, $P_2 x = x$ and $P_1 x = 0$. Therefore, $(P_1 - P_2)x = -x$. This means that $\|P_1 - P_2\| \geq 1$ and we arrive at a contradiction.

■

Lemma 3.12. *Let A be a closed Hermitian operator and let $\lambda_0 \in \pi(A)$. Then, in a sufficiently small ε-neighborhood of the point λ_0, the dimensionality of the defect subspaces $\mathfrak{N}_\lambda$ of the operator A does not depend on λ.*

Proof. Since λ_0 is a point of regular type of the operator A, there exists a number $k_0 = k(\lambda_0) > 0$ such that

$$\|(A - \lambda_0 I)x\| \geq k_0 \|x\| \quad (\forall x \in \mathcal{D}(A)). \tag{3.11}$$

As indicated in Subsection 2.1, the set $\pi(A)$ is open. Hence, a ε-neighborhood of the point λ_0 belongs to $\pi(A)$. Without loss of generality, we can suppose that $\varepsilon < k_0$.

Let $|\lambda - \lambda_0| < \varepsilon$. Assume that

$$\dim \mathfrak{N}_\lambda > \dim \mathfrak{N}_{\lambda_0}. \tag{3.12}$$

Then, by virtue of Lemma 3.10, we obtain $\mathfrak{N}_\lambda \ominus \mathfrak{N}_{\lambda_0} \neq \{0\}$. Thus, there exists a nonzero element $y \in \mathfrak{N}_\lambda$ such that $y \perp \mathfrak{N}_{\lambda_0}$. Since $\mathfrak{N}_{\lambda_0} = \mathcal{H} \ominus \mathcal{R}_A(\lambda_0)$ and, by virtue of Theorem 2.1, $\mathcal{R}_A(\lambda_0)$ is a subspace of the space $\mathcal{H}$, we have $y \in \mathcal{R}_A(\lambda_0)$, i.e., $y = (A - \lambda_0 I)x$ $(x \neq 0)$. Moreover, $y = h + (\lambda - \lambda_0)x$, where $h = (A - \lambda I)x \in \mathcal{R}_A(\lambda)$. Since $y \in \mathfrak{N}_\lambda$, this enables us to conclude that $h \perp y$. Therefore,

$$|\lambda - \lambda_0|^2 \|x\|^2 = \|y - h\|^2 = \|y\|^2 + \|h\|^2, \tag{3.13}$$

where, in view of (3.11),

$$\|y\| \geq k_0 \|x\|, \qquad \|h\| \geq k\|x\| \quad (k = k(\lambda) > 0). \tag{3.14}$$

By inserting (3.14) in (3.13), we get the inequality

$$|\lambda - \lambda_0|^2 \geq k_0^2 + k^2$$

which contradicts the inequality $|\lambda - \lambda_0| < k_0$. If we assume that inequality (3.12) is true, then we obtain a contradiction. Similarly, by replacing λ_0 with λ and vice versa in the relations presented above, we conclude that the opposite inequality is not true. Hence, $\dim \mathfrak{N}_\lambda = \dim \mathfrak{N}_{\lambda_0}$ $(|\lambda - \lambda_0| < \varepsilon)$.

■

Remark. The fact that the operator A is Hermitian is not used in the proof of Lemma 3.12. Therefore, this lemma is also true in the case of an arbitrary closed operator A if its defect subspaces are defined as in the case of Hermitian operators.

Theorem 3.13. *In the upper (lower) half plane, the dimensionality of the defect subspaces* $\mathfrak{N}_\lambda$ *of a closed Hermitian operator does not depend on* λ.

Proof. By virtue of inequality (3.2), all nonreal points are points of regular type of an operator A. Assume that λ and μ are arbitrary fixed points in the upper (lower) half plane. We connect these points by a straight line. In a neighborhood of each point of the segment obtained as a result, the dimensionality of the defect subspaces of the operator A does not depend on λ. Thus, we obtain a covering of this segment by open sets (neighborhoods).

Since any covering of the segment always contains a finite subcovering, we conclude that the defect subspaces $\mathfrak{N}_\lambda$ and $\mathfrak{N}_\mu$ have the same dimensionalities.

■

Let A be a closed Hermitian operator and let $\mathfrak{N}_\lambda$ be its defect subspace. Consider the numbers

$$n_+ = \dim \mathfrak{N}_i \qquad \text{and} \qquad n_- = \dim \mathfrak{N}_{-i} \tag{3.15}$$

In view of Theorem 3.13, we have

$$\dim \mathfrak{N}_\lambda = \begin{cases} n_+ & (\operatorname{Im} \lambda > 0), \\ n_- & (\operatorname{Im} \lambda < 0). \end{cases}$$

The numbers n_+ and n_- defined by equality (3.15) are called the *defect numbers* of the operator A and the ordered pair of these numbers (n_+, n_-) is called the *deficiency index* of the operator A. Thus, the deficiency index of the Hermitian operator A from Example 3.1 is (1, 1).

Note that the defect numbers of an Hermitian operator can be different; moreover, they can take both finite and infinite values. On the other hand, in some cases, we can definitely assert that the defect numbers must coincide.

Theorem 3.14. *If a closed Hermitian operator* A *has at least one real point of regular type, then its defect numbers coincide.*

Proof. If $\lambda_0 \in \pi(A)$ and $\lambda_0 \in \mathbb{R}$, then the defect subspaces of the operator A have the same dimensionality in a certain ε-neighborhood of the point λ_0. By virtue of Theorem 3.13, we conclude that $n_+ = n_-$.

■

3.4. Defect Numbers of a Real Operator. An operator J defined in a space $\mathcal{H}$ is called an operator of *involution* (or conjugation) if, for any x and y from $\mathcal{H}$ we have

$$(Jx, y) = (Jy, x), \qquad J^2 x = x. \tag{3.16}$$

These equalities imply that

$$(Jx, Jy) = (J(Jy), x) = (y, x). \tag{3.17}$$

It follows from equalities (3.16) and (3.17) that the operator J preserves the norm (i.e., $\|Jx\| = \|x\|$). Moreover, it is invertible and such that $J^{-1} = J$. Note that

$$J(\alpha x + \beta y) = \bar{\alpha} Jx + \bar{\beta} Jy \qquad (\{\alpha, \beta\} \subset \mathbb{C}) \tag{3.18}$$

Indeed, for any $h \in \mathcal{H}$, we can write

$$(J(\alpha x + \beta y), h) = (Jh, \alpha x + \beta y) = \bar{\alpha}(Jh, x) + \bar{\beta}\,(Jh, y)$$

$$= \bar{\alpha}(Jx, h) + \bar{\beta}(Jy, h) = (\bar{\alpha} Jx + \bar{\beta} Jy, h),$$

whence we easily arrive at equality (3.18).

The operator J defined by the equality $(Jf)(x) = \overline{f(x)}$ (the operator of transition to the conjugate function) is an example of an operator of involution in the space $L_2(a, b)$.

An operator J defined by the equality

$$Jx = (\bar{x}_1, \bar{x}_2, \ldots) \qquad (x = (x_1, x_2, \ldots))$$

is an operator of this sort defined in the space l_2.

An Hermitian operator A is called a *real operator* with respect to a given involution J if $J\colon \mathcal{D}(A) \to \mathcal{D}(A)$ and $JA = AJ$.

Theorem 3.15. *The defect numbers of a real operator are equal.*

Proof. Let $x \in \mathcal{D}(A)$ and $y \in \mathfrak{N}_\lambda$ $(\operatorname{Im}\lambda \neq 0)$. Since $Jx \in \mathcal{D}(A)$, we have

$$0 = ((A - \lambda I)Jx, y) = (Jy, J(A - \lambda I)Jx). \tag{3.19}$$

In view of equality (3.18), the second equality in (3.16) and the equality $JA = AJ$, we get

$$J(A - \lambda I)Jx = AJ^2 x - \bar{\lambda} J^2 x = (A - \bar{\lambda} I)x. \tag{3.10}$$

By substituting equality (3.20) in (3.19), we get

$$(Jy, (A - \bar{\lambda}I)x) = 0 \qquad (\forall x \in \mathcal{D}(A)).$$

This means that $Jy \in \mathfrak{N}_{\bar{\lambda}}$. Thus, $J\colon \mathfrak{N}_{\lambda} \to \mathfrak{N}_{\bar{\lambda}}$. If we now replace λ by $\bar{\lambda}$, then we obtain $J\colon \mathfrak{N}_{\bar{\lambda}} \to \mathfrak{N}_{\lambda}$. Consequently, the operator J maps the defect subspace $\mathfrak{N}_{\lambda}$ onto the defect subspace $\mathfrak{N}_{\bar{\lambda}}$. Indeed, if $x_{\bar{\lambda}} \in \mathfrak{N}_{\bar{\lambda}}$, then $x_{\lambda} = Jx_{\bar{\lambda}} \in \mathfrak{N}_{\lambda}$. Therefore, we get $x_{\bar{\lambda}} = Jx_{\lambda}$.

Since the operator J is invertible, we conclude that $\dim \mathfrak{N}_{\lambda} = \dim \mathfrak{N}_{\bar{\lambda}}$. Thus, $n_+ = n_-$.

■

Remarks. Theorem 3.13 was proved (in the case of symmetric operators) by von Neumann [1]. Earlier, for a second-order differential operator, the fact that the dimensionality of the defect subspaces in the upper (lower) half planes is stable was established by Weyl. To the case of non-Hermitian operators, Theorem 3.13 was generalized by Krein and Krasnosel'skii [1]. Theorem 3.15 was proved by von Neumann.

4. Symmetric Operators

We now recall some general properties of symmetric operators and, in particular, the von Neumann formulas.

4.1. General Properties. As indicated above, an operator A is called symmetric if it is densely defined and

$$(Ax, y) = (x, Ay) \qquad (\forall \{x, y\} \subset \mathcal{D}(A)). \tag{4.1}$$

It follows from equality (4.1) that if $y \in \mathcal{D}(A)$, then $y \in \mathcal{D}(A^*)$ and $A^*y = Ay$. In other words, the adjoint operator A^* is the extension of the operator A. Any symmetric operator is closable because its adjoint operator is closed. This enables us to assume (without loss of generality) that all symmetric operators under consideration are closed.

Since any symmetric operator A is Hermitian, we can use all concepts and assertions from Section 3. In particular, we can consider the defect subspaces and defect numbers of symmetric operators. It is clear that if $x_{\lambda} \in \mathfrak{N}_{\lambda}$, where $\mathfrak{N}_{\lambda}$ is the defect subspace of a symmetric operator A, then for any x from $\mathcal{D}(A)$, we have $((A - \lambda I)x, x_{\lambda}) = 0$ or

$$(Ax, x_\lambda) = (x, \bar{\lambda} x_\lambda) \quad (\forall x \in \mathcal{D}(A)).$$

This means that $x_\lambda \in \mathcal{D}(A^*)$ and

$$A^* x_\lambda = \bar{\lambda} x_\lambda \quad (x_\lambda \in \mathfrak{N}_\lambda). \tag{4.2}$$

It is obvious that the converse assertion is also true: If $A^* y = \bar{\lambda} y$, then $y \in \mathfrak{N}_\lambda$. Therefore, we obtain

$$\mathfrak{N}_\lambda = \operatorname{Ker}(A^* - \bar{\lambda} I).$$

4.2. Lineal $\mathcal{D}(A^*)$

Theorem 4.1. *Let A be a closed symmetric operator and let* $\operatorname{Im} \lambda \neq 0$. *Then the lineals $\mathcal{D}(A)$, $\mathfrak{N}_\lambda$, and $\mathfrak{N}_{\bar{\lambda}}$ are linearly independent and*

$$\mathcal{D}(A^*) = \mathcal{D}(A) \dotplus \mathfrak{N}_\lambda \dotplus \mathfrak{N}_{\bar{\lambda}}, \tag{4.3}$$

i.e., an arbitrary vector x from $\mathcal{D}(A^*)$ admits a unique representation in the form

$$x = x_0 + x_\lambda + x_{\bar{\lambda}} \quad (x_0 \in \mathcal{D}(A),\ x_\lambda \in \mathfrak{N}_\lambda,\ x_{\bar{\lambda}} \in \mathfrak{N}_{\bar{\lambda}}). \tag{4.4}$$

Furthermore,

$$A^* x = Ax_0 + \bar{\lambda} x_\lambda + \lambda x_{\bar{\lambda}}. \tag{4.5}$$

Proof. Let $x_0 \in \mathcal{D}(A)$, $x_\lambda \in \mathfrak{N}_\lambda$, and $x_{\bar{\lambda}} \in \mathfrak{N}_{\bar{\lambda}}$. Assume that

$$x_0 + x_\lambda + x_{\bar{\lambda}} = 0. \tag{4.6}$$

By applying the operator A^* to both sides of this equality and using the equalities $A^* x_0 = Ax_0$, $A^* x_\lambda = \bar{\lambda} x_\lambda$, and $A^* x_{\bar{\lambda}} = \lambda x_{\bar{\lambda}}$ we obtain

$$Ax_0 + \bar{\lambda} x_\lambda + \lambda x_{\bar{\lambda}} = 0. \tag{4.7}$$

If we multiply equality (4.6) by λ and subtract it from (4.7), then we get

$$(A - \lambda I) x_0 + (\bar{\lambda} - \lambda) x_\lambda = 0.$$

Hence, $(A-\lambda I)x_0=0$ and $x_\lambda=0$ (since $(A-\lambda I)x_0 \in \mathcal{R}_A(\lambda)$ and $x_\lambda \in \mathfrak{N}_\lambda$, these vectors are orthogonal). Further, $\lambda \notin \sigma_p(A)$ (because $\operatorname{Im}\lambda \neq 0$). Thus, we have $x_0=0$. This means that equality (4.6) is true only under the condition that all terms are equal to zero. Therefore, the lineals $D(A)$, $\mathfrak{N}_\lambda$ and $\mathfrak{N}_{\bar\lambda}$ are linearly independent.

Consider the lineal $\mathcal{L}=D(A) \dotplus \mathfrak{N}_\lambda \dotplus \mathfrak{N}_{\bar\lambda}$. Since the lineals $D(A)$, $\mathfrak{N}_\lambda$, and $\mathfrak{N}_{\bar\lambda}$ are contained in $D(A^*)$, we have $\mathcal{L} \subset D(A^*)$. Hence, to prove equality (4.5) it suffices to show that $D(A^*) \subset \mathcal{L}$.

Let $x \in D(A^*)$ and $y=(A^*-\lambda I)x$. Since the operator A acting in the space $\mathcal{H}$ is closed, in view of (1.1) (with $T=A-\lambda I$), we conclude that $\mathcal{H}=\mathcal{R}_A(\lambda) \oplus \mathfrak{N}_\lambda$. Therefore, $y=y_0+y_\lambda$, where $y_0 \in \mathcal{R}_A(\lambda)$ and $y_\lambda \in \mathfrak{N}_\lambda$. Consequently, $y_0=(A-\lambda I)x_0=(A^*-\lambda I)x_0$, where x_0 is a vector from $D(A)$. Consider the vector

$$x_\lambda = \frac{1}{\bar\lambda-\lambda} y_\lambda .$$

By virtue of equality (4.3), we get

$$y_\lambda = (\bar\lambda-\lambda)x_\lambda = (A^*-\lambda I)x_\lambda .$$

Taking into account the expressions for y, y_0, and y_λ, we now rewrite the equality $y=y_0+y_\lambda$ in the form

$$(A^*-\lambda I)x = (A^*-\lambda I)x_0 + (A^*-\lambda I)x_\lambda$$

or

$$(A^*-\lambda I)(x-x_0-x_\lambda) = 0.$$

This means that $x_{\bar\lambda}=x-x_0-x_\lambda \in \mathfrak{N}_{\bar\lambda}$ and, thus, $x=x_0+x_\lambda+x_{\bar\lambda} \in \mathcal{L}$. Hence, the inclusion $D(A^*) \subset \mathcal{L}$ and equality (4.3) are proved. Equality (4.5) immediately follows from equalities (4.4) and (4.2).

■

Corollary 4.2. *A closed symmetric operator is self-adjoint if and only if its defect numbers are equal to zero.*

4.3. Symmetric (Self-Adjoint) Extensions of Symmetric Operators. Let $\tilde A$ be a symmetric, in particular, self-adjoint extension of a symmetric operator A. Since $A \subset \tilde A$ and $\tilde A \subset \tilde A^*$, the operator $\tilde A^*$ is also an extension of the operator A, i.e., $A \subset \tilde A^*$.

Then $\tilde{A}^* \subset A^*$ and $\tilde{A} \subset A^*$. Hence, any symmetric extension $\tilde{A}$ of the operator A satisfies the conditions

$$A \subset \tilde{A} \subset A^*, \qquad A \subset \tilde{A}^* \subset A^* \tag{4.8}$$

In other words, the operators $\tilde{A}$ and $\tilde{A}^*$ are situated "between" the operators A and A^*. Thus, if $\{x, y\} \subset \mathcal{D}(\tilde{A})$, then

$$x = x_0 + x_\lambda + x_{\bar{\lambda}}, \qquad y = y_0 + y_\lambda + y_{\bar{\lambda}}, \tag{4.9}$$

where

$$\{x_0, y_0\} \subset D(A), \qquad \{x_\lambda, y_\lambda\} \subset \mathfrak{N}_\lambda, \qquad \{x_{\bar{\lambda}}, y_{\bar{\lambda}}\} \subset \mathfrak{N}_{\bar{\lambda}}.$$

Since $\tilde{A} \subset A^*$, we have

$$\tilde{A}x = Ax_0 + \bar{\lambda}x_\lambda + \lambda x_{\bar{\lambda}}, \qquad \tilde{A}y = Ay_0 + \bar{\lambda}y_\lambda + \lambda y_{\bar{\lambda}}.$$

In view of the equalities

$$(Ax_0, y_\lambda) = \lambda(x_0, y_\lambda), \qquad (Ax_0, y_{\bar{\lambda}}) = \bar{\lambda}(x_0, y_{\bar{\lambda}}),$$

$$(x_\lambda, Ay_0) = \bar{\lambda}(x_\lambda, y_0), \qquad (x_{\bar{\lambda}}, Ay_0) = \lambda(x_{\bar{\lambda}}, y_0),$$

we obtain

$$0 = (\tilde{A}x, y) - (x, \tilde{A}y) = (\lambda - \bar{\lambda})\left[(x_{\bar{\lambda}}, y_{\bar{\lambda}}) - (x_\lambda, y_\lambda)\right]. \tag{4.10}$$

Therefore, if $\{x, y\} \subset \mathcal{D}(\tilde{A})$, where the vectors x and y are defined by equalities (4.9), then

$$(x_{\bar{\lambda}}, y_{\bar{\lambda}}) = (x_\lambda, y_\lambda). \tag{4.11}$$

In particular, by setting $y = x$, we get

$$\| x_{\bar{\lambda}} \| = \| x_\lambda \|.$$

Theorem 4.3. *Suppose that* A *is a closed symmetric operator,* $\mathfrak{N}_\lambda$ *and* $\mathfrak{N}_{\bar{\lambda}}$ $(\operatorname{Im}\lambda \neq 0)$ *are its defect subspaces, and* $\tilde{A}$ *is a symmetric extension of the operator*

A. Then there exists an isometric operator[3] *V acting from $\mathfrak{N}_\lambda$ into $\mathfrak{N}_{\bar\lambda}$ and such that any pair of vectors x and $\tilde{A}x$ $\left(x \in \mathcal{D}(\tilde{A})\right)$ can be represented in the form*

$$x = x_0 + x_\lambda + Vx_\lambda \quad \textit{and} \quad \tilde{A}x = Ax_0 + \bar{\lambda}x_\lambda + \lambda Vx_\lambda. \tag{4.13}$$

Conversely, if V is an isometric operator acting from $\mathfrak{N}_\lambda$ into $\mathfrak{N}_{\bar\lambda}$, then the operator $\tilde{A}$ defined by equalities (4.13) is a symmetric extension of the operator A.

Proof. Every element x from $\mathcal{D}(\tilde{A})$ admits a unique representation in the form

$$x = x_0 + x_\lambda + x_{\bar\lambda} \quad (x_0 \in \mathcal{D}(A), x_\lambda \in \mathfrak{N}_\lambda, x_{\bar\lambda} \in \mathfrak{N}_{\bar\lambda}). \tag{4.14}$$

Consider an operator V defined by the equality $Vx_\lambda = x_{\bar\lambda}$, where the vectors x_λ and $x_{\bar\lambda}$ are determined by conditions (4.14) (and, thus, $x_\lambda + x_{\bar\lambda} \in \mathcal{D}(\tilde{A})$). Then, in view of equality (4.12), we have $\|Vx_\lambda\| = \|x_\lambda\|$, i.e., V is an isometric operator acting from $\mathfrak{N}_\lambda$ into $\mathfrak{N}_{\bar\lambda}$. Hence, the vectors x and $\tilde{A}x$ $(x \in \mathcal{D}(\tilde{A}))$ can be represented in the form (4.13).

Conversely, let V be an isometric operator acting from $\mathfrak{N}_\lambda$ into $\mathfrak{N}_{\bar\lambda}$ and let $\tilde{A}$ be the operator defined by equalities (4.13). It is clear that $A \subset \tilde{A}$. It remains to prove that the operator $\tilde{A}$ is symmetric. By analogy with the proof of equality (4.11), we can show that

$$(\tilde{A}x, y) - (x, \tilde{A}y) = (\lambda - \bar{\lambda})[(Vx_\lambda, Vy_\lambda) - (x_\lambda, y_\lambda)] = 0$$

for any x and y from $\mathcal{D}(\tilde{A})$ and, therefore, the operator $\tilde{A}$ is symmetric.

■

Relations (4.3), (4.5), and (4.13) are usually called the von Neumann formulas.

Corollary 4.4. *If one defect number of a closed symmetric operator A is equal to zero, then it is impossible to construct any nontrivial symmetric extension of this operator (other than A).*

Proof. Indeed, under the conditions of the corollary, it is impossible to construct an

[3] Recall that an operator V acting from $\mathcal{H}_1$ into $\mathcal{H}_2$ ($\mathcal{H}_1$ and $\mathcal{H}_2$ are Hilbert spaces) is called *isometric* if, for any x and y from $\mathcal{D}(V)$, we can write $(Vx, Vy) = (x, y)$ or, which is the same,

$$\|Vx\| = \|x\| \quad (\forall x \in \mathcal{D}(V)).$$

In particular, if $\mathcal{D}(V) = \mathcal{H}_1$ and $\mathcal{R}(V) = \mathcal{H}_2$, then the isometric operator V is called *unitary*.

isometric operator V (with $\mathcal{D}(V) \neq \{0\}$) acting from one defect subspace of the operator A into the other.

■

A symmetric operator A is called *maximal* if any its symmetric extension acting in the same space as the operator A is trivial.

Thus, it is easy to see that a closed symmetric operator is maximal if and only if at least one of its defect numbers is equal to zero.

Assume that a nontrivial symmetric extension $\tilde{A}$ of the operator A exists and is defined by equalities (4.13). Let us find the defect subspaces of the operator $\tilde{A}$. By virtue of (4.13), we get

$$(\tilde{A} - \lambda I)x = (A - \lambda I)x_0 + (\bar{\lambda} - \lambda)x_\lambda \in \mathcal{R}_A(\lambda) \oplus \mathcal{D}(V),$$

where $\mathcal{D}(V) \subset \mathfrak{N}_\lambda$. Conversely, if $y \in \mathcal{R}_A(\lambda) \oplus \mathcal{D}(V)$, then $y = (\tilde{A} - \lambda I)x$, where x is a vector from $\mathcal{D}(\tilde{A})$. Thus, $\mathcal{R}_{\tilde{A}}(\lambda) = \mathcal{R}_A(\lambda) \oplus \mathcal{D}(V)$. Hence, the defect subspace $\tilde{\mathfrak{N}}_\lambda$ of the operator $\tilde{A}$ is defined by the equality

$$\tilde{\mathfrak{N}}_\lambda = \mathcal{H} \ominus \mathcal{R}_{\tilde{A}}(\lambda) = \mathfrak{N}_\lambda \ominus \mathcal{D}(V).$$

Similarly, we have $\tilde{\mathfrak{N}}_{\bar{\lambda}} = \mathfrak{N}_{\bar{\lambda}} \ominus \mathcal{R}(V)$.

Therefore, if $\mathcal{D}(V) = \mathfrak{N}_\lambda$ (or $\mathcal{R}(V) = \mathfrak{N}_{\bar{\lambda}}$), then $\tilde{\mathfrak{N}}_\lambda = \{0\}$ (or $\tilde{\mathfrak{N}}_{\bar{\lambda}} = \{0\}$) and, thus, one defect number of the operator $\tilde{A}$ is equal to zero. This enables us to conclude that, under the indicated conditions, $\tilde{A}$ is a maximal operator.

In the case where

$$\mathcal{D}(V) = \mathfrak{N}_\lambda \quad \text{and} \quad \mathcal{R}(V) = \mathfrak{N}_{\bar{\lambda}}, \tag{4.15}$$

the defect numbers of the operator $\tilde{A}$ are equal to zero and, hence, the operator $\tilde{A}$ is self-adjoint. Conditions (4.15) yield that the operator V is unitary.

Thus, the extension $\tilde{A}$ of a symmetric operator A defined by equalities (4.13) is self-adjoint if and only if, in equalities (4.13), V is a unitary operator mapping the defect subspace $\mathfrak{N}_\lambda$ onto the defect subspace $\mathfrak{N}_{\bar{\lambda}}$. In this case, the defect subspaces $\mathfrak{N}_\lambda$ and $\mathfrak{N}_{\bar{\lambda}}$ have the same dimensionality.

In other words, the defect numbers n_+ and n_- of the operator A are equal. Hence, we have proved the following theorem:

Theorem 4.5. *Self-adjoint extensions of a symmetric operator A exist if and only if the defect numbers of this operator are equal.*

In the case where the defect numbers of an operator A are different, we can speak only about maximal symmetric extensions of this operator.

4.4. Extensions to a Broader Space. Let A be a symmetric operator in $\mathcal{H}$ with deficiency index (n_+, n_-), where $n_+ \neq n_-$. Let us show that self-adjoint extensions of this operator exist in a broader Hilbert space.

Denote by $\vec{\mathcal{H}} = \mathcal{H} \oplus \mathcal{H}$ the orthogonal sum of the Hilbert spaces $\mathcal{H}$ and $\mathcal{H}$ whose elements are ordered pairs[4] $\langle x_1, x_2 \rangle$, where $\{x_1, x_2\} \subset \mathcal{H}$. In $\vec{\mathcal{H}}$, we define the operations of addition and multiplication by a number $\lambda \in \mathbb{C}$ as

$$\langle x_1, x_2 \rangle + \langle y_1, y_2 \rangle = \langle x_1 + y_1, x_2 + y_2 \rangle \quad \text{and} \quad \lambda \langle x_1, x_2 \rangle = \langle \lambda x_1, \lambda x_2 \rangle.$$

The operation of scalar product is defined by the formula

$$(\langle x_1, x_2 \rangle, \langle y_1, y_2 \rangle)_{\vec{\mathcal{H}}} = (x_1, y_1) + (x_2, y_2), \tag{4.16}$$

where $(\cdot, \cdot)$ is a scalar product in $\mathcal{H}$.

It is clear that $\vec{\mathcal{H}}$ is a Hilbert space with respect to the scalar product (4.16). Consider the set

$$\vec{\mathcal{H}}_1 = \{\langle x, 0 \rangle \mid x \in \mathcal{H}\}$$

which is a subspace of the space $\vec{\mathcal{H}}$ and a linear mapping $T \colon \mathcal{H} \to \vec{\mathcal{H}}_1$ defined by the equality

$$Tx = \langle x, 0 \rangle \quad (x \in \mathcal{H}).$$

It follows from (4.16) that $(Tx, Ty)_{\vec{\mathcal{H}}} = (x, y)$. Thus, the spaces $\mathcal{H}$ and $\vec{\mathcal{H}}_1$ are isometric. Hence, if we identify the elements x and $\langle x, 0 \rangle$ $(x \equiv \langle x, 0 \rangle)$, then the space $\mathcal{H}$ becomes a subspace of the space $\vec{\mathcal{H}}$.

In the space $\vec{\mathcal{H}}$, we consider an operator $\vec{A}$ defined by the formulas

$$\mathcal{D}(\vec{A}) = \{\langle x_1, x_2 \rangle \mid \{x_1, x_2\} \subset \mathcal{D}(A)\}, \tag{4.18}$$

$$\vec{A}\vec{x} = \langle Ax_1, -Ax_2 \rangle \quad (\vec{x} = \langle x_1, x_2 \rangle \in \mathcal{D}(\vec{A})), \tag{4.19}$$

where, as earlier, A is a symmetric operator acting in $\mathcal{H}$. Clearly, the operator $\vec{A}$ is densely defined and, in view of (4.16), we obtain

---

[4] It is always clear whether $\langle x_1, x_2 \rangle$ is an ordered pair or the lineal span of the set $\{x_1, x_2\}$.

$$(\vec{A}\vec{x}, \vec{y})_{\vec{\mathcal{H}}} = (\vec{x}, \vec{A}\vec{y})_{\vec{\mathcal{H}}} \qquad (\forall\ \{\vec{x}, \vec{y}\} \subset \mathcal{D}(\vec{A})).$$

Thus, $\vec{A}$ is a symmetric operator acting in the space $\vec{\mathcal{H}}$. Let us find the defect subspaces $\vec{\mathfrak{N}}_i$ and $\vec{\mathfrak{N}}_{-i}$ of the operator $\vec{A}$. Let $\vec{x} = \langle x_1, x_2 \rangle \in \mathcal{D}(\vec{A})$. Then

$$(\vec{A} - iI)\vec{x} = \langle (A - iI)x_1, (-A - iI)x_2 \rangle,$$

where

$$(A - iI)x_1 \in \mathcal{R}_A(i) \quad \text{and} \quad -(A + iI)x_2 \in \mathcal{R}_A(-i).$$

This enables us to assert that $\mathcal{R}_{\vec{A}}(i) = \mathcal{R}_A(i) \oplus \mathcal{R}_A(-i)$. Then $\vec{\mathfrak{N}}_i = \vec{\mathcal{H}} \ominus \mathcal{R}_{\vec{A}}(i) = \mathfrak{N}_i \oplus \mathfrak{N}_{-i}$.

Similarly, we obtain $\vec{\mathfrak{N}}_{-i} = \mathfrak{N}_{-i} \oplus \mathfrak{N}_i$. By virtue of the equalities presented above, we conclude that the defect subspaces $\vec{\mathfrak{N}}_i$ and $\vec{\mathfrak{N}}_{-i}$ have equal dimensionalities:

$$\dim \vec{\mathfrak{N}}_i = \dim \vec{\mathfrak{N}}_{-i} = \dim \mathfrak{N}_i + \dim \mathfrak{N}_{-i} = n_+ + n_-.$$

The operator $\vec{A}$ defined by equalities (4.18) and (4.19) is a symmetric operator with equal defect numbers. This is why this operator admits a self-adjoint extension $\vec{\vec{A}}$. Moreover, if $x_1 \in \mathcal{D}(A)$, then

$$\vec{\vec{A}}x_1 = \vec{A}x_1 = \vec{A}\langle x_1, 0 \rangle = \langle Ax_1, 0 \rangle = Ax_1.$$

Therefore, the operator $\vec{\vec{A}}$ is also a self-adjoint extension of the operator A. But this extension acts in a broader space than $\mathcal{H}$. In this connection, we speak about self-adjoint extensions of a symmetric operator from the space $\mathcal{H}$ to a broader space $\vec{\mathcal{H}}$.

Theorem 4.6. *For any symmetric operator* A *acting in a Hilbert space* $\mathcal{H}$, *there exists a self-adjoint extension. If the defect numbers of the operator* A *are different, this extension acts in some space* $\vec{\mathcal{H}} \supset \mathcal{H}$.

Proof. The proof follows from the discussion presented above.

■

Remark. For the first time, the description of symmetric and self-adjoint extensions of symmetric operators was given by von Neumann [1]. Formulas (4.5) and (4.13) are called the von Neumann formulas. Krasnosel'skii [1] generalized the von Neumann formulas to the case of self-adjoint extensions of Hermitian (nondensely defined) operators. Later, these formulas were generalized to the case of nonself-adjoint extensions of Hermitian operators (for details, see Kuzhel [7]).

Self-adjoint extensions of symmetric operators to a broader space were considered by Naimark [1, 2].

5. Regular Extensions of Hermitian Operators

In the present section, we introduce and analyze one of the central concepts of the book—the concept of regular extensions of Hermitian operators. For these extensions we establish an analog of the von Neumann formulas frequently used in what follows. We also give the generalization of the Krein formula to the case of regular extensions of Hermitian operators.

5.1. Hermitian Domain of a Linear Operator. Let A be a linear operator acting in a Hilbert space $\mathcal{H}$. The lineal

$$G_A = \{x \in \mathcal{D}(A) \mid (Ax, y) = (x, Ay) \ \forall y \in \mathcal{D}(A)\} \tag{5.1}$$

is called the *Hermitian domain* of the operator A and the operator $A_0 = A|_{G_A}$ is called the *Hermitian part* of A.

It follows from (5.1) that the operator A_0 is Hermitian. Moreover, if the operator A is closed, then it is easy to see that the operator A_0 is also closed.

Proposition 5.1. *Let $A \in N(\mathcal{H})$. Then*

$$G_A = \operatorname{Ker}(A - A^*). \tag{5.2}$$

Proof. If $x \in G_A$, then $(Ax, y) = (x, Ay)$ $(\forall y \in \mathcal{D}(A))$. Consequently, $x \in \mathcal{D}(A^*)$ and $A^*x = Ax$, i.e., $x \in \operatorname{Ker}(A - A^*)$. Conversely, if $A^*x = Ax$, then

$$(Ax, y) = (A^*x, y) = (x, Ay) \quad (\forall y \in \mathcal{D}(A)).$$

Therefore, $x \in G_A$.

■

Proposition 5.2. *The Hermitian parts of N-operators A and A^* are equal.*

Proof. In view of Proposition 5.1, we get

$$G_{A^*} = \operatorname{Ker}(A^* - A^{**}) = \operatorname{Ker}(A - A^*) = G_A.$$

Moreover,

$$(A^*)_0 = A^*|_{G_{A^*}} = A|_{G_A} = A_0.$$

■

Proposition 5.3. *The defect subspace* $\mathfrak{N}_{\bar{\lambda}}$ *of the Hermitian part* A_0 *of the operator* A *is connected with* G_A *by the equality*

$$(A - \lambda I)(\mathfrak{N}_{\bar{\lambda}} \cap \mathcal{D}(A)) = G_A^{\perp} \cap \mathcal{R}(A - \lambda I). \tag{5.3}$$

Proof. Let $x \in \mathfrak{N}_{\bar{\lambda}} \cap \mathcal{D}(A)$. Then, for all $y \in G_A$, we have

$$(y, (A - \lambda I)x) = ((A - \bar{\lambda} I)y, x) = 0.$$

Consequently, $u = (A - \lambda I)x \in G_A^{\perp}$ or, which is the same,

$$(A - \lambda I)(\mathfrak{N}_{\bar{\lambda}} \cap \mathcal{D}(A)) \subset G_A^{\perp} \cap \mathcal{R}(A - \lambda I).$$

Conversely, if $u \in G_A^{\perp} \cap \mathcal{R}(A - \lambda I)$, then $u = (A - \lambda I)x$, where $x \in \mathcal{D}(A)$ and, for any $y \in G_A$, we can write $((A - \lambda I)y, x) = (y, u) = 0$. Therefore, $x \in \mathfrak{N}_{\bar{\lambda}} \cap \mathcal{D}(A)$. Then $u \in (A - \lambda I)(\mathfrak{N}_{\bar{\lambda}} \cap \mathcal{D}(A))$ and equality (5.3) is proved.

■

The corollaries presented below immediately follow from equality (5.3).

Corollary 5.4. *If the lineals* $\mathfrak{N}_{\bar{\lambda}}$ *and* $\mathcal{D}(A)$ *are linearly independent, then the lineals* $G_A^{\perp}$ *and* $\mathcal{R}(A - \lambda I)$ *are also linearly independent. Conversely, if* $\lambda \notin \sigma_p(A)$ *and the lineals* $G_A^{\perp}$ *and* $\mathcal{R}(A - \lambda I)$ *are linearly independent, then the same is true for the lineals* $\mathfrak{N}_{\bar{\lambda}}$ *and* $\mathcal{D}(A)$.

Corollary 5.5. *If* $\overline{G}_A = \mathcal{H}$ *and* $\lambda \notin \sigma_p(A)$, *then the lineals* $\mathfrak{N}_{\bar{\lambda}}$ *and* $\mathcal{D}(A)$ *are linearly independent.*

Corollary 5.6. *If the operator* A *is symmetric, then the lineals* $\mathfrak{N}_{\bar{\lambda}}$ *and* $\mathcal{D}(A)$ *are linearly independent for any nonreal* λ.

Corollary 5.7. *Let A be a closed operator and let $\lambda \in \rho(A)$. Then $\mathfrak{N}_{\bar{\lambda}} \cap \mathcal{D}(A) = (A - \lambda I)^{-1} G_A^{\perp}$.*

5.2. Regular Extensions. An operator A is called a *regular extension* of an Hermitian operator H if $H \subset A_0$, where A_0 is the Hermitian part of the operator A. Obviously, an arbitrary operator A is a regular extension of its Hermitian part A_0.

The collection of all regular extensions of an Hermitian operator H is denoted by $\mathcal{P}(H)$. In particular, $\mathcal{P}(H)$ contains all self-adjoint extensions of the operator H (if these extensions exist).

Proposition 5.8. *An extension A of an Hermitian operator H is regular if and only if*

$$(Hx, y) = (x, Ay) \quad (\forall\, x \in \mathcal{D}(H),\ \forall\, y \in \mathcal{D}(A)). \tag{5.4}$$

In particular, an N-operator A is a regular extension of the operator H if and only if

$$H \subset A \quad \text{and} \quad H \subset A^*. \tag{5.5}$$

Proof. Indeed, if $A \in \mathcal{P}(H)$, then $H \subset A_0$ and, for any $x \in \mathcal{D}(H)$ and $y \in \mathcal{D}(A)$, we have

$$(Hx, y) = (Ax, y) = (x, Ay),$$

i.e., equality (5.4) is true.

Conversely, if $H \subset A$ and equality (5.4) is true, then, for any $x \in \mathcal{D}(H)$ and $y \in \mathcal{D}(A)$,

$$(Ax, y) = (Hx, y) = (x, Ay).$$

Therefore, $x \in \mathcal{D}(A_0)$. Thus, $H \subset A_0$ and, hence, $A \in \mathcal{P}(H)$.

Assume that the N-operator A belongs to $\mathcal{P}(H)$. Then $H \subset A$ and, by virtue of equality (5.4), for any $x \in \mathcal{D}(H)$ and $y \in \mathcal{D}(A)$, we obtain

$$(A^*x, y) = (x, Ay) = (Hx, y),$$

whence $A^*x = Hx$ $(x \in \mathcal{D}(H))$, i.e., $H \subset A^*$. Thus, if the N-operator A belongs to $\mathcal{P}(H)$, then relations (5.5) hold. Conversely, if relations (5.5) are true, then, for $x \in \mathcal{D}(H)$ and $y \in \mathcal{D}(A)$,

$$(Hx, y) = (A^*x, y) = (x, Ay)$$

and, consequently, $A \in \mathcal{P}(H)$.

■

The criterion given by Proposition 5.8 is convenient for many applications.

Note that extensions of an Hermitian operator H are not necessarily regular. Indeed, let $\mathcal{H} = \langle l_1, l_2 \rangle$, $\mathcal{D}(H) = \langle l_1 \rangle$ and $Hl_1 = l_2$. Any extension A of the operator H has the form

$$Al_1 = l_2, \quad Al_2 = al_1 + bl_2,$$

where a and b are complex numbers. It is easy to see that $A \in \mathcal{P}(H)$ only for $a = 1$.

Clearly, for a symmetric (densely defined) operator H the set of its regular extensions $\mathcal{P}(H)$ coincides with the set of its proper extensions (i.e., $\mathcal{P}(H) = \{A \mid H \subset A \subset H^*\}$).

In Theorems 5.9 and 5.10, we show that the fact that an operator A belongs to the set of regular extensions of an Hermitian operator H yields the relationship between the spectra of these operators.

Theorem 5.9. *Let* $A \in \mathcal{P}(H)$. *Then*

$$\sigma_p(H) \subset \sigma_p(A) \subset \sigma_p(H) \cup \sigma_r^*(H), \tag{5.6}$$

$$\sigma_c(H) \subset \sigma_c(A) \subset \sigma_c(H) \cup \sigma_r(H). \tag{5.7}$$

Proof. It is obvious that the first inclusion in (5.6) is true. Let us prove the second inclusion. Let $\lambda \in \sigma_p(A)$ and let x_0 be the corresponding eigenvector. Since $\mathcal{D}(H) \subset \mathcal{D}(A)$, then for any x from $\mathcal{D}(H)$, we can write

$$\big((H - \bar{\lambda} I)x, x_0\big) = (x, (A - \lambda I)x_0) = 0$$

and, hence, $x_0 \perp \mathcal{R}(H - \bar{\lambda} I)$. This means that $\bar{\lambda} \notin \rho(H)$ and $\bar{\lambda} \notin \text{q}(H)$, i.e., $\lambda \notin \rho^*(H)$ and $\lambda \notin \sigma_c^*(H)$. Consequently,

$$\lambda \in \sigma_p^*(H) \cup \sigma_r^*(H) = \sigma_p(H) \cup \sigma_r^*(H)$$

and equality (5.6) is proved.

Let $\lambda \in \sigma_c(H)$. Then there exists a densely defined unbounded operator $B = (H - \lambda I)^{-1}$. Let us show that $\lambda \notin \sigma_p(A)$.

Indeed, assume that $Ax_0 = \lambda x_0$ $(x_0 \neq 0)$. Since $\mathcal{D}(H) \subset \mathcal{D}(A_0)$ and $\bar{\lambda} = \lambda$, for any vector x from $\mathcal{D}(H)$, we obtain $((H-\lambda I)x, x_0) = (x, (A-\lambda I)x_0) = 0$ and, therefore, $x_0 \perp \mathcal{R}(H-\lambda I)$. This contradicts the assumption that the operator B is densely defined.

Thus, $\lambda \notin \sigma_p(A)$ and, hence, the operator $(A-\lambda I)^{-1}$ exists. Moreover, we have

$$(A-\lambda I)^{-1} y = (H-\lambda I)^{-1} y \quad (y \in \mathcal{D}(B)) \tag{5.8}$$

and, consequently, the operator $(A-\lambda I)^{-1}$ is densely defined and unbounded. Therefore, $\lambda \in \sigma_c(A)$. The first inclusion in (5.7) is proved.

Let us now prove the second inclusion in (5.7). In view of the inclusions $\sigma_p(H) \subset \sigma_p(A)$ and $\rho(H) \subset \rho(A)$, for $\lambda \in \sigma_c(A)$, we conclude that $\lambda \notin \sigma_p(H)$ and $\lambda \notin \rho(H)$, i.e., $\lambda \in \sigma_c(H) \cup \sigma_r(H)$.

■

Remark. If $A \in \mathcal{P}(H)$, then, generally speaking, $\sigma_c(H) \neq \sigma_c(A)$. Indeed, let H be an Hermitian operator in $\mathcal{H}$ and let $\lambda \in \sigma_c(H)$. Consider the Hilbert space $\tilde{\mathcal{H}} = \mathcal{H} \oplus \mathcal{H}_1$, where $\mathcal{H}_1$ is a one-dimensional space. On the lineal $\mathcal{D}(A) = \mathcal{D}(H) \oplus \mathcal{H}_1$, we define the operator A as follows:

$$A(x + x_1) = Hx + ax_1 \quad (x \in \mathcal{D}(H),\ x_1 \in \mathcal{H}_1),$$

where $a \neq \lambda$ is a fixed number. It is easy to see that $A \in \mathcal{P}(H)$ and

$$(A-\lambda I)^{-1}(y + y_1) = (H-\lambda I)^{-1} y + \frac{1}{a-\lambda} y_1,$$

where $y \in \mathcal{R}(H-\lambda I)$ and $y_1 \in \mathcal{H}_1$. Therefore, $\lambda \in \sigma_c(A)$. However, if we consider H as an operator in $\tilde{\mathcal{H}}$, then we get $\lambda \notin \sigma_c(H)$ (because, in this case, the operator $(H-\lambda I)^{-1}$ is nondensely defined in $\tilde{\mathcal{H}}$).

Theorem 5.10. *Let H be an Hermitian operator in $\mathcal{H}$ and let $\bar{\lambda} \in \sigma_r(QH)$, where Q is the orthoprojector onto $\overline{\mathcal{D}(H)}$ in $\mathcal{H}$. If the defect subspace $\mathfrak{N}_{\bar{\lambda}}$ of the operator H is nonzero, then there exists the operator $A \in \mathcal{P}(H)$ such that $\lambda \in \sigma_p(A)$ and $\mathcal{H}_\lambda = \mathfrak{N}_{\bar{\lambda}}$, where $\mathcal{H}_\lambda$ is the eigensubspace of the operator A corresponding to the eigenvalue λ.*

Proof. Since $\bar{\lambda} \in \sigma_r(QH)$, we have $\bar{\lambda} \notin \sigma_p(QH)$ and, therefore, by virtue of Proposition 1.3, the lineals $\mathcal{D}(H)$ and $\mathfrak{N}_{\bar{\lambda}}$ are linearly independent.

Consider an operator A defined on the lineal $\mathcal{D}(A) = \mathcal{D}(H) \dotplus \mathfrak{N}_{\bar{\lambda}}$ by the equality

$$A(x + x_{\bar{\lambda}}) = Hx + \lambda x_{\bar{\lambda}} \quad (x \in \mathcal{D}(H),\ x_{\bar{\lambda}} \in \mathfrak{N}_{\bar{\lambda}}).$$

Then it is easy to show that

$$(Ay, x + x_{\lambda}) = (y, A(x + x_{\bar{\lambda}})), \quad (\forall\, y \in \mathcal{D}(H))$$

i.e., $y \in \mathcal{D}(A_0)$. Thus, $\mathcal{D}(H) \subset \mathcal{D}(A_0)$. Moreover,

$$A_0 y = Ay = Hy \quad (y \in \mathcal{D}(H)).$$

Consequently, $A \in \mathcal{P}(H)$ and, obviously, $\mathfrak{N}_{\bar{\lambda}} \subset \mathcal{H}_{\lambda}$.

Let us show that the inverse inclusion is true. Indeed, let $x \in \mathcal{H}_{\lambda}$. Then

$$((H - \lambda I)y, x) = (y, (A - \lambda I)x) = 0 \quad (\forall\, y \in \mathcal{D}(H)),$$

whence we get that $x \in \mathfrak{N}_{\bar{\lambda}}$. Thus, $\mathcal{H}_{\lambda} \subset \mathfrak{N}_{\bar{\lambda}}$ and, hence, $\mathcal{H}_{\lambda} = \mathfrak{N}_{\bar{\lambda}}$.

■

*5.3. **Quasi-Hermitian Operator.*** An operator A from $\mathcal{P}(H)$ is called *quasi-Hermitian* (or a quasi-Hermitian extension of an operator H) if H is a closed Hermitian operator with finite defect numbers.

Theorem 5.11. *If A is a quasi-Hermitian operator, then $\sigma_c(A) \subset \mathbb{R}$.*

Proof. Let $A \in \mathcal{P}(H)$ be a quasi-Hermitian operator, let $\lambda \notin \sigma_p(A)$ $(\operatorname{Im}\lambda \neq 0)$, let $y = (A - \lambda I)^{-1}x$, and let $\mathfrak{N}_{\lambda}$ be the defect subspace of the operator H. Since $\mathcal{H} = \mathcal{R}(H - \lambda I) \oplus \mathfrak{N}_{\lambda}$, we can write

$$x = (A - \lambda I)y = x_1 + x_2 \quad (x_1 \in \mathcal{R}(H - \lambda I),\ x_2 \in \mathfrak{N}_{\lambda}),$$

where $x_1 = (H - \lambda I)y_1$, $x_2 = (A - \lambda I)y_2$, and $y_1 + y_2 = y$.

Thus,

$$(A - \lambda I)^{-1}x = (H - \lambda I)^{-1}P_1 x + (A - \lambda I)^{-1}P_2 x,$$

where P_1 and P_2 are the orthoprojectors onto $\mathcal{R}(H - \lambda I)$ and $\mathfrak{N}_{\lambda}$, respectively. By using the inequality

$$\|(H-\lambda I)^{-1}\| \leq \frac{1}{|\operatorname{Im}\lambda|}$$

and the fact that the operator $(A-\lambda I)^{-1}P_2$ is finite-dimensional, we conclude that the operator $(A-\lambda I)^{-1}$ is bounded, i.e., $\lambda \notin \sigma_c(A)$.

■

Theorem 5.12. *Let* $A \in \mathcal{P}(H)$, *where* H *is an Hermitian operator with deficiency index* (m, n), *let* $\lambda \in \sigma_p(A)$, *and let* $\mathcal{H}_\lambda$ *be the corresponding eigensubspace of the operator* A. *Then*

$$\dim \mathcal{H}_\lambda \leq \begin{cases} n & (\operatorname{Im}\lambda > 0), \\ m & (\operatorname{Im}\lambda < 0). \end{cases}$$

Proof. Let $\lambda \in \sigma_p(A)$ $(\operatorname{Im}\lambda \neq 0)$ and $x \in \mathcal{H}_\lambda$. Then, for any $y \in \mathcal{D}(H)$, we obtain

$$((H-\bar{\lambda}I)y, x) = (y, (A-\lambda I)x) = 0$$

and, hence, $x \in \mathfrak{N}_{\bar{\lambda}}$, i.e., $\mathcal{H}_\lambda \subset \mathfrak{N}_{\bar{\lambda}}$. Thus, if $\operatorname{Im}\lambda > 0$, then

$$\dim \mathcal{H}_\lambda \leq \dim \mathfrak{N}_{\bar{\lambda}} = \dim \mathfrak{N}_{-i} = n.$$

Similarly, $\dim \mathcal{H}_\lambda \leq m$ whenever $\operatorname{Im}\lambda < 0$.

■

*5.4. **Operator*** $T_{\mu\lambda}$. Suppose that H is an Hermitian operator acting in $\mathcal{H}$, an operator A is a regular extension of the operator H, and $\mathfrak{N}_A(\lambda) = \mathcal{R}_A(\lambda) \ominus \mathcal{R}_H(\lambda)$, where $\mathcal{R}_T(\lambda) = \mathcal{R}(T-\lambda I)$ $(T \in \{A, H\})$. Clearly, $\mathfrak{N}_A(\lambda) \subset \mathfrak{N}_\lambda$, where $\mathfrak{N}_\lambda$ is the defect subspace of the operator H. It is also obvious that the equality $\mathfrak{N}_A(\lambda) = \mathfrak{N}_\lambda$ is true for $\lambda \in \rho(A)$.

Let $\lambda \notin \sigma_p(A)$. We introduce an operator $T_{\mu\lambda}\colon \mathcal{R}_A(\lambda) \to \mathcal{R}_A(\mu)$ defined by the equality

$$T_{\mu\lambda} = (A-\mu I)(A-\lambda I)^{-1} \tag{5.9}$$

Theorem 5.13. *If* $A \in \mathcal{P}(H)$ *and* $\lambda \notin \sigma_p(A)$, *then*

$$T_{\mu\lambda}\colon \mathcal{R}_A(\lambda) \cap \mathfrak{N}_{\bar{\mu}} \to \mathfrak{N}_{\bar{\lambda}}. \tag{5.10}$$

In particular, if $\lambda \in \rho(A)$, *then*

$$T_{\mu\lambda} : \mathfrak{N}_{\bar{\mu}} \to \mathfrak{N}_{\bar{\lambda}} \quad \textit{and} \quad T^*_{\mu\lambda} : \mathfrak{N}_{\mu} \to \mathfrak{N}_{\lambda}. \tag{5.11}$$

Proof. Let $f \in \mathcal{R}_A(\lambda) \cap \mathfrak{N}_{\bar{\mu}}$, i.e., $f = (A - \lambda I)\varphi$ $(\varphi \in \mathcal{D}(A))$ and

$$(f, (H - \bar{\mu} I)x) = 0 \quad (\forall x \in \mathcal{D}(H)). \tag{5.12}$$

In view of equality (5.9), we get

$$T_{\mu\lambda} f = (A - \mu I)\varphi = f + (\lambda - \mu)\varphi.$$

Therefore, for $x \in \mathcal{D}(H)$, we can write

$$(T_{\mu\lambda} f, (H - \bar{\lambda} I)x) = (f, (H - \bar{\lambda} I)x) + (\lambda - \mu)(\varphi, (H - \bar{\lambda} I)x)), \tag{5.13}$$

where, by virtue of (5.12),

$$(f, (H - \bar{\lambda} I)x) = (\mu - \lambda)(f, x). \tag{5.14}$$

Further, since $x \in G_A$, we have

$$(\varphi, (H - \bar{\lambda} I)x) = (\varphi, Hx) - \lambda(\varphi, x) = (A\varphi, x) - \lambda(\varphi, x) = (f, x). \tag{5.15}$$

Hence, it follows from (5.14) and (5.15) that, for any $x \in \mathcal{D}(H)$, the right-hand side of relation (5.13) is equal to zero. Thus, $T_{\mu\lambda} f \in \mathfrak{N}_{\bar{\lambda}}$ and we arrive at (5.10).

The first relation in (5.11) follows from (5.10) for $y \in \rho(A)$. To prove the second relation we note that if $f \in \mathfrak{N}_{\mu}$ and $x \in \mathcal{D}(H)$, then

$$(T^*_{\mu\lambda} f, (H - \lambda I)x) = (f, (H - \mu I)x) = 0$$

and, therefore, $T^*_{\mu\lambda} f \in \mathfrak{N}_{\lambda}$.

■

Corollary 5.14. *Let* $A \in \mathcal{P}(H)$ *and let* $\{\mu, \lambda\} \subset \rho(A)$. *Then*

$$\dim \mathfrak{N}_{\mu} = \dim \mathfrak{N}_{\lambda} \quad \textit{and} \quad \dim \mathfrak{N}_{\bar{\mu}} = \dim \mathfrak{N}_{\bar{\lambda}}, \tag{5.16}$$

where $\mathfrak{N}_{\alpha}$ $(\alpha \in \{\mu, \lambda, \bar{\mu}, \bar{\lambda}\})$ *are the defect subspaces of the operator* H.

Proof. If $\{\mu, \lambda\} \subset \rho(A)$, then the operator $T_{\mu\lambda}$ is completely invertible and $T_{\mu\lambda}^{-1} = T_{\lambda\mu}$. It follows from Theorem 5.13 that

$$T_{\mu\lambda}: \mathfrak{N}_{\bar{\mu}} \to \mathfrak{N}_{\bar{\lambda}} \quad \text{and} \quad T_{\mu\lambda}^{-1}: \mathfrak{N}_{\bar{\lambda}} \to \mathfrak{N}_{\bar{\mu}}.$$

Hence, the second equality in (5.16) is true.

By analogy, in view of the fact that $\left(T_{\mu\lambda}^{*}\right)^{-1} = T_{\lambda\mu}^{*}$, we establish the first equality in (5.16).

■

5.5. Analog of the Von Neumann Formulas

Theorem 5.15. *Suppose that an operator A is a regular extension of an Hermitian operator H and λ is a nonreal number which does not belong to the point spectrum of the operator A ($\lambda \notin \sigma_p(A)$). Then an arbitrary vector x from $\mathcal{D}(A)$ can be represented in the form*

$$x = x_0 + x_\lambda + \Phi x_\lambda, \tag{5.17}$$

where $x_0 \in \mathcal{D}(H)$, $x_\lambda \in \mathcal{D}(\Phi)$, and Φ is a linear operator acting from $\mathfrak{N}_\lambda$ into $\mathfrak{N}_{\bar{\lambda}}$ and such that $-1 \notin \sigma_p(\Phi)$. Furthermore, the operator A acts according to the formula

$$A(x_0 + x_\lambda + \Phi x_\lambda) = Hx_0 + \bar{\lambda} x_\lambda + \lambda \Phi x_\lambda \tag{5.18}$$

and the lineals $\mathcal{D}(H)$ and $(\Phi + I)\,\mathcal{D}(\Phi)$ are linearly independent.

Proof. Let $y \in \mathcal{R}_A(\lambda)$. Then $y = y_0 + y_1$, where $y_0 \in \mathcal{R}_H(\lambda)$, $y_1 \in \mathfrak{N}_\lambda$, and, hence,

$$y_0 = (H - \lambda I)x_0 = (A - \lambda I)x_0 \quad (x_0 \in \mathcal{D}(H)),$$

$$y_1 = (A - \lambda I)x_1 \quad (x_1 \in \mathcal{D}(A)).$$

Therefore, we have

$$y = (A - \lambda I)x = (A - \lambda I)(x_0 + x_1) \quad (x \in \mathcal{D}(A)).$$

Since $\lambda \notin \sigma_p(A)$, we get $x = x_0 + x_1$. Let $T_\lambda = T_{\bar{\lambda}\lambda} = (A - \bar{\lambda} I)(A - \lambda I)^{-1}$. Then

$$T_\lambda y = (A - \bar{\lambda} I)x \quad \text{and} \quad y = (A - \lambda I)x.$$

These equalities imply that

$$x = \frac{1}{\lambda - \bar{\lambda}}(T_\lambda y - y) \quad \text{and} \quad Ax = \frac{1}{\lambda - \bar{\lambda}}(\lambda T_\lambda y - \bar{\lambda} y). \tag{5.19}$$

Since $y = y_0 + y_1$, we have

$$x = \frac{1}{\lambda - \bar{\lambda}}(T_\lambda y_0 - y_0) + \frac{1}{\bar{\lambda} - \lambda} y_1 + \frac{1}{\lambda - \bar{\lambda}} T_\lambda y_1. \tag{5.20}$$

Further, we have $y_0 = (H - \lambda I)x_0$ ($x_0 \in \mathcal{D}(H)$) and $y_1 \in \mathfrak{N}_\lambda$. Consequently,

$$\frac{1}{\lambda - \bar{\lambda}}(T_\lambda y_0 - y_0) = \frac{1}{\lambda - \bar{\lambda}}\left[(H - \bar{\lambda} I)x_0 - (H - \lambda I)x_0\right] = x_0, \tag{5.21}$$

$$\frac{1}{\bar{\lambda} - \lambda} y_1 = x_\lambda \in \mathfrak{N}_\lambda, \quad \frac{1}{\lambda - \bar{\lambda}} T_\lambda y_1 = -T_\lambda x_\lambda = \Phi x_\lambda, \tag{5.22}$$

where $\Phi = -T_\lambda\big|_{\mathfrak{N}_\lambda \cap \mathcal{D}(T_\lambda)}$. By virtue of Theorem 5.13, Φ is a linear operator acting from $\mathfrak{N}_\lambda$ into $\mathfrak{N}_{\bar{\lambda}}$. Thus, in view of relations (5.21) and (5.22), the vector x can be represented in the form (5.17).

By using the second equality in (5.19), we obtain the following representation of the vector Ax:

$$Ax = \frac{1}{\lambda - \bar{\lambda}}\left(\lambda T_\lambda y_0 - \bar{\lambda} y_0\right) + \frac{\bar{\lambda}}{\bar{\lambda} - \lambda} y_1 + \frac{\lambda}{\lambda - \bar{\lambda}} T_\lambda y_1, \tag{5.23}$$

where

$$\lambda T_\lambda y_0 - \bar{\lambda} y_0 = \lambda (H - \bar{\lambda} I)x_0 - \bar{\lambda}(H - \lambda I)x_0 = (\lambda - \bar{\lambda})Ax_0 \tag{5.24}$$

and, hence, by virtue of (5.22)–(5.24), the vector Ax can be represented in the form (5.18).

Let us show that the lineals $\mathcal{D}(H)$ and $(\Phi + I)\,\mathcal{D}(\Phi)$ are linearly independent. Indeed, assume that $x_0' \neq 0$, $x_\lambda \neq 0$, and

$$\alpha x_0 + \beta(\Phi + I)x_\lambda = 0 \quad (x_0 \in \mathcal{D}(H),\ x_\lambda \in \mathcal{D}(\Phi)). \tag{5.25}$$

By applying the operator A to both sides of equality (5.25) and using equality (5.18), we obtain

$$\alpha H x_0 + \beta(\lambda \Phi x_\lambda + \bar{\lambda} x_\lambda) = 0. \tag{5.26}$$

Therefore, in view of (5.25) and (5.26), we get

$$\alpha(H - \lambda I)x_0 + \beta(\bar{\lambda} - \lambda)x_\lambda = 0.$$

This equality implies that $\lambda = \beta = 0$ (since $\operatorname{Im}\lambda \neq 0$, $\lambda \notin \sigma_p(H)$, $(H - \lambda I)x_0 \in \mathcal{R}_H(\lambda)$, and $x_\lambda \perp \mathcal{R}_H(\lambda)$).

■

The following theorem shows that the converse assertion is also true.

Theorem 5.16. *Assume that* H *is an Hermitian operator,* A *is the operator defined by (5.18) with the domain of definition (5.17), where the operator* Φ *satisfies the conditions of Theorem 5.15. Then the operator* A *is a regular extension of the operator* H, $\lambda \notin \sigma_p(A)$, *and*

$$\bar{\lambda} \in \sigma_p(A) \Leftrightarrow 0 \in \sigma_p(\Phi). \tag{5.27}$$

Proof. Since the lineals $\mathcal{D}(H)$ and $(I + \Phi)\mathcal{D}(\Phi)$ are linearly independent and $-1 \notin \sigma_p(\Phi)$, the operator A is correctly defined by relations (5.17) and (5.18). If $y \in \mathcal{D}(H)$, then $Ay = Hy$ and it is easy to see that

$$(Ax, y) - (x, Ay) = -(x_\lambda, (H - \lambda I)y) - (\Phi x_\lambda, (H - \bar{\lambda} I)y) = 0$$

for any $x \in \mathcal{D}(A)$.

Hence, by virtue of Proposition 5.8, we conclude that A is a regular extension of the operator H.

Assume that $Ax = \lambda x$, where $x = x_0 + x_\lambda + \Phi x_\lambda$. Then, in view of equality (5.18), we have $Hx_0 + \bar{\lambda} x_\lambda = \lambda x_0 + \lambda x_\lambda$, i.e., $(H - \lambda I)x_0 + (\bar{\lambda} - \lambda)x_\lambda = 0$. Since the operator H is Hermitian, the vectors $(H - \lambda I)x_0$ and x_λ are orthogonal, and $\operatorname{Im}\lambda \neq 0$, we conclude that $x_0 = x_\lambda = 0$. Thus, $x = 0$ and, hence, $\lambda \notin \sigma_p(A)$.

Similarly, we show that if $Ax = \bar{\lambda} x$ $(x \neq 0)$, then $(H - \bar{\lambda} I)x_0 = (\bar{\lambda} - \lambda)\Phi x_\lambda$, whence it follows that $x_0 = \Phi x_\lambda = 0$. Note that $x_\lambda \neq 0$ (since $x = 0$) and, consequently, $0 \in \sigma_p(\Phi)$.

Analogously, if $\Phi x_\lambda = 0$ $(x_\lambda \neq 0)$, then $Ax_\lambda = \bar{\lambda} x_\lambda$. Theorem 5.16 is proved. ■

5.6. Generalization of the Krein Resolvent Formula. Let H be a symmetric operator acting in a Hilbert space $\mathcal{H}$. Assume that its deficiency index (r, r) is finite ($r < \infty$) and that $\mathring{A}$ and A are arbitrary self-adjoint extensions of the operator H. These extensions are called *relatively simple* if

$$\mathcal{D}(A) \cap \mathcal{D}(\mathring{A}) = \mathcal{D}(H). \tag{5.28}$$

Krein showed [1] that, for any λ from the set

$$\rho(A, \mathring{A}) = \rho(A) \cap \rho(\mathring{A})$$

the resolvents R_λ and $\mathring{R}_\lambda$ of relatively simple self-adjoint extensions A and $\mathring{A}$ of the operator H satisfy the equality

$$R_\lambda = \mathring{R}_\lambda - \sum_{k, m = 1}^{r} (\cdot, g_m(\lambda)) p_{mk}(\lambda) g_k(\bar{\lambda}), \tag{5.29}$$

where $\{g_k(\lambda)\}_1^r$ and $\{g_k(\bar{\lambda})\}_1^r$ are bases in the defect subspaces $\mathfrak{N}_\lambda$ and $\mathfrak{N}_{\bar{\lambda}}$ of the operator H, respectively. Moreover, the matrix function $P(\lambda) = \| p_{mk}(\lambda) \|_{m,k=1}^r$ is invertible and, for any λ and μ from $\rho(A, \mathring{A})$, we have

$$P^{-1}(\lambda) = P^{-1}(\mu) + (\lambda - \mu) G(\bar{\mu}, \lambda), \tag{5.30}$$

where

$$G(\bar{\mu}, \lambda) = \| (g_k(\bar{\mu}), g_m(\lambda)) \|_{m,k=1}^r. \tag{5.31}$$

In what follows, we show that the resolvent formula (5.29) remains true in a much more general case, namely, when we do not assume that the operator H is densely defined and its regular extensions $\mathring{A}$ and A are self-adjoint. However, in the general case, the relationship between the matrices $P(\lambda)$ and $P(\mu)$ is much more complicated [see, relations (5.49) and (5.52)].

Thus, let $\mathring{A}$ and A be regular extensions of an Hermitian operator H and we do not assume that the defect numbers of the operator H are finite or equal. Let us show that

$$(\mathring{R}_\lambda - R_\lambda)x = \begin{cases} = 0 & (x \in \mathfrak{M}_\lambda), \\ \in \mathfrak{N}_{\bar{\lambda}} & (x \in \mathfrak{N}_{\bar{\mu}},\ \mu \neq \lambda). \end{cases} \tag{5.32}$$

for $\lambda \in \rho(A, \mathring{A})$, where $\mathfrak{M}_\lambda = \mathcal{R}(H - \lambda I)$.

Indeed, let $y = R_\lambda x$, where $x \in \mathfrak{M}_\lambda$. Then $x = (A - \lambda I)y$, where $y \in \mathcal{D}(H)$. Hence, $x = (H - \lambda I)y = (\mathring{A} - \lambda I)y$. Therefore,

$$(\mathring{R}_\lambda - R_\lambda)x = 0 \quad \text{as} \quad x \in \mathfrak{M}_\lambda.$$

If $x \in \mathfrak{N}_{\bar{\mu}}$, where $\mu \neq \lambda$, then, by virtue of Theorem 5.13, we have

$$g(\bar{\lambda}) = T_{\mu\lambda}x \in \mathfrak{N}_{\bar{\lambda}}. \tag{5.33}$$

Moreover,

$$T_{\mu\lambda} = (A - \mu I)(A - \lambda I)^{-1} = I + (\lambda - \mu)R_\lambda. \tag{5.34}$$

and, hence, by using relations (5.33) and (5.34), we can write

$$R_\lambda x = \frac{g(\bar{\lambda}) - x}{\lambda - \mu}. \tag{5.35}$$

Similarly, we obtain

$$\mathring{R}_\lambda x = \frac{\mathring{g}(\bar{\lambda}) - x}{\lambda - \mu}, \tag{5.36}$$

where $\mathring{g}(\bar{\lambda})$ is a vector from $\mathfrak{N}_{\bar{\lambda}}$. Consequently, in view of (5.35) and (5.36), the vector $(\mathring{R}_\lambda - R_\lambda)x$ belongs to $\mathfrak{N}_{\bar{\lambda}}$.

Consider the operator

$$\Gamma_\lambda = \mathring{R}_\lambda - R_\lambda \quad (\lambda \in \rho(\mathring{A}, A)). \tag{5.37}$$

According to relation (5.32), this operator is equal to zero on $\mathfrak{M}_\lambda$ and, for $\mu \neq \lambda$, it maps $\mathfrak{N}_{\bar{\mu}}$ into $\mathfrak{N}_{\bar{\lambda}}$.

Let $\{\lambda, \mu\} \subset \rho(\mathring{A}, A)$. Substituting the values of R_λ and R_μ from (5.37) in the functional equation

$$R_\lambda - R_\mu = (\lambda - \mu) R_\lambda) R_\mu$$

and taking into account the equality

$$\mathring{R}_\lambda - \mathring{R}_\mu = (\lambda - \mu) \mathring{R}_\lambda \mathring{R}_\mu,$$

we obtain

$$\Gamma_\lambda - \Gamma_\mu = (\lambda - \mu)\left[\mathring{R}_\lambda \Gamma_\mu + \Gamma_\lambda \mathring{R}_\mu - \Gamma_\lambda \Gamma_\mu\right].$$

In view the relation

$$\mathring{T}_{\mu\lambda} = (\mathring{A} - \mu I)(\mathring{A} - \lambda I)^{-1} = I + (\lambda - \mu)\mathring{R}_\lambda,$$

the last equality can be rewritten in the form

$$\Gamma_\lambda \mathring{T}_{\lambda\mu} = \mathring{T}_{\mu\lambda}\Gamma_\mu + (\mu - \lambda)\Gamma_\lambda \Gamma_\mu. \tag{5.38}$$

By using the relation

$$\mathring{T}_{\lambda\mu} - (\mu - \lambda)\Gamma_\mu = I + (\mu - \lambda)\mathring{R}_\mu - (\mu - \lambda)\left(\mathring{R}_\mu - R_\mu\right) = T_{\lambda\mu}, \tag{5.39}$$

one can show that equality (5.38) takes the form

$$\Gamma_\lambda T_{\lambda\mu} = \mathring{T}_{\mu\lambda}\Gamma_\mu. \tag{5.40}$$

By using the fact that the operator $T_{\lambda\mu}$ is invertible, relation (5.39), and the equality $T_{\lambda\mu}^{-1} = \left[\mathring{T}_{\lambda\mu} + (\lambda - \mu)\Gamma_\mu\right]^{-1}$, we can rewrite relation (5.40) as

$$\Gamma_\lambda = \mathring{T}_{\mu\lambda}\Gamma_\mu\left[\mathring{T}_{\lambda\mu} + (\lambda - \mu)\Gamma_\mu\right]^{-1}.$$

The last formula makes it possible to find the operator Γ_λ for all $\lambda \in \rho(\overset{\circ}{A}, A)$, provided that we know the form of this operator at some fixed point $\mu \in \rho(\overset{\circ}{A}, A)$.

In what follows, we assume that H is an Hermitian operator with deficiency index (r, r) $(r < \infty)$ and $\{\lambda, \mu\} \subset \rho(\overset{\circ}{A})$, where $\overset{\circ}{A}$ is a fixed regular extension of the operator H.

We also assume that μ $(\operatorname{Im} \mu \neq 0)$ is fixed and choose a basis $\{g_k(\bar{\mu})\}_{k=1}^r$ in the defect subspace $\mathfrak{N}_{\bar{\mu}}$ of the operator H.

Since the operator $\overset{\circ}{T}_{\mu\lambda}$ is invertible and, by virtue of Theorem 5.13, maps $\mathfrak{N}_{\bar{\mu}}$ onto $\mathfrak{N}_{\bar{\lambda}}$, we conclude that the system of vectors $\{g_k(\bar{\lambda})\}_{k=1}^r$, where

$$g_k(\bar{\lambda}) = \overset{\circ}{T}_{\mu\lambda}\, g_k(\bar{\mu}) \qquad (k = 1, \dots, r) \tag{5.41}$$

can be used as a basis in $\mathfrak{N}_{\bar{\lambda}}$. Then the operator Γ_λ mapping the entire space $\mathfrak{N}_{\bar{\mu}}$ into the defect subspace $\mathfrak{N}_{\bar{\lambda}}$ can be represented in the form

$$\Gamma_\lambda = \sum_{k=1}^{r} c_k(\cdot)\, g_k(\bar{\lambda}), \tag{5.42}$$

where $c_k(\cdot)$ are linear functionals depending on the parameter λ. In view of equality (5.32) and the fact that the system of vectors (5.41) is linearly independent, these functionals $c_k(\cdot)$ are equal to zero on vectors from $\mathfrak{M}_\lambda$.

Hence, $c_k(\cdot) = (\cdot, h_k(\lambda))$, where $h_k(\lambda) \in \mathfrak{N}_\lambda$. Therefore, we have

$$h_k(\lambda) = \sum_{m=1}^{r} \overline{p_{mk}(\lambda)}\, g_m(\lambda),$$

where $\{g_m(\lambda)\}_{m=1}^r$ is a basis in the subspace $\mathfrak{N}_\lambda$ and, thus,

$$c_k(\cdot) = \sum_{m=1}^{r} p_{mk}(\lambda)(\cdot, g_m(\lambda)). \tag{5.43}$$

By inserting (5.43) in (5.42), we obtain

$$\Gamma_\lambda = \sum_{k,m=1}^{r} (\cdot, g_m(\lambda))\, p_{mk}(\lambda)\, g_k(\bar{\lambda}). \tag{5.44}$$

By virtue of (5.37), this equality can be rewritten in the form (5.29). Thus, for regular extensions, the resolvent formula has the same form as in the case of self-adjoint extensions.

To establish the properties of the matrix function $P(\lambda) = \left\| p_{mk}(\lambda) \right\|_{m,k}^{r}$, we consider an $r \times r$ matrix

$$M = \left\| \left(\Gamma_\lambda \overset{\circ}{T}_{\lambda\mu}\, g_\alpha(\mu),\, g_\beta(\bar{\lambda}) \right) \right\|.$$

In view of equality (5.38), it can be represented as

$$M = \left\| \left(\overset{\circ}{T}_{\mu\lambda}\, \Gamma_\mu\, g_\alpha(\mu),\, g_\beta(\bar{\lambda}) \right) \right\| + (\mu - \lambda) \left\| (\Gamma_\lambda \Gamma_\mu\, g_\alpha(\mu),\, g_\beta(\bar{\lambda})) \right\|. \tag{5.45}$$

On the other hand, by virtue of equality (5.44), we get

$$M = \left\| \left(\overset{\circ}{T}_{\alpha\mu}\, g_\alpha(\mu),\, g_m(\lambda) \right) \right\| P(\lambda) \cdot G(\bar{\lambda}), \tag{5.46}$$

where $G(\bar{\lambda}) = \left\| (g_k(\bar{\lambda}), g_m(\bar{\lambda})) \right\|$ is the Gramm matrix of basis vectors $\left\{ g_k(\bar{\lambda}) \right\}_{k=1}^{r}$. Since $\overset{\circ}{T}{}^{*}_{\lambda\mu}$ maps $\mathfrak{N}_\lambda$ into $\mathfrak{N}_\mu$, we have

$$\overset{\circ}{T}{}^{*}_{\lambda\mu}\, g_m(\lambda) = \sum_{j=1}^{r} t_{mj}\, g_j(\mu). \tag{5.47}$$

Moreover,

$$\left\| \left(\overset{\circ}{T}_{\lambda\mu}\, g_\alpha(\mu),\, g_m(\lambda) \right) \right\| = \left\| \sum_{j=1}^{r} \left(g_\alpha(\mu), g_j(\mu) \right) t^{*}_{jm} \right\| = G(\mu) T^{*}, \tag{5.48}$$

where $t^{*}_{jm} = \bar{t}_{mj}$ and the elements of the matrix $T = \| t_{mj} \|$ depend on λ and μ. Inserting (5.48) in (5.46), we obtain

$$M = G(\mu)\, T^{*} P(\lambda)\, G(\bar{\lambda}). \tag{5.49}$$

To transform the first term on the right-hand side of (5.45), we note that relations (5.41) and (5.44) imply that

$$\mathring{T}_{\mu\lambda}\Gamma_\mu g_\alpha(\mu) = \sum_{k,m=1}^{r} (g_\alpha(\mu), g_m(\mu))\, p_{mk}(\mu)\, g_k(\bar{\lambda}).$$

Then

$$\left\|\left(\mathring{T}_{\mu\lambda}\Gamma_\mu g_\alpha(\mu), g_\beta(\bar{\lambda})\right)\right\| = G(\mu)P(\mu)G(\bar{\lambda}). \tag{5.50}$$

Similarly, by using (5.44), we obtain

$$\Gamma_\lambda\Gamma_\mu g_\alpha(\mu) = \sum_{k,m=1}^{r} (g_\alpha(\mu), g_m(\mu))\, p_{mk}(\mu)\Gamma_\lambda g_k(\bar{\mu})$$

$$= \sum_{k,m,s,j=1}^{r} (g_\alpha(\mu), g_m(\mu))\, p_{mk}(\mu)\big(g_k(\bar{\mu}), g_j(\lambda)\big)p_{js}(\lambda)g_s(\bar{\lambda}).$$

This yields

$$\left\|(\Gamma_\lambda\Gamma_\mu g_\alpha(\mu), g_\beta(\bar{\lambda}))\right\| = G(\mu)P(\mu)G(\bar{\mu},\lambda)P(\lambda)G(\bar{\lambda}), \tag{5.51}$$

where

$$G(\bar{\mu},\lambda) = \left\|\big(g_k(\bar{\mu}), g_j(\lambda)\big)\right\|. \tag{5.52}$$

We substitute (5.49)–(5.51) in (5.45) and multiply the equality obtained as a result from the left and from the right by $G^{-1}(\mu)$ and $G^{-1}(\bar{\lambda})$, respectively. This gives

$$T^* P(\lambda) = P(\mu) + (\mu-\lambda)P(\mu)G(\bar{\mu},\lambda)P(\lambda). \tag{5.53}$$

Thus, the values of the matrix functions $P(\lambda)$ and $P(\mu)$ for any two points λ and μ from the set $\rho(\mathring{A}, A)$ satisfy equality (5.53).

Let us determine the structure of the matrix T under the additional assumption that $\{\lambda, \bar{\mu}\} \subset \rho(\mathring{A})$. Under this condition, by virtue of (5.41), we can write

$$g_k(\lambda) = \mathring{T}_{\mu\bar{\lambda}}\, g_k(\bar{\mu}) \quad \text{and} \quad g_n(\mu) = \mathring{T}_{\mu\bar{\mu}}\, g_n(\bar{\mu}). \tag{5.54}$$

By multiplying equality (5.47) from the right by the vector $g_n(\mu)$, we get

$$\left\| \left(g_k(\lambda), \overset{\circ}{T}_{\lambda\mu}\, g_n(\mu) \right) \right\| = T\,G(\mu)\,. \tag{5.55}$$

Hence, in view of the equality $\overset{\circ}{T}_{\lambda\mu} = I + (\mu - \lambda)\, \overset{\circ}{R}_\mu$, we obtain

$$\overset{\circ}{T}_{\lambda\mu} g_n(\mu) = g_n(\mu) + (\mu - \lambda)\, \overset{\circ}{R}_\mu g_n(\mu). \tag{5.56}$$

By applying the operator $\overset{\circ}{T}_{\bar\mu\mu} = I + (\mu - \bar\mu)\, \overset{\circ}{R}_\mu$ to both sides of the second inequality (5.54), we get

$$g_n(\bar\mu) = \overset{\circ}{T}_{\bar\mu\mu} g_n(\mu) = g_n(\mu) + (\mu - \bar\mu)\, \overset{\circ}{R}_\mu\, g_n(\mu),$$

whence

$$\overset{\circ}{R}_\mu g_n(\mu) = \frac{g_n(\bar\mu) - g_n(\mu)}{\mu - \bar\mu}.$$

By using this equality, we rewrite relation (5.56) in the form

$$T_{\lambda\mu} g_n(\mu) = \frac{\mu - \lambda}{\bar\mu - \mu} g_n(\mu) + \frac{\mu - \lambda}{\mu - \bar\mu} g_n(\bar\mu). \tag{5.57}$$

Taking into account (5.55) and (5.57), we arrive at the relation

$$T = \left[\frac{\mu - \bar\lambda}{\mu - \bar\mu} G(\lambda, \mu) + \frac{\bar\mu - \bar\lambda}{\bar\mu - \mu} G(\lambda, \bar\mu) \right] G^{-1}(\mu). \tag{5.58}$$

Finally, consider the case where the operator $\overset{\circ}{A}$ is self-adjoint. Let us show that, under this condition, T is the identity matrix. Indeed, if $\overset{\circ}{A}{}^* = \overset{\circ}{A}$, then $\overset{\circ}{T}{}^*_{\lambda\mu} = \overset{\circ}{T}_{\bar\lambda\bar\mu}$. Hence, it follows from (5.47) that

$$\overset{\circ}{T}_{\bar\lambda\bar\mu} g_m(\lambda) = \sum_{j=1}^{r} t_{mj} g_j(\mu), \tag{5.59}$$

where, in view of equality (5.54),

$$\overset{\circ}{T}_{\bar\lambda\bar\mu} g_m(\lambda) = \overset{\circ}{T}_{\bar\lambda\bar\mu}\, \overset{\circ}{T}_{\mu\bar\lambda}\, g_m(\bar\mu) = \overset{\circ}{T}_{\mu\bar\mu}\, g_m(\bar\mu) = g_m(\mu)\,. \tag{5.60}$$

Therefore, by using (5.59) and (5.60), we conclude that $T = E$.

5.7. Relatively Simple Extensions. As in the case of self-adjoint extensions, the regular extensions A and $\mathring{A}$ of an Hermitian operator H are called relatively simple if they satisfy equality (5.28).

Let us show that regular relatively simple extensions of an operator H satisfy the equality

$$\operatorname{Ker}\Gamma_\lambda = \mathfrak{M}_\lambda. \tag{5.61}$$

Indeed, it follows from (5.32) that $\mathfrak{M}_\lambda \subset \operatorname{Ker}\Gamma_\lambda$. Let us prove that, in the considered case, the inverse inclusion is also true. Let $\Gamma_\lambda = \mathring{R}_\lambda x - R_\lambda x = 0$. Then the vector $y = R_\lambda x = \mathring{R}_\lambda x$ belongs to the domains of definition of the operators A and $\mathring{A}$. Since A and $\mathring{A}$ are relatively simple extensions of the operator H, we have $y \in \mathcal{D}(H)$. Hence,

$$x = (\mathring{A} - \lambda I)y = (H - \lambda I)y \in \mathfrak{M}_\lambda$$

and equality (5.61) is proved.

If H is an Hermitian operator with deficiency index (r, r) $(r < \infty)$, then the invertibility of the matrix $P(\lambda)$ for $\lambda \in \rho(\mathring{A}, A)$ follows from equalities (5.44) and (5.61) and the fact that the system of vectors $\{g_k(\bar{\lambda})\}_{k=1}^{r}$ is linearly independent. Therefore, equality (5.53) can be rewritten as

$$P^{-1}(\lambda) = P^{-1}(\mu)T^* + (\lambda - \mu)G(\bar{\mu}, \lambda). \tag{5.62}$$

In particular, if $\mathring{A}^* = \mathring{A}$, then, as shown above, $T = E$. Hence, in this case equality (5.62) coincides with equality (5.30).

In conclusion, we note that if the operators $\mathring{A}$ and A are self-adjoint, then $\mathring{R}_\lambda^* = \mathring{R}_{\bar{\lambda}}$ and $R_\lambda^* = R_{\bar{\lambda}}$ and, hence, $\Gamma_\lambda^* = \Gamma_{\bar{\lambda}}$. Taking into account the expression for the operator Γ_λ [equality (5.44)], we conclude that self-adjoint extensions of the Hermitian operator H satisfy the equality

$$P^*(\lambda) = P(\bar{\lambda}). \tag{5.63}$$

The converse assertion is true in the following sense: If equality (5.63) is true and one of the extensions $\mathring{A}$ or A of the operator H is self-adjoint, then the second extension is also self-adjoint.

Remarks. The notion of regular extensions of Hermitian operators was introduced and studied for the first time by A. Kuzhel in [2] (see also [7]).

The von Neumann formulas were generalized to the case of self-adjoint Hermitian operators by Krasnosel'skii [1]. Later, these formulas, under different restrictions, were generalized to various classes of nonself-adjoint extensions of Hermitian operators (see, e.g., Tsekanovskii and Shmulyan [1]). A. Kuzhel and Rudenko [1, 2] extended the von Neumann formulas to the case of regular extensions of Hermitian operators. Note that any linear operator A is a regular extensions of its Hermitian part A_0 and, thus, the von Neumann formulas can actually be applied to any linear operator whose point spectrum does not fill the entire plane $(\sigma_p(A) \neq \mathbb{C})$.

The Krein[5] formula (Krein [1]) plays an important role in the description of self-adjoint extensions of symmetric operators, in the construction of models in the theory of extensions to broader space, and in scattering theory (see, e.g., Albeverio, Gesztesy, Høegh-Krohn, and Holden [1], Koshmanenko [1], Pavlov and Shushkov [1], Adamjan and Pavlov [1]). A. Kuzhel generalized the Krein formula to the case of regular extensions of Hermitian operators [3].

6. Dissipative Extensions of Hermitian Operators

We now recall some general properties of dissipative operators. In particular, we show that any dissipative extension of an Hermitian operator is, at the same time, a regular extension of this operator (an analog of the Phillips lemma). We establish the conditions under which a dissipative operator is maximal. By using the von Neumann formulas, we prove the condition of dissipativity of linear operators.

6.1. Dissipative Operator. An operator A acting in a Hilbert space $\mathcal{H}$ is called *dissipative* if

$$\operatorname{Im}(Ax, x) \geq 0 \qquad (\forall x \in \mathcal{D}(A)). \tag{6.1}$$

In particular, an operator A with $\mathcal{D}(A^*) = \mathcal{D}(A)$ is dissipative if and only if

$$\operatorname{Im} A = \frac{A - A^*}{2i} \geq 0. \tag{6.2}$$

Note that, unlike the commonly used approach, in the general case, we do not assume that the dissipative operator is densely defined.

[5] In the work of Naimark [3] which appeared earlier than the work of Krein, the same formula was obtained in a somewhat different form.

Let us now formulate several propositions whose proofs can be found in the monograph by A. Kuzhel [7].

Proposition 6.1. *If* A *is a dissipative operator and* $\operatorname{Im}\lambda < 0$, *then* $\lambda \notin \sigma_p(A)$ *and*

$$\|(A-\lambda I)^{-1}\| \leq \frac{1}{|\operatorname{Im}\lambda|}. \tag{6.3}$$

Proposition 6.2. *If* A *is a dissipative operator and* $\operatorname{Im}\lambda < 0$, *then the lineals* $\mathcal{R}(A-\lambda I)$ *and* $\mathcal{D}(A)^{\perp}$ *are linearly independent.*

Proposition 6.3. If A *is a dissipative N-operator with nonempty resolvent set and* $\operatorname{Im}\lambda < 0$, *then* $\lambda \in \rho(A)$.

Proposition 6.4. *If a closed dissipative operator* A *has at least one regular point in the lower half plane, then all points of this half plane belong to the resolvent set* $\rho(A)$.

Proposition 6.5. (Phillips lemma). *If* A *is a dissipative extension of a symmetric operator* H, *then* $H \subset A \subset H^*$.

Proposition 6.6. (generalization of the Phillips lemma). *Any dissipative extension* A *of an Hermitian operator* H *is a regular extension of the operator* H *(i.e.,* $A \in \mathcal{P}(A)$*).*

Proposition 6.7. (analog of the Sz.-Nagy–Foias theorem on triangulation of contractions). *Let* $\mathcal{H}_1$ *be an invariant subspace of a bounded dissipative operator* A *acting in a Hilbert space* $\mathcal{H}$ ($\mathcal{D}(A) = \mathcal{H}$) *and let* $\mathcal{H}_2 = \mathcal{H} \ominus \mathcal{H}_1$. *Then the operator* A *can be represented in the form*

$$A = \begin{pmatrix} A_1 & Q_{A_1} L Q_{A_2} \\ 0 & A_2 \end{pmatrix}, \tag{6.4}$$

where $A_1 = A|_{\mathcal{H}_1}$ *and* $A_2 = P_2 A|_{\mathcal{H}_2}$ *are, respectively, dissipative operators acting in the subspaces* $\mathcal{H}_1$ *and* $\mathcal{H}_2$, P_2 *is the orthoprojector onto* $\mathcal{H}_2$ *in* $\mathcal{H}$, $Q_{A_k} = (\operatorname{Im} A_k)^{1/2}$, *and* L ($\|L\| \leq 2$) *is a linear operator which maps the lineal* $Q_{A_2}\mathcal{H}_2$ *into the subspace* $\mathcal{H}_1$.

Proposition 6.8. *Suppose that the operator A is defined in the space $\mathcal{H} = \mathcal{H}_1 \oplus \mathcal{H}_2$ by equality (6.4), where A_1 and A_2 are bounded dissipative operators acting in the subspaces $\mathcal{H}_1$ and $\mathcal{H}_2$, respectively. Moreover, assume that $L : \mathcal{H}_2 \to \mathcal{H}_1$ is a linear operator and $\|L\| \leq 2$. Then the operator A is dissipative.*

Proposition 6.9. *Any densely defined dissipative operator is closable.*

Proof. Let A be a dissipative operator acting in $\mathcal{H}$, let $\overline{\mathcal{D}(A)} = \mathcal{H}$, and let $x_n \to 0$ and $Ax_n \to y$, where $\{x_n\}$ is a sequence of elements from $\mathcal{D}(A)$. Note that, for any $x \in \mathcal{D}(A)$ and $\alpha \in \mathbb{C}$, we have

$$\operatorname{Im}(A(x+\alpha x_n), x+\alpha x_n) \geq 0.$$

Therefore, for $n \to \infty$, we obtain

$$\operatorname{Im}(Ax, x) + \operatorname{Im}[\alpha(y, x)] \geq 0 \qquad (\forall\, \alpha \in \mathbb{C}).$$

This is possible only under the condition that $(y, x) = 0$ $(\forall\, x \in \mathcal{D}(A))$. Hence, $y = 0$. Thus, the operator A is closed.

■

Note that, in the case of nondensely defined dissipative operators, Proposition 6.9 is not always true. Indeed, consider the space l_2 and an operator A defined on the lineal

$$\mathcal{D}(A) = \left\{ x \in l_2 \,\middle|\, x = (0, x_1, x_2, \dots),\ \sum_{k=1}^{\infty} |x_k| < \infty \right\}$$

by the equality

$$A(0, x_1, x_2, \dots) = \left(\sum_{k=1}^{\infty} x_k, 0, 0, \dots \right).$$

Clearly, $(Ax, x) = 0$ for all $x \in \mathcal{D}(A)$ and, it is easy to see that the operator A is both dissipative and Hermitian.

By analogy with Example 1.1, we can show that the operator A is neither closed nor closable.

6.2. Condition of Dissipativity of Linear Operators in Terms of the von Neumann Formulas. Let A be a dissipative extension of an Hermitian operator H and let λ

($\operatorname{Im}\lambda < 0$) be a fixed number. The fact that $\lambda \notin \sigma_p(A)$ and Proposition 1.6 imply that $A \in \mathcal{P}(H)$ and, hence, in view of Theorem 5.15, we conclude that the vectors x and Ax can be represented in the form

$$x = x_0 + x_\lambda + \Phi x_\lambda, \qquad Ax = Hx_0 + \bar{\lambda} x_\lambda + \lambda \Phi x_\lambda, \qquad x \in \mathcal{D}(A), \tag{6.5}$$

where Φ is a linear operator acting from $\mathfrak{N}_\lambda$ and $\mathfrak{N}_{\bar{\lambda}}$. By using the equalities

$$(Hx_0, x_\lambda) = \lambda(x_0, x_\lambda) \quad \text{and} \quad (Hx_0, \Phi x_\lambda) = \bar{\lambda}(x_0, \Phi x_\lambda),$$

we get

$$(Ax, x) = (Hx_0, x_0) - i \operatorname{Im}\lambda \left[\|x_\lambda\|^2 - \|\Phi x_\lambda\|^2 \right]$$

$$+ 2 \operatorname{Re}\left[\lambda(x_0, x_\lambda) + \lambda(\Phi x_\lambda, x_0) + \lambda(\Phi x_\lambda, x_\lambda)\right] + \operatorname{Re}\lambda \left[\|x_\lambda\|^2 + \|\Phi x_\lambda\|^2 \right].$$

This yields

$$\operatorname{Im}(Ax, x) = |\operatorname{Im}\lambda| \left[\|x_\lambda\|^2 - \|\Phi x_\lambda\|^2 \right]. \tag{6.7}$$

Thus, if A is a dissipative extension of the operator H, then, by virtue of (6.7), the operator Φ is a contractive operator acting from $\mathfrak{N}_\lambda$ to $\mathfrak{N}_{\bar{\lambda}}$.

Conversely, if A is a regular extension of the Hermitian operator H, $\operatorname{Im}\lambda < 0$, $\lambda \notin \sigma_p(A)$, and the operator Φ in equalities (6.5) is a contraction, then, in view of (6.7), we easily conclude that

$$\operatorname{Im}(Ax, x) \geq 0 \qquad (\forall x \in \mathcal{D}(A)).$$

Therefore, the operator A is dissipative.

Note that, in some works, the authors consider a dissipative operator A_λ defined, for fixed λ ($\operatorname{Im}\lambda < 0$), on the lineal $\mathcal{D}(A_\lambda) = \mathcal{D}(H) \dotplus \mathfrak{N}_\lambda$ by the equality

$$A_\lambda(x_0 + x_\lambda) = Hx_0 + \bar{\lambda} x_\lambda, \tag{6.8}$$

where H is a symmetric operator and $\mathfrak{N}_\lambda$ is its defect subspace. The operator A_λ defined above is a special case of the operator A defined by equalities (6.5) (with $\Phi = 0$).

The operator A_λ defined by equality (6.8) is also meaningful for an Hermitian operator H. Indeed, by virtue of Corollary 3.4, the lineals $\mathcal{D}(H)$ and $\mathfrak{N}_\lambda$ are linearly independent and, thus, one can define the operator A_λ on the lineal $\mathcal{D}(A_\lambda) = \mathcal{D}(H) \dotplus \mathfrak{N}_\lambda$ by equality (6.8). Clearly, in this case, the operator A_λ is also a dissipative extension of the operator H.

6.3. Maximal Dissipative Operators. A dissipative operator A is called *maximal* if all its dissipative extensions are trivial (i.e., coincide with A).

Theorem 6.10. *A closed dissipative operator A acting in $\mathcal{H}$ is a maximal dissipative operator if and only if the equality*

$$\mathcal{R}(A - \lambda I) = \mathcal{H} \tag{6.9}$$

holds for all λ $(\operatorname{Im}\lambda < 0)$.

Proof. Let A be a maximal closed dissipative operator acting in $\mathcal{H}$. In view of inequality (6.3), an arbitrary point λ from the lower half plane is a point of regular type for the operator A and, hence, it follows from Theorem 2.1 that the lineal $\mathcal{R}(A - \lambda I)$ is closed. Assume that $\mathcal{R}(A - \lambda I) \neq \mathcal{H}$ and $\mathcal{H}_\lambda = \mathcal{H} \ominus \mathcal{R}(A - \lambda I)$. Let us show that the lineals $\mathcal{D}(A)$ and $\mathcal{H}_\lambda$ are linearly independent. Indeed, if $x \in \mathcal{D}(A) \cap \mathcal{H}_\lambda$, then

$$(Ax, x) = \lambda (x, x)$$

and, therefore,

$$\operatorname{Im}(Ax, x) = (\operatorname{Im}\lambda)\, \|x\|^2.$$

This is possible only for $x = 0$ (since $\operatorname{Im}\lambda < 0$). Thus, the lineals $\mathcal{D}(A)$ and $\mathcal{H}_\lambda$ are linearly independent.

Consider an extension B of the operator A defined on the lineal

$$\mathcal{D}(B) = \mathcal{D}(A) \dotplus \mathfrak{N}_\lambda$$

by the equality

$$B(x + y) = Ax + \bar{\lambda} y \quad (x \in \mathcal{D}(A),\ y \in \mathcal{H}_\lambda). \tag{6.10}$$

Then, by using the equality $(Ax, y) = \lambda (x, y)$, we obtain

$$(B(x + y),\ x + y) = (Ax, x) + 2\operatorname{Re}[\lambda (x, y)] + \bar{\lambda}\, \|y\|^2$$

and, therefore,

$$\operatorname{Im}(B(x+y),\ x+y) = \operatorname{Im}(Ax,x) + (\operatorname{Im}\bar{\lambda})\,\|y\|^2 \geq 0,$$

i.e., the operator B is a nontrivial dissipative extension of the operator A. We arrive at a contradiction. This means that equality (6.9) is true.

Conversely, let A be a dissipative operator and let equality (6.9) be true. Assume that a nontrivial dissipative extension B of the operator A exists and $y = (B-\lambda I)x$, where $x \in \mathcal{D}(B)$ and $x \notin \mathcal{D}(A)$. It follows from (6.9) that $\mathcal{D}(A)$ contains vector x_0 such that

$$y = (A-\lambda I)x_0 = (B-\lambda I)x.$$

Hence, $(B-\lambda I)(x-x_0) = 0$, where $x - x_0 \neq 0$ and we conclude that $\lambda \in \sigma_p(B)$. This contradicts Proposition 6.1. Thus, under condition (6.9), the operator A is a maximal dissipative operator.

■

Corollary 6.11. *A closed dissipative operator A is maximal if and only if all points from the lower half plane belong to its resolvent set.*

Proof. Indeed, by virtue of Proposition 6.1, a point λ $(\operatorname{Im}\lambda < 0)$ may belong only to the residual spectrum of the dissipative operator A. For maximal dissipative operators, this situation is impossible by virtue of equality (6.9).

■

Corollary 6.12. *Any closed dissipative operator A is either maximal or can be extended to a maximal dissipative operator.*

Proof. If an operator A is not maximal, then, in view of Theorem 6.10, we get $\mathcal{R}(A-\lambda I) \neq \mathcal{H}$. Hence, as shown above, the operator B defined by equality (6.10) is a dissipative extension of the operator A. Moreover,

$$\mathcal{R}(B-\lambda I) = \mathcal{R}(A-\lambda I) \oplus \mathcal{H}_\lambda = \mathcal{H}$$

and, thus, the operator B is a maximal dissipative extension of the operator A.

■

Corollary 6.13. *Any closed Hermitian operator can be extended to a maximal dissipative operator.*

Proof. It suffices to use Corollary 6.12 because any Hermitian operator is dissipative.

■

Note that, in view of Theorem 6.10, the operator A_λ defined by equality (6.8) is a maximal dissipative extension of the symmetric operator H. This is also true in the case where H is a closed Hermitian operator.

6.4. Operator Adjoint to a Dissipative Operator. If A is a bounded dissipative operator or a dissipative operator with $\mathcal{D}(A) = \mathcal{D}(A^*)$, then it is easy to show that the operator $-A^*$ is also dissipative. As follows from the results of Phillips [1], the indicated property of the operator $-A^*$ remains valid if $\mathcal{D}(A) \neq \mathcal{D}(A^*)$.

A. Kuzhel [7] proved the dissipativity of the operator $-A^*$ by using the properties of the auxiliary operator $T_{\mu\lambda}$. Here, we present another proof of this assertion based on the use of the generalized von Neumann formulas.

Theorem 6.14. *If A is a closed maximal dissipative operator, then the operator $-A^*$ is also a maximal dissipative operator.*

Proof. Assume that an operator A satisfies the conditions of the theorem and λ is a fixed point in the lower half plane. Then, by virtue of Propositions 6.2, 6.10, and 6.11, we conclude that $\lambda \in \rho(A)$ and the operator A is densely defined. If, in this case, A_0 is the Hermitian part of the operator A, then $A \in \mathcal{P}(A_0)$ and, hence, in view of Theorem 5.15, the vectors x and Ax $(x \in \mathcal{D}(A))$ admit the representation

$$x = x_0 + x_\lambda + \Phi x_\lambda, \qquad Ax = A_0 x_0 + \bar{\lambda} x_\lambda + \lambda \Phi x_\lambda, \tag{6.11}$$

where $x_0 \in \mathcal{D}(A_0)$, $x_\lambda \in \mathcal{D}(\Phi)$, and the operator Φ acts from $\mathfrak{N}_\lambda$ into $\mathfrak{N}_{\bar{\lambda}}$. Here, $\mathfrak{N}_\lambda$ and $\mathfrak{N}_{\bar{\lambda}}$ are the defect subspaces of the operator A_0. Moreover, since the operator A is maximal, we have

$$\mathcal{H} = \mathcal{R}(A - \lambda I) = \mathcal{R}(A_0 - \lambda I) \oplus \mathcal{D}(\Phi),$$

whence it follows that $\mathcal{D}(\Phi) = \mathfrak{N}_\lambda$.

Further, since $\bar{\lambda} \in \rho(A^*)$, by using Theorem 5.15, we conclude that the vectors y and $A^* y$ $(y \in \mathcal{D}(A^*))$ admit the representation

$$y = y_0 + y_{\bar{\lambda}} + \Phi^* y_{\bar{\lambda}}, \qquad A^* y = A_0 y_0 + \lambda y_{\bar{\lambda}} + \bar{\lambda} \Phi^* y_{\bar{\lambda}}, \tag{6.12}$$

where Φ^* maps $\mathfrak{N}_{\bar{\lambda}}$ into $\mathfrak{N}_\lambda$. By analogy with the proof of equality (6.7), we can show that

$$\operatorname{Im}(A^* y, y) = |\operatorname{Im}\lambda| \left[\|\Phi^* y_{\bar{\lambda}}\|^2 - \|y_{\bar{\lambda}}\|^2 \right]. \tag{6.13}$$

Since the operator A is dissipative, it is easy to see that, as shown in Subsection 6.2, the operator Φ is a contraction from $\mathfrak{N}_\lambda$ into $\mathfrak{N}_{\bar{\lambda}}$. Therefore, Φ^* is a contraction from $\mathfrak{N}_{\bar{\lambda}}$ into $\mathfrak{N}_\lambda$ and, thus, by virtue of (6.13), we obtain

$$\operatorname{Im}(A^* y, y) \le 0 \qquad (\forall\, y \in \mathcal{D}(A^*)). \tag{6.14}$$

Hence, $\operatorname{Im}(-A^* y, y) \ge 0$.

■

*6.5. **Other Results.*** In numerous works, meaningful model operators are constructed for a broad class of dissipative operators, which enables one to reduce the problem of investigation of an abstract dissipative operator to the problem of investigation of its model in a certain functional space. In particular, this method makes it possible to establish a criterion of completeness for the collection of root vectors for a some class of dissipative operators. In the case of bounded operators, similar results can be found in Lifshits [1]. If A is a simple dissipative operator with finite-dimensional (or nuclear) imaginary component

$$A_J = \frac{A - A^*}{2i},$$

then

$$\sum_{k=1}^{N} \operatorname{Im}\mu_k \le \operatorname{tr} A_J, \tag{6.15}$$

where $\{\mu_k\}_{k=1}^N$ $(N \le \infty)$ are all (nonreal) eigenvalues of the operator A and $\operatorname{tr} A_J$ is the trace of the operator A_J. (An operator A is called simple if it does not induce a self-adjoint operator in any nontrivial subspace.) Further, the collection of root vectors of an operator A is complete if and only if relation (6.15) is true as an equality.

Similar results for the case of unbounded dissipative operators can be found in A. Kuzhel [7].

Self-adjoint dilations are very important objects for the investigation of dissipative operators. Recall that a linear operator B acting in a Hilbert space $\tilde{\mathcal{H}}$ is called a *dila-*

tion of a bounded linear operator A acting in a Hilbert space $\mathcal{H}$ if $\mathcal{H}$ is a subspace of the space $\tilde{\mathcal{H}}$ and

$$A^n = PA^n|_{\mathcal{H}} \quad (\forall n \in \mathbb{N}), \tag{6.16}$$

where P is the orthoprojector onto $\mathcal{H}$ in $\tilde{\mathcal{H}}$.

Condition (6.16) is equivalent to the following condition:

$$(A - \lambda I)^{-1} = P(B - \lambda I)^{-1}|_{\mathcal{H}} \quad (\lambda \in \rho(A) \cap \rho(B)).$$

The last condition can be regarded as a definition of dilations for unbounded operators.

The construction of self-adjoint dilations of maximal dissipative operators and the relevant bibliography can be found in A. Kuzhel [7].

Remarks. In some works (see, e.g., Phillips [1] and Balakrishan [1]), an operator L is called dissipative if

$$\operatorname{Re}(Lx, x) \le 0 \quad (\forall x \in \mathcal{D}(L)). \tag{6.17}$$

Clearly, an operator L is dissipative in a sense of (6.17) if and only if the operator $A = -iL$ is dissipative in a sense of the definition from Subsection 6.1.

An operator A is called *accumulative* if

$$\operatorname{Im}(Ax, x) \le 0 \quad (\forall x \in \mathcal{D}(A)).$$

Thus, if A is an accumulative operator, then $-A$ is a dissipative operator.

7. Accretive Operators

In the present section, we consider general properties of accretive operators.

7.1. Definition and Properties. A linear (not only densely defined) operator A is called *accretive* if

$$\operatorname{Re}(Ax, x) \ge 0 \quad (\forall x \in \mathcal{D}(A)). \tag{7.1}$$

In particular, the operator A with $\mathcal{D}(A^*) = \mathcal{D}(A)$ is accretive if and only if

$$\operatorname{Re} A = \frac{A + A^*}{2} \geq 0.$$

It follows from (6.1) and (7.1) that an operator A is accretive if and only if the operator iA is dissipative. This is why many properties of accretive operators immediately follow from the properties of dissipative operators.

At the same time, in what follows, we show that some properties of dissipative operators are not necessarily true for accretive operators (e.g, the Phillips lemma and its generalization). Conversely, some properties of accretive operators are not necessarily properties of dissipative operators.

Consider the following properties similar for accretive and dissipative operators:

Proposition 7.1. *Let* A *be an accretive operator and* $\operatorname{Re}\lambda < 0$. *Then*

(i) $\lambda \notin \sigma_p(A)$ *and*

$$\|(A - \lambda I)^{-1}\| \leq \frac{1}{|\operatorname{Re}\lambda|}; \tag{7.2}$$

(ii) the lineals $\mathcal{R}(A - \lambda I)$ and $\mathcal{D}(A)^{\perp}$ *are linearly independent.*

Proposition 7.2. *If inequality (7.2) is true for all negative* λ, *then the operator* A *is accretive.*

Proof. Let $\lambda < 0$, $x \in \mathcal{D}(A)$, and $y = (A - \lambda I)x$. Then, by using inequality (7.2), we obtain

$$\|x\| = \|(A - \lambda I)^{-1} y\| \leq \frac{1}{|\lambda|}\|y\|$$

and, hence, $|\lambda|\,\|x\| \leq \|(A - \lambda I)x\|$. By squaring both sides of the last inequality, we obtain

$$\lambda^2\|x\|^2 \leq \|Ax\|^2 + 2|\lambda|\operatorname{Re}(Ax, x) + \lambda^2\|x\|^2,$$

i.e.,

$$\|Ax\|^2 + 2|\lambda|\operatorname{Re}(Ax, x) \geq 0 \tag{7.3}$$

Since $|\lambda|$ is an arbitrary positive number, inequality (7.3) holds only under the condition that $\mathrm{Re}\,(Ax,x) \geq 0$.

■

Proposition 7.3. *If a closed accretive operator has at least one regular point in the half plane* $\mathrm{Re}\,\lambda < 0$, *then it is densely defined (and, therefore, it is an N-operator).*

Proof. Let λ $(\mathrm{Re}\,\lambda < 0)$ be a regular point of a closed accretive operator A acting in the space $\mathcal{H}$. Then $\mathcal{R}(A - \lambda I) = \mathcal{H}$ and, hence, according to statement (ii) in Proposition 7.1, we get $\mathcal{D}(A)^{\perp} = \{0\}$.

■

Proposition 7.4. *Any densely defined accretive operator is closable.*

Proof. If A is an accretive operator, then the operator iA is dissipative. Thus, it suffices to apply Proposition 6.9.

■

As an important property of dissipative operators, one can mention the fact that any dissipative extension of an Hermitian operator is a regular extension of this operator (and, therefore, these extensions satisfy the von Neumann formulas). For accretive extensions, this assertion is not true. In other words an accretive extension of an Hermitian operator H is not always a regular extension of this operator.

Indeed, consider the space $\mathcal{H} = \langle l_1, l_2 \rangle$ with orthonormal basis $\{l_1, l_2\}$ and an Hermitian operator H acting in this space according to the formula

$$Hl_1 = al_1 + bl_2 \quad (a \in \mathcal{R},\ b \in \mathbb{C}), \qquad \mathcal{D}(H) = \langle l_1 \rangle.$$

Any extension A of the operator H is defined by the equalities

$$Al_1 = Hl_1 \quad \text{and} \quad Al_2 = pl_1 + ql_2, \tag{7.4}$$

where p and q are arbitrary fixed complex numbers.

Let $x = x_1 l_1 + x_2 l_2$. Then

$$(Ax, x) = a|x_1|^2 + px_2\bar{x}_1 + bx_1\bar{x}_2 + q|x_2|^2. \tag{7.5}$$

This means that the operator A is dissipative if and only if $p = \bar{b}$ and $\mathrm{Im}\,q \geq 0$. Under these conditions, the operator A is a regular extension of the operator H. Thus, according to Proposition 6.6, the dissipative operator A belongs to $\mathcal{P}(H)$.

On the other hand, if $a = q = 1$ and $b = p = i$, then, in view of equality (7.5), we obtain

$$\operatorname{Re}(Ax, x) = \|x\|^2 \geq 0.$$

Therefore, in the considered case, the operator A defined by equalities (7.4) is an accretive extension of the operator H but $A \notin \mathcal{P}(H)$ (since $p \neq \bar{b}$).

7.2. Maximal Accretive Operators. An accretive operator A is called *maximal accretive* (or, according to Kato [1], *m*-accretive) if all its accretive extensions are trivial (i.e., coincide with A).

Taking into account the relationship between accretive and dissipative operators, we arrive at the following assertions:

Proposition 7.5. *A closed accretive operator A acting in $\mathcal{H}$ is maximal if and only if the equality $\mathcal{R}(A - \lambda I) = \mathcal{H}$ holds for all λ $(\operatorname{Re}\lambda < 0)$.*

Corollary 7.6. *A closed accretive operator A is maximal if and only if all points from the left open half plane belong to the resolvent set of this operator.*

Theorem 7.7. *If A is a closed maximal accretive operator, then the operator A^* is also maximal accretive operator.*

Proof. Indeed, if an operator A satisfies the condition of the theorem, then the operator $T = iA$ is a maximal dissipative operator. To complete the proof, it remains to use the equalities $A = -iT$ and $A^* = -i(-T^*)$ and Theorem 6.10.

■

7.3. Sectorial (θ-Sectorial) Operators. If A is an accretive operator, then its number domain

$$M(A) = \{(Ax, x) \mid x \in \mathcal{D}(A), \ \|x\| = 1\}$$

belongs to the right half plane $\operatorname{Re}\lambda \geq 0$, i.e., to the sector $|\arg\lambda| \leq \pi/2$. In this connection, it is natural to introduce the following concept (Kato [1]): an accretive operator is called *sectorial (or θ-sectorial)* if its number domain is a subset of the sector

$$|\arg\lambda| \leq \theta < \frac{\pi}{2}. \tag{7.6}$$

For $\theta = 0$, sectorial operators turn into (nonnegative) Hermitian operators.

If $\theta \neq 0$, then the fact that A is a sectorial operator means that

$$\operatorname{Re}(Ax, x) \geq \cot\theta \, |\operatorname{Im}(Ax, x)| \tag{7.7}$$

for any $x \in \mathcal{D}(A)$.

Note that any θ-sectorial operator is an accretive operator. Therefore, as in Subsection 7.2, we can introduce the notion of a *maximal* θ*-sectorial operator.*

Theorem 7.8. *The point and continuous spectra of a* θ*-sectorial operator belong to sector (7.6). Furthermore, if* λ *does not belong to sector (7.6), then*

$$\|(A - \lambda I)^{-1}\| \leq \frac{1 + \cot\theta}{|\operatorname{Im}\lambda|\cot\theta - \operatorname{Re}\lambda}. \tag{7.8}$$

Proof. If λ does not belong to sector (7.6), then, obviously, $\lambda \notin \sigma_p(A)$ and we have

$$\operatorname{Re}\lambda < |\operatorname{Im}\lambda|\cot\theta. \tag{7.9}$$

Let $x \in \mathcal{R}(A - \lambda I)$ and $y = (A - \lambda I)^{-1}x$. Hence, $x = Ay - \lambda y$ and we can write

$$(x, y) = (Ay, y) - \lambda\|y\|^2. \tag{7.10}$$

Further, since the operator A is θ-sectorial, by using relations (7.7) and (7.10), we obtain

$$\begin{aligned} \operatorname{Re}(x, y) + (\operatorname{Re}\lambda)\|y\|^2 &\geq \cot\theta \, |\operatorname{Im}(x, y) + (\operatorname{Im}\lambda)\|y\|^2| \\ &\geq |\operatorname{Im}\lambda| \, \|y\|^2 - |\operatorname{Im}(x, y)| \\ &\geq \cot\theta \, [\, |\operatorname{Im}\lambda| \, \|y\|^2 - |\operatorname{Im}(x, y)|\,], \end{aligned}$$

whence, by virtue of inequality (7.9) and the inequalities

$$\operatorname{Re}(x, y) \leq \|x\| \, \|y\| \quad \text{and} \quad |\operatorname{Im}(x, y)| \leq \|x\| \, \|y\|,$$

it follows that

$$\|y\| \leq \frac{1 + \cot\theta}{|\operatorname{Im}\lambda|\cot\theta - \operatorname{Re}\lambda}\|x\|. \tag{7.11}$$

Inequality (7.8) follows from (7.11) and the equality $y = (A - \lambda I)^{-1} x$. Therefore, $\lambda \notin \sigma_c(A)$.

■

If A is a θ-sectorial operator, then the operator $T = (I - A)(I + A)^{-1}$ is contractive. Moreover,

$$\|x\|^2 - \|Tx\|^2 \geq 2\cot\theta\,|\operatorname{Im}(Tx, x)|. \tag{7.12}$$

Indeed, if $x \in \mathcal{D}(T)$, then

$$x = (I + A)y \quad \text{and} \quad Tx = (I - A)y \quad (y \in \mathcal{D}(A)) \tag{7.13}$$

and, hence, in view of inequality (7.7),

$$\|x\|^2 - \|Tx\|^2 \geq 4\cot\theta\,|\operatorname{Im}(Ay, y)|. \tag{7.14}$$

It follows from (7.13) that

$$(Tx, x) = \|y\|^2 - \|Ay\|^2 - 2i\operatorname{Im}(Ay, y).$$

Therefore,

$$|\operatorname{Im}(Tx, x)| = 2\,|\operatorname{Im}(Ay, y)| \tag{7.15}$$

and, by using (7.14) and (7.15), we arrive at inequality (7.12).

Conversely, if the operator T satisfies condition (7.12) and $-1 \notin \sigma_p(T)$, then the operator $A = (I - T)(I + T)^{-1}$ is θ-sectorial.

Condition (7.12) is equivalent to the condition

$$\|T\sin\theta \pm iI\cos\theta\| \leq 1. \tag{7.16}$$

For this reason, the extensions of operators satisfying condition (7.12) or (7.16) were extensively studied. In this connection, one can mention the works of Arlinskii [1, 2] and Arlinskii and Tsekanovskii [1].

*7.4. **Accretive Operators and Contracting Semigroups.*** Let $\{T(s)\}_{s \geq 0}$ be a one-parameter continuous semigroup of contractions in a Hilbert space $\mathcal{H}$. An operator A defined by the equality

$$Ax = \lim_{s \to 0} \frac{T(s) - I}{s} x, \tag{7.17}$$

is called the *generator* (or the *infinitesimal operator*) of the semigroup $\{T(s)\}$. The set $\mathcal{D}(A)$ of elements x of the space $\mathcal{H}$ for which limit (7.17) exists is dense in $\mathcal{H}$ and the operator A is closed. The operator A defined as indicated above uniquely determines the semigroup $\{T(s)\}$ and $-A$ is a maximal accretive operator.

Thus, $1 \in \rho(A)$ because $-1 \in \rho(-A)$ and, hence, we can consider the operator

$$T = (A + I)(A - I)^{-1} \qquad (\mathcal{D}(T) = \mathcal{H})$$

which is contractive in $\mathcal{H}$. The operator T thus defined is called the *cogenerator* of the semigroup $\{T(s)\}$.

According to the equality $A = (T + I)(T - I)^{-1}$, the cogenerator T uniquely defines the corresponding generator and, hence, the semigroup $\{T(s)\}$. Moreover, if T is a contraction in $\mathcal{H}$ and $1 \notin \sigma_p(T)$, then there exists a continuous semigroup of contractions $\{T(s)\}$ whose cogenerator coincides with T. Therefore, any m-accretive operator A generates a continuous semigroup of contractions whose generator coincides with the operator $-A$.

The proofs of all results obtained in this subsection can be found in the monograph by Foias and Szökefalvi-Nagy [1].

Remarks. Accretive and θ-sectorial extensions were studied in the works by Phillips [1], Grendal [1], Kuzhel and Rotkevich [1], Evans and Knowles [1], Arlinskii [1, 2], Arlinskii and Tsekanovskii [1], Malamud [2], Storozh [2], Kato [1], etc.

2. Regular extensions with restrictions

In Chapter 2, we consider extensions of Hermitian (in particular, symmetric) operators with various restrictions, namely, self-adjoint bound-preserving extensions of semibounded symmetric operators with justification of the corresponding von Neumann hypothesis by the Friedrichs method and Krein method, regular extensions of Hermitian operators commuting with a certain collection of unitary operators, etc. We give an improved version of the proof of the Krein theorem on self-adjoint extensions of Hermitian contractions and prove the Ando–Nishio criterion for the existence of self-adjoint bound-preserving extensions of semibounded Hermitian operators. We define and study canonical dissipative extensions of Hermitian operators; these extensions are essentially used in Chapter 3 for the construction and investigation of spaces of boundary values of Hermitian operators with different defect numbers.

1. Self-Adjoint Bound-Preserving Extensions of Semibounded Symmetric Operators. Theorem and Hypothesis of Von Neumann

Recall the von Neumann hypothesis about self-adjoint bound-preserving extensions of semibounded operators.

1.1. Semibounded Operators. A symmetric operator A is called *semibounded from below (from above)* if there exists a real number c such that, for any $x \in \mathcal{D}(A)$, the following inequality holds:

$$(Ax, x) \geq c(x, x) \quad ((Ax, x) \leq c(x, x)). \tag{1.1}$$

In this case, the number c is called a *lower (upper) bound* of the operator A.

If A is an operator semibounded from above, then the operator $-A$ is semibounded from below. For this reason, in what follows, we consider only operators semibounded from below.

If A is an operator semibounded from below, then the number

$$m(A) = \inf_{x \in \mathcal{D}'(A)} \frac{(Ax, x)}{(x, x)} \qquad (\mathcal{D}'(A) = \mathcal{D}(A) \setminus \{0\}) \tag{1.2}$$

is called *the greatest lower bound* of this operator.

Proposition 1.1. *If* $\lambda < m(A)$, *then* $\lambda \in \pi(A)$.

Proof. Indeed, if $\lambda < m(A)$ and $x \in \mathcal{D}(A)$, then

$$(m(A) - \lambda) \|x\|^2 \le (Ax, x) - \lambda(x, x) \le \|(A - \lambda I)x\| \, \|x\|.$$

Hence, for $x \in \mathcal{D}(A)$, we have

$$\|(A - \lambda I)x\| \ge k \|x\|,$$

where $k = m(A) - \lambda > 0$.

■

Corollary 1.2. *The defect numbers of a semibounded operator are equal.*

Proof. The proof follows from Proposition 1.1 and Theorem 3.14 (Chapter 1).

■

1.2. Theorem and Hypothesis of Von Neumann. Let $\tilde{A}$ be a symmetric (in particular, self-adjoint) extension of a symmetric operator A semibounded from below. Then, obviously, $m(\tilde{A}) \le m(A)$. The self-adjoint extensions $\tilde{A}$ of the operator A for which $m(\tilde{A}) = m(A)$ are especially important for theoretical and practical applications. In this case, $\tilde{A}$ is called a bound-preserving extension of the operator A. Below in this section, we always consider bound-preserving extensions.

Von Neumann [1] proved the following theorem:

Theorem 1.3. *If* A *is a closed symmetric operator semibounded from below, then for any* $\varepsilon > 0$, *there exists a self-adjoint extension* $\tilde{A}$ *of the operator* A *with the greatest lower bound* $m(\tilde{A}) = m(A) - \varepsilon$.

Proof. Denote $\lambda = m(A) - \varepsilon$. Let us show that the domain of definition $\mathcal{D}(A)$ and the defect subspace $\mathfrak{N}_\lambda$ of the operator A are linearly independent. Indeed, if $h \in$

$\mathcal{D}(A) \cap \mathfrak{N}_\lambda$, then $((A-\lambda I)h,h)=0$, i.e., $(Ah,h)=\lambda(h,h)$. Since $\lambda < m(A)$, the last equality is possible only for $h=0$.

Consider the operator $\tilde{A}$ defined on the lineal $\mathcal{D}(\tilde{A}) = \mathcal{D}(A) \dotplus \mathfrak{N}_\lambda$ by the equality

$$\tilde{A}(x_0+x_\lambda) = Ax_0+\lambda x_\lambda \quad (x_0 \in \mathcal{D}(A),\ x_\lambda \in \mathfrak{N}_\lambda). \tag{1.3}$$

It is easy to see that $\tilde{A}$ is a symmetric extension of the operator A. Moreover, if $x = x_0+x_\lambda$, then

$$(\tilde{A}x,x) = (\tilde{A}x_0,x_0)+(Ax_0,x_\lambda)+\lambda(x_\lambda,x_0)+\lambda(x_\lambda,x_\lambda).$$

Since $(Ax_0,x_0) \geq \lambda(x_0,x_0)$ and $(Ax_0,x_\lambda)=\lambda(x_0,x_\lambda)$, we get

$$(\tilde{A}x,x) \geq \lambda[(x_0,x_0)+(x_0,x_\lambda)+(x_\lambda,x_0)+(x_\lambda,x_\lambda)],$$

i.e.,

$$(\tilde{A}x,x) \geq \lambda(x,x) \quad (\forall x \in \mathcal{D}(\tilde{A})).$$

Therefore, λ is a lower bound of the operator $\tilde{A}$. Taking into account the equality $(\tilde{A}x_\lambda,x_\lambda)=\lambda(x_\lambda,x_\lambda)$, we conclude that $m(\tilde{A})=\lambda$, whence $m(\tilde{A})=m(A)-\varepsilon$.

Let us now show that the defect numbers of the operator $\tilde{A}$ are equal to zero. Indeed, if $\tilde{x}_i = \tilde{\mathfrak{N}}_i$, where $\tilde{\mathfrak{N}}_i = \mathcal{H} \ominus \mathcal{R}_{\tilde{A}}(i)$ is the defect subspace of the operator $\tilde{A}$, then for all $x_0 \in \mathcal{D}(A)$ and $x_\lambda \in \mathfrak{N}_\lambda$, by virtue of equality (1.3), we obtain

$$((\tilde{A}-iI)(x_0+x_\lambda),\tilde{x}_i) = ((A-iI)x_0+(\lambda-i)x_\lambda,\tilde{x}_i) = 0. \tag{1.4}$$

In particular, for $x_0=0$, we get

$$(x_\lambda,\tilde{x}_i)=0 \quad (\forall x_\lambda \in \mathfrak{N}_\lambda).$$

This means that $\tilde{x}_i \perp \mathfrak{N}_\lambda$, i.e., $\tilde{x}_i \in \overline{\mathcal{R}_A(\lambda)}$. Since the operator A is closed and $\lambda \in \pi(A)$, in view of Theorem 2.1 (Chapter 1), we have $\overline{\mathcal{R}_A(\lambda)} = \mathcal{R}_A(\lambda)$. Thus, $\tilde{x}_i \in \mathcal{R}_A(\lambda)$. Then $\tilde{x}_i=(A-\lambda I)y$, where y is some vector from $\mathcal{D}(A)$.

On the other hand, for $x_\lambda=0$, we can rewrite equality (1.4) in the form

$$((A-iI)x_0,\tilde{x}_i) = 0 \quad (\forall x_0 \in \mathcal{D}(A)).$$

Hence, $\tilde{x}_i \in \mathfrak{N}_i$, where $\mathfrak{N}_i$ is a defect subspace of the operator A. Then

$$\| \tilde{x}_i \|^2 = ((A - iI)y + (i - \lambda)y, \tilde{x}_i)$$

$$= (i - \lambda)(y, \tilde{x}_i) = (i - \lambda)\left[(y, Ay) - \lambda(y, y)\right]. \tag{1.5}$$

Since $(y, Ay) - \lambda(y, y) \geq 0$ and $\lambda \in \mathbb{R}$, equality (1.5) holds only if $(y, Ay) = \lambda(y, y)$. Therefore, the right-hand side of equality (1.5) is equal to zero. This yields $\tilde{x}_i = 0$ and, thus, $\tilde{\mathfrak{N}}_i = \{0\}$. By virtue of Corollary 1.2, we get $\tilde{\mathfrak{N}}_{-i} = \{0\}$. Thus, the operator A has the deficiency index $\langle 0, 0 \rangle$. By taking (if necessary) the closure of the operator $\tilde{A}$, we obtain the operator $\bar{\tilde{A}}$ semibounded from below with the greatest lower bound $m(\bar{\tilde{A}}) = m(A) - \varepsilon$.

■

The constructive nature of the von Neumann theorem makes it especially convenient for applications.

According to the von Neumann theorem, for any symmetric operator A semibounded from below, there exist a self-adjoint extension $\tilde{A}$ whose greatest lower bound is arbitrarily close to the greatest lower bound of the operator A. Moreover, von Neumann advanced a hypothesis that there exists a self-adjoint extension $\tilde{A}$ of the operator A with the greatest lower bound $m(\tilde{A}) = m(A)$.

Later, different approaches to the proof of this statement were proposed by Stone [1], Friedrichs [1], Freudenthal [1], Krein [2], and Phillips [1]. In what follows, we consider the most frequently used Friedrichs method and the Krein method. Note that it will be shown below that the proofs of the theorems of the Krein method require certain correction.

2. Proof of the Von Neumann Hypothesis (Friedrichs Method)

The method proposed by Friedrichs [1] for the proof of the von Neumann hypothesis is well known and described in detail in literature (see remarks at the end of this section). This is why we recall only the basic steps of this method.

2.1. Auxiliary Hilbert Space $\mathcal{D}_0$. Let A be a symmetric operator semibounded from below and acting in $\mathcal{H}$. In what follows, without loss of generality, we assume that $m(A) = 1$. Thus,

$$(Ax, x) \geq (x, x) \quad (x \in \mathcal{D}(A)). \tag{2.1}$$

On the lineal $\mathcal{D}(A)$, we introduce scalar product $(\cdot, \cdot)_0$ and norm $\|\cdot\|_0$ according to the equalities

$$(x, y)_0 = (Ax, y) \text{ and } \|x\|_0 = \sqrt{(x, x)_0} \quad (\{x, y\} \subset \mathcal{D}(A)). \tag{2.2}$$

It is clear that $\|x\|_0 \geq \|x\|$. By completing the lineal $\mathcal{D}(A)$ in the norm $\|\cdot\|_0$, we obtain the Hilbert space $\mathcal{D}_0 \subset \mathcal{H}$, the norm and scalar product in which are defined by equalities (2.2).

2.2. Friedrichs' Proof of the Von Neumann Hypothesis

Theorem 2.1. *An arbitrary semibounded symmetric operator A can be extended to a certain semibounded self-adjoint operator $\tilde{A}$ with the same bound as the operator A.*

Proof. Without loss of generality, we can assume that A is a closed symmetric operator semibounded from below and such that $m(A) = 1$. Consider the operator $\tilde{A}$ defined as follows:

$$\mathcal{D}(\tilde{A}) = \mathcal{D}(A^*) \cap \mathcal{D}_0, \quad \tilde{A}x = A^*x \quad (x \in \mathcal{D}(\tilde{A})), \tag{2.3}$$

where $\mathcal{D}_0$ is the auxiliary Hilbert space constructed above. It is obvious that $A \subset \tilde{A}$.

Let $\{x, y\} \subset \mathcal{D}(\tilde{A})$. Since $\mathcal{D}(\tilde{A}) \subset \mathcal{D}_0$, there exist sequences $\{x_n\}$ and $\{y_n\}$ of elements from $\mathcal{D}(A)$ that converge in the norm of the space $\mathcal{H}$ (as well as in the norm of the space $\mathcal{D}_0$) to some elements x and y, respectively. Furthermore,

$$(\tilde{A}x, y_n) = (A^*x, y_n) = (x, Ay_n) = (x, y_n)_0.$$

Passing to the limit as $n \to \infty$ in this equality, we obtain $(\tilde{A}x, y) = (x, y)_0$. Similarly, by using the equality $(x_n, \tilde{A}y) = (x_n, y)_0$, we get $(x, \tilde{A}y) = (x, y)_0$. Thus,

$$(\tilde{A}x, y) = (x, \tilde{A}y) \quad (\forall \{x, y\} \subset \mathcal{D}(\tilde{A})),$$

i.e., $\tilde{A}$ is a symmetric operator and, moreover,

$$(\tilde{A}x, x) = (x, x)_0 \geq (x, x) \quad (\forall x, \in \mathcal{D}(\tilde{A})).$$

Hence, $\tilde{A}$ has the same lower bound as the operator A. Let us show that

$$\mathcal{R}(\tilde{A}) = \mathcal{H}. \tag{2.4}$$

Indeed, let $y \in \mathcal{H}$. On the space $\mathcal{D}_0$, consider the functional $f(\cdot) = (\cdot, y)$ and assume that $\|x_n - x\|_0 \to 0$. Then $\|x_n - x\| \to 0$ and, therefore,

$$\lim_{n\to\infty} f(x_n) = \lim_{n\to\infty} (x_n, y) = (x, y) = f(x).$$

This means that the functional f is continuous on $\mathcal{D}_0$. Hence, by virtue of the Riesz theorem, this functional can be represented in the form $f(\cdot) = (\cdot, h)_0$, where $h \in \mathcal{D}_0$. Furthermore, for any x from $\mathcal{D}(A)$, we have

$$(Ax, h) = (x, h)_0 = f(x) = (x, y).$$

Consequently, $h \in \mathcal{D}(A^*)$ and $y = A^*h$. This yields $h \in \mathcal{D}(\tilde{A})$ and $y = A^*h = \tilde{A}h \in \mathcal{R}(\tilde{A})$, which proves the inclusion $\mathcal{H} \subset \mathcal{R}(\tilde{A})$ and, hence, equality (2.4) is true.

In view of (2.4), the defect subspace $\mathfrak{N}_0 = \mathcal{H} \ominus \mathcal{R}(\tilde{A})$ of the operator $\tilde{A}$ coincides with $\{0\}$ and, hence, taking the equality $m(\tilde{A}) = 1$ and Proposition 1.1 into account, we conclude that $0 \in \pi(\tilde{A})$. Hence, by virtue of Lemma 3.12 (Chapter 1), the defect numbers of the operator $\tilde{A}$ are equal to zero, i.e., the operator $\tilde{A}$ is self-adjoint.

■

The self-adjoint operator $\tilde{A}$ constructed in the proof of Theorem 2.1 is called the *Friedrichs extension* of the semibounded symmetric operator A.

Remarks. There are many methods for the construction of the Friedrichs extension of a semibounded symmetric operator (see, e.g., Dunford and Schwartz [1], Reed and Simon [1], Riesz and Szökefalvi-Nagy [1], etc.). In particular, in the monograph of Reed and Simon [1], one can find various important examples of application of this extension.

Here, we have used (with certain modification) the version of the proof of the von Neumann hypothesis presented in monograph of Dunford and Schwartz [1] (see also Freudenthal [1]).

Also note the nonstandard method for the construction of the Friedrichs extension proposed by Prokai and Sebestyen [1]. In this work, the auxiliary Hilbert space $\mathcal{R}_0$ was defined as a completion of the lineal $\mathcal{R}(A)$ in the norm generated by the scalar product

$$(Ax, Ay)_0 = (Ax, y) \quad (\{x, y\} \subset \mathcal{D}(A)),$$

and the Friedrichs extension of the operator A was represented in the form $\tilde{A} = Q^* Q^{**}$, where the operator $Q : \mathcal{D}(A) \to \mathcal{R}_0$ is densely defined in $\mathcal{H}$ by the equality

$$Qx = Ax \qquad (x \in \mathcal{D}(A)).$$

3. Self-Adjoint Norm-Preserving Extensions of Hermitian Contractions

In this section, we propose a counterexample (Subsection 3.2) to demonstrate the necessity of introducing corrections in the proof of the von Neumann hypothesis by the Krein method and present the required modifications.

3.1. Preliminary Comments. The Friedrichs method described above enables one to construct a (unique) bound-preserving self-adjoint extension $\tilde{A}$ of a semibounded symmetric operator A. Note that the Friedrichs extension is very important for applications.

Unlike the Friedrichs method, the Krein method enables one to describe all bound-preserving self-adjoint extensions of a semibounded operator. The Krein method is based on the reduction of the problem of bound-preserving self-adjoint extensions of semibounded (nonnegative) symmetric operators to the problem of norm-preserving self-adjoint extensions of Hermitian contractions.

In this section, we present a modified version of the theorem on extensions of Hermitian contractions and prove it by a method slightly different from the commonly used methods. Our corrections simplify the proof of this theorem and make it more precise. These results enable us to describe all self-adjoint extensions of nonnegative symmetric operators (Section 5). The necessity of these corrections is discussed in Subsection 3.2.

3.2. Counterexample. As stated above, Krein [2] reduced the problem of description of bound-preserving self-adjoint extensions of nonnegative symmetric operators to the problem of description of norm-preserving self-adjoint extensions of Hermitian contractions. In this connection, Krein [2] formulates the theorem on self-adjoint extensions of Hermitian contractions in the following form (see also Akhiezer and Glazman [1] and Riesz and Szökefalvi-Nagy [1]):

Theorem (Krein [2]). *Any Hermitian bounded operator S has at least one self-adjoint extension with the same norm as the operator S.*

To prove this theorem, we first construct the lineal $\mathfrak{N} \subset \mathcal{H}$ and then consider the factor spaces $\mathcal{H}/\mathfrak{N}$ and $\mathcal{D}(S)/\mathfrak{N}$. However, in the general case, $\mathfrak{N} \not\subset \mathcal{D}(S)$ and, moreover, there are cases where $\mathcal{D}(S) \cap \mathfrak{N} = \{0\}$. Thus, under the conditions of this theorem we cannot consider the factor space $\mathcal{D}(S)/\mathfrak{N}$ and, hence, all subsequent constructions are meaningless.

We illustrate this observation by the following example: Let $\mathcal{H}=\langle e_1, e_2\rangle$ be a two-dimensional space with orthonormal basis $\{e_1, e_2\}$. Consider an operator S defined on the lineal $\mathcal{D}(S)=\langle e_1\rangle$ by the equality

$$Se_1 = ae_1 + be_2 \quad (a \in \mathbb{R}, b \in \mathbb{C}, a^2 + |b|^2 = 1).$$

Thus, the operator S is Hermitian and $\|S\| = 1$.

We also consider the operator $S_1 = PS$, where P is the orthoprojector onto $\mathcal{D}(S)$ in $\mathcal{H}$ and a functional f defined by the equality

$$f(x) = (Sx, y) \quad (x \in \mathcal{D}(S)),$$

where y is an arbitrary fixed element from $\mathcal{H}$. This functional can be represented in the form $(Sx, y) = (x, h)$, where $h \in \mathcal{D}(S)$ and $\|h\| \le \|y\|$. Thus, we can consider the operator $\tilde{S}_1 : \mathcal{H} \to \mathcal{D}(S)$ defined by the equality $\tilde{S}_1 y = h$. It is clear that $\|\tilde{S}_1\| \le 1$ and $S_1 \subset \tilde{S}_1$.

In our case, $h = (y, Se_1)e_1$ and, hence,

$$\tilde{S}_1 y = (y, Se_1)e_1 \quad (\forall y \in \mathcal{H}).$$

To extend the operator $S_2 = (I-P)S$ to $\mathcal{H}$, we consider the scalar product

$$(x, y)_0 = (x, y) - (\tilde{S}_1 x, \tilde{S}_1 y)$$

and the lineal $\mathfrak{N} = \{x \in \mathcal{H} \mid (x, x)_0 = 0\}$. Note that

$$(x, x)_0 = (x, x) - |(x, Se_1)|^2$$

and, thus, $\mathfrak{N} = \langle Se_1\rangle$. Hence, for $b \neq 0$, we get $\mathfrak{N} \cap \mathcal{D}(S) = \{0\}$.

Thus, the proof of Krein's theorem should be made more precise. In Subsection 3.3, we show that these difficulties can be eliminated if we consider strong Hermitian contractions.

3.3. Self-Adjoint Norm-Preserving Extensions of Hermitian Contractions. Let S be an Hermitian contraction acting in a Hilbert space $\mathcal{H}$ (i.e., the operator S is Hermitian and $\|S\| \le 1$). Without loss of generality, we can assume that the domain of definition $\mathcal{D}(S)$ of the operator S is a subspace of the space $\mathcal{H}$. Furthermore, $\mathcal{D}(S) \neq \mathcal{H}$

and $\|S\| = 1$. In what follows, the contraction S is called *strong* if

$$\|Sx\| < \|x\| \qquad (\forall\, x \in \mathcal{D}(S),\ x \neq 0). \tag{3.1}$$

Theorem 3.1. *Any strong Hermitian contraction S can be extended to a self-adjoint operator $\tilde{S}$ with the same norm.*

Proof. Consider the operators

$$S_1 = PS \quad \text{and} \quad S_2 = QS, \tag{3.2}$$

where P and $Q = I - P$ are, respectively, the orthoprojectors onto the subspaces $\mathcal{D}(S)$ and $\mathcal{H} \ominus \mathcal{D}(S)$ in $\mathcal{H}$. Thus, $\mathcal{D}(S_1) = \mathcal{D}(S_2) = \mathcal{D}(S)$ and, for $x \in \mathcal{D}(S)$, we have

$$Sx = S_1 x + S_2 x \quad \text{and} \quad \|Sx\|^2 = \|S_1 x\|^2 + \|S_2 x\|^2. \tag{3.3}$$

Consider an extension $\tilde{S}_1$ of the operator S_1, which can be constructed in the following way:

For any fixed $y \in \mathcal{H}$, we consider the functional $\varphi(\cdot) = (S\,\cdot, y)$ defined in the Hilbert space $\mathcal{D}(S)$. This functional is bounded and, by virtue of (3.1), we conclude that

$$|\varphi(x)| < \|y\|\,\|x\| \qquad (x \neq 0,\ y \neq 0). \tag{3.4}$$

According to the Riesz theorem, we can represent this functional in the form $\varphi(x) = (x, h)$, where h is a uniquely defined element of the space $\mathcal{D}(S)$ and, moreover,

$$\|h\| = \|\varphi\| \leq \|y\|. \tag{3.5}$$

Consider an operator $\tilde{S}_1 : \mathcal{H} \to \mathcal{D}(S)$ defined by the equality $\tilde{S}_1 y = h$. Therefore, $\mathcal{D}(\tilde{S}_1) = \mathcal{H}$ and

$$(Sx, y) = (x, \tilde{S}_1 y) \qquad (x \in \mathcal{D}(S),\ y \in \mathcal{H}). \tag{3.6}$$

Moreover, by using (3.5), we obtain $\|\tilde{S}_1 y\| = \|h\| \leq \|y\|$. Let us show that

$$\|\tilde{S}_1 y\| < \|y\| \qquad (\forall\, y \in \mathcal{H},\ y \neq 0), \tag{3.7}$$

i.e., that $\tilde{S}_1$ is a strong contraction in $\mathcal{H}$. Indeed, assume that, for some $y \neq 0$,

$$\|\tilde{S}_1 y\| = \|y\|. \tag{3.8}$$

By substituting $\tilde{S}_1 y$ for x in (3.6) and taking into account relations (3.1) and (3.8), we get

$$\|y\|^2 = \|\tilde{S}_1 y\|^2 = (S\tilde{S}_1 y, y) < \|\tilde{S}_1 y\|\,\|y\|^2.$$

This contradiction proves inequality (3.7).

Let us show that $\tilde{S}_1$ is an extension of the operator S_1. Indeed, if $\{x, y\} \subset \mathcal{D}(S)$, then

$$(x, \tilde{S}_1 y) = (Sx, y) = (x, PSy) = (x, S_1 y).$$

Hence, $\tilde{S}_1 y = S_1 y$ $(\forall y \in \mathcal{D}(S))$ and, thus, $S_1 \subset \tilde{S}_1$.

We now construct an extension $\tilde{S}_2$ of the operator S_2 satisfying the condition

$$\|\tilde{S}_1 x\|^2 + \|\tilde{S}_2 x\|^2 < \|x\|^2 \quad (x \in \mathcal{H}). \tag{3.9}$$

For this purpose, in $\mathcal{H}$, we consider a bilinear form $(\cdot, \cdot)_0$ defined by the equality

$$(x, y)_0 = (x, y) - (\tilde{S}_1 x, \tilde{S}_1 y) \quad (\{x, y\} \subset \mathcal{H}).$$

Since $\tilde{S}_1$ is a strong contraction, we have

$$(x, x)_0 = \|x\|^2 - \|\tilde{S}_1 x\|^2 \geq 0$$

and the equality $(x, x)_0 = 0$ holds only for $x = 0$. Thus, the bilinear form $(\cdot, \cdot)_0$ is a scalar product in $\mathcal{H}$ and, therefore, the space $\mathcal{H}$ is a pre-Hilbert space with scalar product $(\cdot, \cdot)_0$. The norm $\|\cdot\|_0$ in the space $\mathcal{H}$ is defined by the equality

$$\|x\|_0^2 = \|x\|^2 - \|\tilde{S}_1 x\|^2 \tag{3.10}$$

and, obviously,

$$\|x\|_0 \leq \|x\| \quad (\forall x \in \mathcal{H}). \tag{3.11}$$

If the space $\mathcal{H}$ is not complete with respect to the norm $\|\cdot\|_0$, then, clearly, its completion with respect to this norm $\hat{\mathcal{H}} \supset \mathcal{H}$ is a Hilbert space.

Let $x \in \mathcal{D}(S)$. Then, in view of inequality (3.1) and the second equality in (3.3), we can write

$$\|S_2 x\|^2 = \|Sx\|^2 - \|S_1 x\|^2 \leq \|x\|^2 - \|\tilde{S}_1 x\|^2 = \|x\|_0^2.$$

By virtue of (3.11), we obtain

$$\|S_2x\|_0 \le \|S_2x\| \le \|x\|_0. \tag{3.12}$$

Thus, the operator S_2 is bounded (moreover, it is a contraction) in the space $\hat{\mathcal{H}}$ and, hence, we can continuously extend this operator to the closure $\hat{\mathcal{D}}(S)$ of the lineal $\mathcal{D}(S)$ in the norm $\|\cdot\|_0$. As a result, we obtain the contraction $\hat{S}_2$ defined in $\hat{\mathcal{D}}(S)$.

If a sequence $\{x_n\}$ converges to an element x in the norm $\|\cdot\|$ (in the norm $\|\cdot\|_0$), then we write

$$x_n \to x \quad (x_n \overset{0}{\to} x).$$

Assume that $x_n \overset{0}{\to} x$, where $\{x_n\}$ is a sequence of elements from $\mathcal{D}(S)$. Then, in view of the second inequality in (3.12), the sequence $\{S_2x_n\}$ is fundamental in the space $\mathcal{H}$ and, thus, converges to an element $h \in \mathcal{H}$.

It follows from the first inequality in (3.12) that $h = \hat{S}_2x$. Thus, if $x_n \overset{0}{\to} x$, then $S_2x_n \to \hat{S}_2x$. In particular, if $x \in \mathcal{D}(S)$, then $S_2x = \hat{S}_2x$ and, therefore, $S_2 \subset \hat{S}_2$.

Equalities (3.2) imply the following relation:

$$(S_2x_n, y) = 0 \quad (\{x_n, y\} \subset \mathcal{D}(S)).$$

By passing in this equality to the limit as $n \to \infty$, we obtain

$$(\hat{S}_2x, y) = 0 \quad (\forall y \in \mathcal{D}(S))$$

and, hence, the operator $\hat{S}_2$ maps the lineal $\hat{\mathcal{D}}(S)$ into the subspace $\mathcal{H} \ominus \mathcal{D}(S)$.

We also note that if we replace x by x_n in (3.12) and pass to the limit as $x_n \overset{0}{\to} x$, then we get

$$\|\hat{S}_2x\|_0 \le \|\hat{S}_2x\| \le \|x\|_0. \tag{3.13}$$

Consider an operator $B: \hat{\mathcal{H}} \to \mathcal{H} \ominus \mathcal{D}(S)$ defined in the space $\hat{\mathcal{H}}$ by the equality

$$Bx = \hat{S}_2\hat{P}x \quad (x \in \hat{\mathcal{H}}),$$

where $\hat{P}$ is the orthoprojector onto the subspace $\hat{\mathcal{D}}(S)$ in $\hat{\mathcal{H}}$. The operator B is a contraction in $\hat{\mathcal{H}}$ (as a product of the contractions $\hat{S}_2$ and $\hat{P}$). Denote by $\tilde{S}_2$ the restriction of the operator B to $\mathcal{H}$: $\tilde{S}_2 = B|_{\mathcal{H}}$. By using (3.11) and (3.13), for any $x \in \mathcal{H}$, we obtain

$$\| \tilde{S}_2 x \| = \| \hat{S}_2 \hat{P} x \| \leq \| \hat{P} x \|_0 \leq \| x \|_0 \leq \| x \|, \tag{3.14}$$

i.e., $\| \tilde{S}_2 \| \leq 1$. Further, if $x \in \mathcal{D}(S)$, then

$$\tilde{S}_2 x = \hat{S}_2 x = S_2 x$$

and, thus, $\tilde{S}_2$ is an extension of the operator S_2 in the space $\mathcal{H}$.

Consider the operator $T = \tilde{S}_1 + \tilde{S}_2$. Since, for any x from the space $\mathcal{H}$, the vectors $\tilde{S}_1 x$ and $\tilde{S}_2 x$ are orthogonal, we get

$$\| Tx \|^2 = \| \tilde{S}_1 x \|^2 + \| \tilde{S}_2 x \|^2. \tag{3.15}$$

By using (3.14), we conclude that

$$\| \tilde{S}_2 x \|^2 \leq \| x \|_0^2 = \| x \|^2 - \| \tilde{S}_1 x \|^2. \tag{3.16}$$

Equalities (3.15) and (3.16) imply that $\| Tx \| \leq \| x \|$. Moreover, if $x \in \mathcal{D}(S)$, then

$$Tx = S_1 x + S_2 x = Sx.$$

Hence, the contraction T is an extension of the contraction S (however, generally speaking, $T^* \neq T$).

Let us show that T^* is an extension of the operator S. Indeed, if $x \in \mathcal{D}(S)$ and $y \in \mathcal{H}$, then it follows from equality (3.6) that

$$(T^* x, y) = (x, Ty) = (x, \tilde{S}_1 y + \tilde{S}_2 y) = (x, \tilde{S}_1 y) = (Sx, y),$$

whence it follows that $T^* x = Sx$ $[\forall x \in \mathcal{D}(S)]$. This means that $S \subset T^*$. Note that, in view of Proposition 5.8 (Chapter 1), the relations $S \subset T$ and $S \subset T^*$ imply that T is a regular extension of the operator S.

Consider the operator $\tilde{S} = \frac{1}{2}(T + T^*)$. This operator is a self-adjoint extension of the operator S and, for any x from $\mathcal{H}$, we have

$$\| \tilde{S} x \| \leq \frac{1}{2}\left(\| Tx \| + \| T^* x \| \right) \leq \| x \|.$$

Thus, the extension $\tilde{S}$ of the operator S is a contraction in $\mathcal{H}$. This completes the proof of the theorem.

■

Corollary 3.2. *Let S be a strong Hermitian contraction and let $\tilde{S}$ be its self-adjoint extension with the same norm. Assume that, for any $x \in \mathcal{D}(S)$,*

$$SCx = \theta CSx, \qquad \theta \in \{-1, 1\}, \tag{3.17}$$

where C is a bounded self-adjoint operator. Then the equality

$$\tilde{S}Cy = \theta C\tilde{S}y \tag{3.18}$$

holds for all $y \in \mathcal{H}$.

Proof. It follows from equality (3.17) that $C\colon \mathcal{D}(S) \to \mathcal{D}(S)$. This means that the contraction $\tilde{S}_1 : \mathcal{H} \to \mathcal{D}(S)$ constructed in Theorem 3.1 satisfies the relations

$$\tilde{S}_1 C\colon \mathcal{D}(S) \to \mathcal{D}(S) \quad \text{and} \quad C\tilde{S}_1 : \mathcal{H} \to \mathcal{D}(S). \tag{3.19}$$

Equalities (3.6) and (3.17) imply that

$$(x, \tilde{S}_1 Cy) = (Sx, Cy) = (CSx, y) = \theta(SCx, y)$$

$$= \theta(Cx, \tilde{S}_1 y) = (x, \theta C\tilde{S}_1 y) \quad (x \in \mathcal{D}(S),\ y \in \mathcal{H}).$$

By virtue of (3.19), this yields

$$\tilde{S}_1 C = \theta C\tilde{S}_1. \tag{3.20}$$

Let us show that the operator C is bounded in the norm $\|\cdot\|_0$. Indeed, in view of (3.10) and (3.20), we have

$$\|Cx\|_0^2 = \|Cx\|^2 - \|\tilde{S}_1 Cx\|^2 = (C^2x, x) - (\tilde{S}_1 C^2x,\ \tilde{S}_1 x)$$

$$= (C^2x, x)_0 \le \|C^2x\|_0^2\, \|x\|_0.$$

Therefore,

$$\|Cx\|_0^2 \le \|C^2x\|_0 \|x\|_0. \tag{3.21}$$

Let us show that the following inequality holds for all $n \in \mathbb{N}$:

$$\| Cx \|_0^{2^n} \le \| C^{2^n} x \|_0 \| x \|_0^{2^n - 1}. \tag{3.22}$$

According to (3.21), relation (3.22) is true for $n = 1$. Suppose that it is true for some $n \in \mathbb{N}$. Consider the operator $C_1 = C^{2^n}$.

In view of (3.20), we obtain $\tilde{S}_1 C_1 = C_1 \tilde{S}_1$. This means that relation (3.21) remains true if we replace C by C_1. Thus,

$$\| C^{2^n} x \|_0^2 \le \| C^{2^{n+1}} x \|_0 \| x \|_0. \tag{3.23}$$

If we now square both sides of inequality (3.22) and take into account inequality (3.23), then we get

$$\| Cx \|_0^{2^{n+1}} \le \| C^{2^{n+1}} x \|_0 \| x \|^{2^{n+1} - 1}.$$

Thus, inequality (3.22) is true for all $n \in \mathbb{N}$.

Since

$$\| C^{2^n} x \|_0 \le \| C^{2^n} x \| \le \| C^{2^n} \| \, \| x \|,$$

inequality (3.22) implies that

$$\| Cx \|_0 \le \| C^{2^n} \|^{2^{-n}} \| x \|^{2^{-n}} \| x \|_0^{1 - 2^{-n}}.$$

By passing in this inequality to the limit as $n \to \infty$ and using the equality

$$\lim_{n \to \infty} \| C^{2^n} \|^{2^{-n}} = \| C \|,$$

we obtain $\| Cx \|_0 \le \| C \| \, \| x \|_0$. Hence, the operator C is bounded in the norm $\| \cdot \|_0$ and $\| C \|_0 \le \| C \| < \infty$. Moreover, in view of relations (3.10) and (3.20), it is easy to see that $(Cx, y)_0 = (x, Cy)_0$ for any x and y from $\mathcal{H}$. Hence, the continuous extension of the operator C to the entire space $\hat{\mathcal{H}}$ is a bounded self-adjoint operator and we denote it by $\hat{C}$.

Since $C \colon \mathcal{D}(S) \to \mathcal{D}(S)$, we get $\hat{C} \colon \mathcal{D}(\hat{S}) \to \mathcal{D}(\hat{S})$. Therefore, $\hat{C}\hat{P} = \hat{P}\hat{C}$, where $\hat{P}$ is the orthoprojector onto $\hat{\mathcal{D}}(S)$ in $\hat{\mathcal{H}}$.

Consider the operator $S_2 = S - S_1$. According to the proof of Theorem 3.1, $\tilde{S}_1|_{\mathcal{D}(S)} = S_1$. Hence, it follows from equality (3.20) that

$$S_2 Cx = \theta CS_2 x \qquad (x \in \mathcal{D}(S)).$$

Since the operator S_2 can be continuously extended to the closure $\hat{\mathcal{D}}(S)$ of the lineal $\mathcal{D}(S)$ in the space $\hat{\mathcal{H}}$ (see Theorem 3.1), this equality implies that

$$\hat{S}_2 \hat{C} x = \theta \hat{C} \hat{S}_2 x \qquad (x \in \hat{\mathcal{D}}(S))$$

or, which is the same,

$$\hat{S}_2 \hat{C} \hat{P} x = \theta \hat{C} \hat{S}_2 \hat{P} x \qquad (x \in \hat{\mathcal{H}}).$$

Taking into account the facts that the operator $\hat{C}$ commutes with $\hat{P}$ and the operator $\tilde{S}_2$ is a restriction of the operator $\hat{S}_2 \hat{P}$ to $\mathcal{H}$, we obtain

$$\tilde{S}_2 Cx = \theta C \tilde{S}_2 x \qquad (x \in \mathcal{H}).$$

This and equality (3.20) yield

$$TC = \theta CT \quad \text{and} \quad T^* C = \theta CT^*,$$

where $T = \tilde{S}_1 + \tilde{S}_2$. Thus, the operator $\tilde{S} = \frac{1}{2}(T + T^*)$ satisfies condition (3.18).

The proof of Corollary 3.2 is completed.

■

3.4. Description of Self-Adjoint Norm-Preserving Extensions of Hermitian Contractions. The following lemma is important for the description of extensions of Hermitian contractions:

Lemma 3.3. *Assume that B is a bounded nonnegative operator acting in $\mathcal{H}$ and $\mathcal{H}_1$ is a subspace of the space $\mathcal{H}$. Let $\mathcal{M}$ be a set of all bounded self-adjoint operators C such that*

$$C \le B \quad \textit{and} \quad \mathcal{H}_1 \subset \operatorname{Ker} C. \tag{3.24}$$

Then the set $\mathcal{M}$ contains a maximal operator $\hat{C}$ and this operator admits a representation

$$\hat{C} = \sqrt{B}Q\sqrt{B},$$

where Q is the orthoprojector onto the subspace $\mathcal{H} \ominus (\sqrt{B}\mathcal{H}_1)$ in $\mathcal{H}$.

Proof. If $C \in \mathcal{M}$, then, by virtue of (3.24), for any $x \in \mathcal{H}$ and $x_1 \in \mathcal{H}_1$, we have $(Cx, x_1) = 0$. Hence,

$$(Cx, x) = (C(x - x_1), x - x_1)$$

and, therefore,

$$(Cx, x) \leq (B(x - x_1), x - x_1) = \|\sqrt{B}(x - x_1)\|^2. \tag{3.25}$$

Note that the left-hand side of (3.25) does not depend on x_1. Consequently,

$$(Cx, x) \leq \inf_{x_1 \in \mathcal{H}_1} \|\sqrt{B}(x - x_1)\|^2. \tag{3.26}$$

Denote by $P = I - Q$ the orthoprojector onto $\overline{\sqrt{B}\mathcal{H}_1}$ in $\mathcal{H}$. The equalities

$$Q\sqrt{B}x_1 = 0 \quad \text{and} \quad P\sqrt{B}x_1 = \sqrt{B}x_1 \tag{3.27}$$

imply that

$$\|\sqrt{B}(x - x_1)\|^2 = \|P\sqrt{B}(x - x_1) + Q\sqrt{B}x\|^2 = \|P\sqrt{B}(x - x_1)\|^2 + \|Q\sqrt{B}x\|^2$$

It is obvious that

$$\inf_{x_1 \in \mathcal{H}_1} \|P\sqrt{B}(x - x_1)\|^2 = 0.$$

Hence, in view of (3.26), for any $x \in \mathcal{H}$, we have

$$(Cx, x) \leq \|Q\sqrt{B}x\|^2 = (\sqrt{B}Q\sqrt{B}x, x) = (\hat{C}x, x),$$

i.e., $C \leq \hat{C}$. Moreover, $\hat{C}^* = \hat{C}$ and, by using (3.27), we get

$$\hat{C}x_1 = \sqrt{B}Q\sqrt{B}x_1 = 0 \quad (x_1 \in \mathcal{H}_1).$$

Consequently, $\mathcal{H}_1 \subset \operatorname{Ker} \hat{C}$ and, thus, $\hat{C} \in \mathcal{M}$.

Denote by $\mathcal{P}_0(S)$ the set of all self-adjoint norm-preserving extensions of an Hermitian contraction S.

Let $\tilde{S}_0$ be a fixed operator from $\mathcal{P}_0(S)$. We choose an arbitrary operator $\tilde{S} \in \mathcal{P}_0(S)$ and consider the operators

$$C = \tilde{S} - \tilde{S}_0, \quad B_- = I + \tilde{S}_0, \quad \text{and} \quad B_+ = I - \tilde{S}_0. \tag{3.28}$$

Since $-I \leq \tilde{S} \leq I$, we have

$$-B_- \leq C \leq B_+. \tag{3.29}$$

Further, the inclusions $S \subset \tilde{S}_0$ and $S \subset \tilde{S}$ imply that

$$Cx = 0 \quad (\forall x \in \mathcal{D}(S)). \tag{3.30}$$

At the same time, if a self-adjoint operator C satisfies conditions (3.29) and (3.30), then the operator $\tilde{S} = C + \tilde{S}_0$ belongs to the set $\mathcal{P}_0(S)$. Indeed, by virtue of (3.29) and (3.30), for $x \in \mathcal{D}(S)$, we get $Sx = \tilde{S}_0 x = \tilde{S}x$ and $-I \leq \tilde{S} \leq I$. Hence, $S \subset \tilde{S}$ and $\|\tilde{S}\| = 1$, i.e., $\tilde{S} \in \mathcal{P}_0(S)$.

Denote by Q_- and Q_+ the orthoprojectors onto the subspaces

$$\mathcal{H} \ominus \left(\sqrt{B_-}\,\mathcal{D}(S)\right) \quad \text{and} \quad \mathcal{H} \ominus \left(\sqrt{B_+}\,\mathcal{D}(S)\right), \tag{3.31}$$

respectively. Here, the nonnegative operators B_- and B_+ are defined by relations (3.28). Consider the operators

$$\hat{C}_- = \sqrt{B_-}\,Q_-\sqrt{B_-} \quad \text{and} \quad \hat{C}_+ = \sqrt{B_+}\,Q_+\sqrt{B_+}. \tag{3.32}$$

By using these operators, we introduce self-adjoint operators

$$S_- = \tilde{S}_0 - \hat{C}_- \quad \text{and} \quad S_+ = \tilde{S}_0 + \hat{C}_+. \tag{3.33}$$

Relations (3.31) and (3.32) and the property of nonnegativity of the operators B_+ and B_- immediately imply that

$$\hat{C}_- x = \hat{C}_+ x = 0 \;\; (x \in \mathcal{D}(S)), \quad -B_+ \leq \hat{C}_- \leq B_-, \quad \text{and} \quad -B_- \leq \hat{C}_+ \leq B_+. \tag{3.34}$$

Therefore, the operators $-\hat{C}_-$ and $\hat{C}_+$ satisfy relations (3.29) and (3.30) and, hence, the self-adjoint operators S_- and S_+ belong to the set $\mathcal{P}_0(S)$.

Theorem 3.4. *A self-adjoint operator T is an extension of a strong Hermitian contraction S with the same norm if and only if*

$$S_- \leq T \leq S_+. \tag{3.35}$$

Proof. Let $T \in \mathcal{P}_0(S)$ and $C = T - \tilde{S}_0$. Clearly, the operator C satisfies conditions (3.29) and (3.30). Hence, in view of Lemma 3.3, we have $C \leq \hat{C}_+$. Therefore, equalities (3.33) imply that $T = \tilde{S}_0 + C \leq S_+$.

By using the same reasoning as above and taking into account the inequality $-C \leq B_-$, we conclude that $-C \leq \hat{C}_-$. Thus, $T = \tilde{S}_0 + C \geq \tilde{S}_0 - \hat{C}_- = S_-$. This means that if $T \in \mathcal{P}_0(S)$, then condition (3.35) is satisfied.

Conversely, assume that the self-adjoint operator T satisfies condition (3.35). Consider the operator $C = T - \tilde{S}_0$. By virtue of (3.33) and (3.35), we obtain

$$-\hat{C}_- \leq C \leq \hat{C}_+. \tag{3.36}$$

Let $W = C + \hat{C}_-$. In view of (3.36), we have

$$0 \leq W \leq \hat{C}_- + \hat{C}_+.$$

This relation and the first equality in (3.34) imply that $Wx = 0$ $(x \in \mathcal{D}(S))$ and, hence, $Cx = 0$ for all $x \in \mathcal{D}(S)$. At the same time, it follows from second and third relations in (3.34) and equality (3.36) that $-B_- \leq C \leq B_+$. Thus, the operator $C = T - \tilde{S}_0$ satisfies conditions (3.29) and (3.30). Therefore, as shown above, the operator $T = C + \tilde{S}_0$ belongs to $\mathcal{P}_0(S)$.

■

Note that the operators S_- and S_+ does not depend on choice of the operator $\tilde{S}_0$. Indeed, let $\tilde{S}'$ be another operator from $\mathcal{P}_0(S)$ and let S'_- and S'_+ be the operators defined by the equalities similar to equalities (3.33), namely,

$$S'_- = \tilde{S}' - \tilde{C}_- \quad \text{and} \quad S'_+ = \tilde{S}' - \tilde{C}_+.$$

Since $\{S_-, S'_-\} \subset \mathcal{P}_0(S)$, in view of Theorem 3.4, we obtain $S'_- \leq S_-$ and $S_- \leq S'_-$. Thus, $S'_- = S_-$. Similarly, we have $S'_+ = S_+$.

3.5. Condition under Which the Operator $\tilde{S}_0$ Coincides with S_-. As above, we assume that $\tilde{S}_0 \in \mathcal{P}_0(S)$. In the space $\mathcal{H}$, we consider the B_--scalar product and the relevant B_--norm:

$$(x, y)_- = (B_- x, y) \quad \text{and} \quad \|x\|_- = \sqrt{(x, x)_-} \quad \{x, y\} \subset \mathcal{H},$$

where the operator B_- is defined by equality (3.28).

Note that the kernel of operator B_- may differ from the zero set and, hence, the equality $\|x\|_- = 0$ may hold for $x \neq 0$.

Indeed, assume that $\mathcal{H} = \langle e_1, e_2 \rangle$ is a two-dimensional space with orthonormal basis $\{e_1, e_2\}$ and the operator S is defined by the equality $Se_1 = e_2$ $(\mathcal{D}(S) = \langle e_1 \rangle)$. Then there exists a unique self-adjoint norm-preserving extension $\tilde{S}_0$ of the operator S defined by the formulas $\tilde{S}_0 e_1 = e_2$ and $\tilde{S}_0 e_2 = e_1$. In this case,

$$B_-(e_1 - e_2) = e_1 - e_2 + \tilde{S}_0(e_1 - e_2) = 0$$

and, thus, $\operatorname{Ker} B_- \neq \{0\}$.

A lineal $\mathcal{L} \subset \mathcal{H}$ is called *B_--dense* in $\mathcal{H}$ if

$$\inf_{x \in \mathcal{L}} \|y - x\|_- = 0 \quad (\forall y \in \mathcal{H}). \tag{3.37}$$

Suppose that $\mathcal{L}$ is a B_--dense lineal in $\mathcal{H}$ and an element $y \in \mathcal{H}$ satisfies the condition

$$(x, y)_- = 0 \quad (\forall x \in \mathcal{L}). \tag{3.38}$$

Let us show that, in this case, $y \in \operatorname{Ker} B_-$. Indeed, by virtue of (3.38),

$$\|y - x\|_-^2 = \|y\|_-^2 + \|x\|_-^2 \geq \|y\|_-^2$$

and, therefore, equality (3.37) is possible only under the condition

$$\|y\|_-^2 = (B_- y, y) = 0,$$

whence, in view of the inequality $B_- \geq 0$, we conclude that $B_- y = 0$, i.e., $y \in \operatorname{Ker} B_-$.

Theorem 3.5. *The operator $\tilde{S}_0$ coincides with the operator S_- if and only if the lineal $\mathcal{D}(S)$ is B_--dense in $\mathcal{H}$.*

Proof. Let $\tilde{S}_0 = S_-$. Then it follows from (3.33) that $\hat{C}_- = 0$. Hence, by using the second equality in (3.32), we get

$$0 = (\hat{C}_- x, x) = \|Q_- \sqrt{B_-} x\|^2 \quad (\forall x \in \mathcal{H}), \tag{3.39}$$

i.e., $Q_- \sqrt{B_-} = 0$. Then $P_- \sqrt{B_-} = \sqrt{B_-}$, where $P_- = I - Q_-$ is the orthoprojector onto the subspace $\overline{\sqrt{B_-}\mathcal{D}(S)}$ in the space $\mathcal{H}$.

Consequently, for any $y \in \mathcal{H}$, we obtain

$$\begin{aligned} 0 = \|\sqrt{B_-}y - P_-\sqrt{B_-}y\|^2 &= \inf_{x \in \mathcal{D}(S)} \|\sqrt{B_-}y - \sqrt{B_-}x\|^2 \\ &= \inf_{x \in \mathcal{D}(S)} (B_-(y-x), y-x) = \inf_{x \in \mathcal{D}(S)} \|y-x\|_-^2 \end{aligned} \tag{3.40}$$

and, thus, the lineal $\mathcal{D}(S)$ is B_--dense in $\mathcal{H}$.

Conversely, let $\mathcal{D}(S)$ be a B_--dense lineal in $\mathcal{H}$, i.e.,

$$\inf_{x \in \mathcal{D}(S)} \|y - x\|_- = 0 \quad (\forall y \in \mathcal{H}). \tag{3.41}$$

Thus, by virtue of (3.40) and (3.41), for any $y \in \mathcal{H}$, we can write

$$\|Q_- \sqrt{B_-}\, y\| = \|\sqrt{B_-}\, y - P_- \sqrt{B_-}\, y\| = \inf_{x \in \mathcal{D}(S)} \|y - x\|_- = 0.$$

In view of (3.39), this yields

$$(\hat{C}_- y, y) = 0 \quad (\forall y \in \mathcal{H}),$$

Hence, $\hat{C}_- = 0$ and, consequently, $\tilde{S}_0 = S_-$.

■

Remarks. The Krein theorem on the existence of self-adjoint extensions of Hermitian contractions with the same norm was studied in detail by Akhiezer and Glazman [1] and Riesz and Szökefalvi–Nagy [1]. The counterexample from Subsection 3.2 was constructed by A. Kuzhel [6]. Corollary 3.2 was proved by Phillips [2].

4. Normal Norm-Preserving Extensions of Hermitian Contractions

In the present section, the results obtained in Subsections 3.4 and 3.5 concerning the description of self-adjoint norm-preserving extensions of Hermitian contractions are generalized to the case of normal extensions of Hermitian contractions.

4.1. Regular Normal Extensions. Let S be a closed Hermitian contraction acting in a Hilbert space $\mathcal{H}$. Without loss of generality, we can assume that $\|S\| = 1$. Denote by $\mathcal{P}_1(S)$ the set of all regular normal norm-preserving extensions of the operator S, namely,

$$\mathcal{P}_1(S) = \{T \in \mathcal{P}(S) \,|\, T^*T = TT^*,\ \|T\| = 1,\ \mathcal{D}(T) = \mathcal{H}\}.$$

Any operator T from $\mathcal{P}_1(S)$ can be represented in the form

$$T = R + iK \quad \left(R = \frac{T+T^*}{2},\ K = \frac{T-T^*}{2i}\right),$$

where the self-adjoint operators R and K are commuting (i.e., $RK = KR$; sometimes, we simply write $R \cup T$) and $R^2 + K^2 \le I$.

By virtue of Proposition 5.1 (Chapter 1), we obtain

$$T^*x = Tx \quad (\forall x \in G_T),$$

where G_T is an Hermitian domain of the operator T.

This equality implies that $K|_{G_T} = 0$ and, hence, the Hermitian part of the operator T coincides with $R|_{G_T}$:

$$T_0 = T|_{G_T} = R|_{G_T}.$$

Since $S \subset T_0$, it follows from the last equality that $S \subset R$.

Let $T_1 = R_1 + iK$ be a fixed operator from $\mathcal{P}_1(S)$ and let

$$\mathcal{P}(S; K) = \{T \in \mathcal{P}_1(S) \,|\, T = R + iK\}. \tag{4.1}$$

In particular, $\mathcal{P}(S; 0) = \mathcal{P}_0(S)$ is the collection of all self-adjoint extensions of the operator S with the same norm (these extensions are studied in Section 3).

We take an arbitrary operator $T = R + iK$ from $\mathcal{P}(S; K)$ and consider the operator

$$C = T - T_1 = R - R_1. \tag{4.2}$$

Since $S \subset R$ and $S \subset R_1$, the equality $Cx = 0$ is true for all $x \in \mathcal{D}(S)$. Moreover, $K \cup R$ and $K \cup R_1$ and, hence, $K \cup C$. In addition,

$$(R_1 + C)^2 = R^2 \leq I - K^2. \tag{4.3}$$

Thus,

$$-B \leq R_1 + C \leq B, \quad \text{where} \quad B = (I - K^2)^{1/2}.$$

Consequently,

$$-B_- \leq C \leq B_+ \quad (B_- = B + R_1, \ B_+ = B - R_1). \tag{4.4}$$

Conversely, if the operator C satisfies inequalities (4.4) and commutes with K, then (4.3) follows from (4.4) and, therefore, $R^2 + K^2 \leq I$. But then

$$\| Tx \|^2 = \| Rx \|^2 + \| Kx \|^2 \leq \| x \|^2,$$

i.e., $\| T \| \leq 1$. Further, if $C|_{\mathcal{D}(S)} = 0$, then

$$T^* x = Tx = Sx \quad (\forall x \in \mathcal{D}(S))$$

and, thus, $T \in \mathcal{P}(S; K)$.

Denote by Q_- and Q_+ the orthoprojectors onto the subspaces

$$\mathcal{H} \ominus (\sqrt{B_-}\,\mathcal{D}(S)) \quad \text{and} \quad \mathcal{H} \ominus (\sqrt{B_+}\,\mathcal{D}(S)), \tag{4.5}$$

respectively, and consider the operators

$$\hat{C}_- = \sqrt{B_-}\, Q_- \sqrt{B_-} \quad \text{and} \quad \hat{C}_+ = \sqrt{B_+}\, Q_+ \sqrt{B_+}. \tag{4.6}$$

By using the operators $\hat{C}_-$ and $\hat{C}_+$, we define the following self-adjoint operators:

$$R_- = R_1 - \hat{C}_- \quad \text{and} \quad R_+ = R_1 + \hat{C}_+. \tag{4.7}$$

Theorem 4.1. *A normal operator* $T = R + iK$ *belongs to* $\mathcal{P}(S; K)$ *if and only if* $\mathcal{D}(S) \subset \operatorname{Ker} K$ *and*

$$R_- \le R \le R_+. \tag{4.8}$$

Proof. Let $T = R + iK \in \mathcal{P}(S, K)$. Then $S \subset T_0$ and, in view the equality $K|_{G_T} = 0$, we get $\mathcal{D}(S) \subset \operatorname{Ker} K$. Moreover, it is easy to see that $R = R_1 + C$, where the operator C is equal to zero on $\mathcal{D}(S)$ and satisfies condition (4.4). Thus, $C \le B_+$. Hence, by virtue of Lemma 3.3, we conclude that $C \le \hat{C}_+$.

Similarly, since $-C \le B_-$, we have $-C \le \hat{C}_-$. Therefore, $-\hat{C}_- \le C \le \hat{C}_+$ and, consequently,

$$R_1 - \hat{C}_- \le R_1 + C \le R_1 + \hat{C}_+,$$

i.e., $R_- \le R \le R_+$.

Conversely, assume that $\mathcal{D}(S) \subset \operatorname{Ker} K$ and inequalities (4.8) are true. First, we show that $S \subset T$.

Let $W = R - R_-$. Then, in view of (4.8),

$$0 \le W \le R_+ - R_-.$$

It follows from (4.5) and (4.6) that $\hat{C}_- x = \hat{C}_+ x = 0 \quad (x \in \mathcal{D}(S))$. This implies that $R_- x = R_+ x = R_1 x$ for all $x \in \mathcal{D}(S)$. Therefore,

$$0 \le (Wx, x) \le ((R_+ - R_-)x, x) = 0.$$

By virtue of the inequality $W \ge 0$, this yields

$$Wx = 0 \quad (\forall x \in \mathcal{D}(S)).$$

Under the condition of the theorem, $K|_{\mathcal{D}(S)} = 0$ Hence, for $x \in \mathcal{D}(S)$,

$$Tx = Rx = R_1 x = Sx,$$

i.e., $S \subset T$. Similarly, we prove that $S \subset T^*$. Thus, in view of Proposition 5.8 (Chapter 1), the operator T is a regular extension of the operator S.

In this case, since $C = R - R_1$, we get $C|_{\mathcal{D}(S)} = 0$ and, in view of (4.7) and (4.8), $-\hat{C}_- \le C \le \hat{C}_+$. Thus, it follows from (4.6) that inequalities (4.4) are true. As indicated above, these inequalities imply that $\|T\| \le 1$. Therefore, $T = R + iK \in \mathcal{P}(S; K)$.

■

Proposition 4.2. *The operators* $T_- = R_- + iK$ *and* $T_+ = R_+ + iK$ *belong to the set* $\mathcal{P}(S; K)$.

Proof. Indeed, let $\mathcal{H}_1 = \overline{\sqrt{B_-}\mathcal{D}(S)}$ and $\mathcal{H}_2 = \mathcal{H} \ominus \mathcal{H}_1$. Since $K \cup \sqrt{B_-}$ and $\mathcal{D}(S) \subset \operatorname{Ker} K$, we get $\mathcal{H}_1 \subset \operatorname{Ker} K$. Consequently, $K: \mathcal{H} \to \mathcal{H}_2$. In view of the fact that Q_- is the orthoprojector onto $\mathcal{H}_2$ in $\mathcal{H}$, we conclude that $Q_- \cup K$. Hence, $K \cup \hat{C}_-$ and, clearly, $K \cup R_-$. Moreover, $\| T_- \| \le 1$. Therefore, $T_- \in \mathcal{P}(S; K)$. Similarly, it is easy to see that $T_+ \in \mathcal{P}(S; K)$.

■

4.2. Condition under Which the Operator $R_1 = \operatorname{Re} T_1$ ***Coincides with*** R_-. Let $T_1 = R_1 + iK \in \mathcal{P}(S; K)$. Since $R_1^2 + K^2 \le 1$, we get $R_1^2 \le B^2$, where, as earlier, $B = (I - K^2)^{1/2}$. In view of the fact that the operators R_1 and B are commuting, we obtain $R_1 \ge -B$. Hence, the operator $B_- = B + R_1$ is nonnegative.

In the space $\mathcal{H}$, we consider the B_--scalar product and the B_--norm defined by the formulas

$$(x, y)_- = (B_- x, y) \quad \text{and} \quad \| x \|_- = \sqrt{(x, x)_-} \qquad (\{x, y\} \subset \mathcal{H}).$$

Note that the kernel of the operator B_- may differ from the zero set and, therefore, the B_--scalar product can be degenerate.

Thus, consider the operator S in the two-dimensional space $\mathcal{H} = \langle e_1, e_2 \rangle$ with an orthonormal basis $\{e_1, e_2\}$ defined by the formulas

$$Se_1 = e_2 \quad \text{and} \quad \mathcal{D}(S) = \langle e_1 \rangle.$$

There exists a unique regular extension T_1 of the operator S with the same norm defined by the equalities $T_1 e_1 = e_2$ and $T_1 e_2 = e_1$. It is clear that this extension is self-adjoint. In this case, $B_- = I + T_1$ $(T_1 = R_1)$ and, consequently, $B_-(e_1 - e_2) = 0$.

Note that we can formulate sufficient conditions for the invertibility of the operator B_- (e.g., $1 \notin \sigma_p(T_1^* T_1)$). However, in what follows, we do not take into account conditions of this sort and the case $\operatorname{Ker} B_- \ne \{0\}$ is not excluded.

A lineal $\mathcal{L}$ in $\mathcal{H}$ is called *B_--dense* in the space $\mathcal{H}$ if

$$\inf_{x \in \mathcal{L}} \| y - x \|_- = 0 \qquad (\forall y \in \mathcal{H}). \tag{4.9}$$

By analogy with Subsection 3.5, it is not difficult to show that if the lineal $\mathcal{L}$ is B_--dense in $\mathcal{H}$ and

$$(x, y)_- = 0 \qquad (\forall x \in \mathcal{L}), \tag{4.10}$$

then the element y belongs to $\operatorname{Ker} B_-$.

Theorem 4.3. *If $T_1 = R_1 + iK \in \mathcal{P}(S; K)$, then the operator R_1 coincides with the operator R_- if and only if the lineal $\mathcal{D}(S)$ is B_--dense in $\mathcal{H}$*

Proof. Let $R_1 = R_-$. This means that, by virtue of (4.7), $\hat{C}_- = 0$. Further, by repeating of the proof of Theorem 3.5, we show that Theorem 4.3 is true.

■

Note that the operator R_- does not depend on choice of the operator T_1. Indeed, assume that $T = R + iK$ is another operator from $\mathcal{P}(S; K)$ and the operator $\tilde{R}_-$, by analogy with (4.7), is defined by the equality $\tilde{R}_- = R - \hat{\tilde{C}}_-$. According to Theorem 4.1, we conclude that $R_- \leq \tilde{R}_-$ and $\tilde{R}_- \leq R_-$. This means that $\tilde{R}_- = R_-$. Thus, Theorem 4.3 is true for any extension T from $\mathcal{P}(S; K)$.

Remark. Our presentation in Section 4 is based on the results obtained by A. Kuzhel and Rotkevich in [1, 2].

5. Krein's Proof of the Von Neumann Hypothesis

In Section 5, we use the results presented in Section 3 to describe self-adjoint bound-preserving extensions of symmetric operators by the Krein method.

5.1. Extensions of Nonnegative Symmetric Operators. Let A be a closed symmetric operator acting in a Hilbert space $\mathcal{H}$ and semibounded from below. Without loss of generality, we can assume that its greatest lower bound $m(A)$ is equal to zero. Thus, $(Ax, x) \geq 0$ for all $x \in \mathcal{D}(A)$.

By virtue of the generalized Schwartz inequality[6]

$$|(Ax, y)|^2 \leq (Ax, x)(Ay, y) \qquad \{x, y\} \subset \mathcal{D}(A),$$

it is easy to see that

$$\operatorname{Ker} A = \{x \in \mathcal{D}(A) \mid (Ax, x) = 0\}.$$

[6] See, e.g., Riesz and Szökefalvi-Nagy [1].

In the subspace $\operatorname{Ker} A$, the operator A induces a self-adjoint (zero) operator. Therefore, without loss of generality, we can restrict ourselves to the case of positive symmetric operators, i.e., operators satisfying the condition

$$(Ax, x) > 0 \qquad (x \in \mathcal{D}(A),\ x \neq 0). \tag{5.1}$$

Theorem 5.1. *Any positive symmetric operator can be extended to a nonnegative self-adjoint operator.*

Proof. Assume that A is a symmetric operator acting in $\mathcal{H}$ and such that condition (5.1) is satisfied. Without loss of generality, we can assume that this operator is closed. In view of Proposition 1.1, the number -1 is a point of regular type for the operator A. This means that the lineal $(I+A)\mathcal{D}(A)$ is closed.

Consider the operator

$$S = (I-A)(I+A)^{-1} \qquad (\mathcal{D}(S) = (I+A)\mathcal{D}(A)). \tag{5.2}$$

Let us show that S is a strong Hermitian contraction in $\mathcal{H}$. Indeed, let $\{x, y\} \subset \mathcal{D}(S)$. Then

$$x = (I+A)u \quad \text{and} \quad y = (I+A)v, \tag{5.3}$$

where u and v are elements of the lineal $\mathcal{D}(A)$. In this case, by virtue of (5.2), we get

$$Sx = (I-A)u \quad \text{and} \quad Sy = (I-A)v. \tag{5.4}$$

Hence, in view of (5.3) and (5.4), we can write

$$(Sx, y) = (u, v) - (Au, Av) = (x, Sy) \quad (\forall \{x, y\} \subset \mathcal{D}(S)),$$

i.e., the operator S is Hermitian. Further, by using (5.1), we obtain

$$\begin{aligned} \|Sx\|^2 &= \|u - Au\|^2 = \|u\|^2 - 2(Au, u) + \|Au\|^2 \\ &< \|u\|^2 + 2(Au, u) + \|Au\|^2 = \|u + Au\|^2 = \|x\|^2. \end{aligned} \tag{5.5}$$

Thus, $\|Sx\| < \|x\|$ for $x \neq 0$ and, consequently, the operator S is a strong Hermitian contraction.

Let $\tilde{S}$ be an arbitrary self-adjoint extension of the operator S with the same norm. Let us show that $-1 \notin \sigma_p(\tilde{S})$. Indeed, assume that the equality $(\tilde{S} + I)y = 0$ is true for some $y \in \mathcal{H}$.

Then, in view of (5.2)–(5.4), the element $x = (I + A)u$ ($u \in \mathcal{D}(A)$) belongs to $\mathcal{D}(S)$ and $Sx = (I - A)u$. This yields

$$2(u, y) = ((S + I)x, y) = ((\tilde{S} + I)x, y) = (x, (\tilde{S} + I)y) = 0$$

and, hence, $y \perp \mathcal{D}(A)$. Since $\overline{\mathcal{D}(A)} = \mathcal{H}$, we obtain $y = 0$. Thus, $-1 \notin \sigma_p(\tilde{S})$. This enables us to introduce the operator

$$\tilde{A} = (I - \tilde{S})(I + \tilde{S})^{-1} \quad (\mathcal{D}(\tilde{A}) = (I + \tilde{S})\mathcal{H}). \tag{5.6}$$

We show that $\tilde{A}$ is a nonnegative self-adjoint extension of the operator A.

Let $\{u, v\} \in \mathcal{D}(\tilde{A})$. Then

$$u = (I + \tilde{S})x \quad \text{and} \quad v = (I + \tilde{S})y, \tag{5.7}$$

where x and y are some elements from $\mathcal{H}$. In this case,

$$\tilde{A}u = (I - \tilde{S})x \quad \text{and} \quad \tilde{A}v = (I - \tilde{S})y. \tag{5.8}$$

Hence,

$$(\tilde{A}u, v) = (x, y) - (\tilde{S}x, \tilde{S}y) = (u, \tilde{A}v) \tag{5.9}$$

and, thus, the operator $\tilde{A}$ is Hermitian. Moreover, it follows from (5.9) that

$$(\tilde{A}u, u) = \|x\|^2 - \|\tilde{S}x\|^2 \geq 0. \tag{5.10}$$

This means that the operator $\tilde{A}$ is nonnegative.

Let us show that $A \subset \tilde{A}$. Indeed, if $u \in \mathcal{D}(A)$, then, by virtue of (5.3) and (5.4), we obtain

$$x = u + Au \in \mathcal{D}(S) \quad \text{and} \quad \tilde{S}x = Sx = u - Au, \tag{5.11}$$

whence

$$u = \frac{1}{2}(I + \tilde{S})x \in \mathcal{D}(\tilde{A}).$$

Then, by using (5.7), (5.8), and (5.11), we get

$$\tilde{A}u = \frac{1}{2}(I - \tilde{S})x = \frac{1}{2}(I - S)x = Au$$

and, hence, $A \subset \tilde{A}$. It remains to show that $\tilde{A}^* = \tilde{A}$.

Let $x \in \mathcal{H}$ and $u = (I+\tilde{S})x$. Then, by virtue of (5.7) and (5.8), we conclude that $(\tilde{A}+I)u = 2x$, i.e.,

$$\mathcal{R}(\tilde{A}+I) = \mathcal{H}. \tag{5.12}$$

In this case, the inequality $\tilde{A} \geq 0$ implies that $-1 \in \pi(\tilde{A})$, whence, in view of (5.12), we get $-1 \in \rho(\tilde{A})$. Thus, the defect numbers of the symmetric operator $\tilde{A}$ are equal to zero and, therefore, $\tilde{A}$ is a self-adjoint operator.

■

Let $\mathcal{P}_+(A)$ be the set of all nonnegative self-adjoint extensions of a positive symmetric operator A and let, as earlier (Section 3), $\mathcal{P}_0(S)$ be the set of all self-adjoint norm-preserving extensions of a strong Hermitian contraction S. It follows from the proof of Theorem 5.1 that relation (5.6) associates every operator $\tilde{S}$ from $\mathcal{P}_0(S)$ with an operator $\tilde{A} \in \mathcal{P}_+(A)$. Conversely, if $\tilde{A} \in \mathcal{P}_+(A)$, then the operator $\tilde{S}$ defined by relation (5.2) (with $\tilde{A}$ substituted for A) belongs to $\mathcal{P}_0(S)$. Hence, by virtue of Theorem 3.4, we conclude that the operators

$$A_\mu = (I-S_-)(I+S_-)^{-1} \quad \text{and} \quad A_M = (I-S_+)(I+S_+)^{-1}, \tag{5.13}$$

are elements of the set $\mathcal{P}_+(A)$. Here, S_- and S_+ are the operators defined by equalities (3.33).

The operators A_μ and A_M are called the *hard* and *soft* extensions of a positive symmetric operator A, respectively. Generally speaking, $A_\mu \neq A_M$. In the case where $A_\mu = A_M$, it is obvious that the operator A has only one nonnegative self-adjoint extension.

5.2. Conditions under Which the Hard Extension A_μ Coincides with the Friedrichs Extension. In Section 2, we constructed the Friedrichs extension of a symmetric operator A with lower bound $m(A) = 1$. If, instead of A, we consider the nonnegative operator $A - I$, then, clearly, we can speak about the Friedrichs extension of an arbitrary nonnegative symmetric operator (and, in general, about the Friedrichs extension of an arbitrary symmetric operator semibounded from below). In this connection, it is reasonable to ask: Is it possible to establish a relationship between the Friedrichs extension $\tilde{A}$ of a positive symmetric operator A and its hard extension A_μ?

Theorem 5.2. *The hard extension A_μ of a positive symmetric operator A coincides with the Friedrichs extension $\tilde{A}$ of this operator.*

Proof. Let $\tilde{A}$ be the Friedrichs extension of a positive symmetric operator A. Then the operator $\tilde{B} = I + \tilde{A}$ is the Friedrichs extension of the operator $B = I + A$ such that

$$(Bx, x) \geq (x, x) \quad (x \in \mathcal{D}(B) = \mathcal{D}(A)).$$

All construction in Section 2 are true for the operator B. In particular, we can introduce the relevant scalar product

$$(x, y)_0 = (Bx, y) = ((I + A)x, y)$$

and construct the lineal $\mathcal{D}_0$ such that

$$\mathcal{D}(A) = \mathcal{D}(B) \subset \mathcal{D}_0 \subset \mathcal{H}.$$

Moreover, the lineal $\mathcal{D}(A)$ is dense in $\mathcal{D}_0$ both in the original norm and in the norm $\|\cdot\|_0$. In our case,

$$\|x\|_0^2 = (x, x)_0 = ((I + A)x, x) \quad (x \in \mathcal{D}(A)).$$

This equality remains true for all $x \in \mathcal{D}(\tilde{A})$, i.e.,

$$\|x\|_0^2 = (x, x)_0 = ((I + \tilde{A})x, x) \quad (x \in \mathcal{D}(\tilde{A})). \tag{5.14}$$

Consider the operators

$$S = (I - A)(I + A)^{-1} \quad \text{and} \quad \tilde{S}_0 = (I - \tilde{A})(I + \tilde{A})^{-1}.$$

As shown in Theorem 5.1, the operator S is a strong Hermitian contraction in the space $\mathcal{H}$ and $\tilde{S}_0$ is a norm-preserving self-adjoint extension of the operator S. Since

$$\mathcal{H} = \mathcal{D}(\tilde{S}_0) = (I + \tilde{A})\,\mathcal{D}(\tilde{A}),$$

any element y of the space $\mathcal{H}$ can be represented in the form $y = (I + \tilde{A})h$, where $h \in \mathcal{D}(\tilde{A})$. Then $\tilde{S}_0 y = (I - \tilde{A})h$ and, hence, in view of (3.28),

$$h = \frac{1}{2}(y + \tilde{S}_0 y) = \frac{1}{2} B_- y. \tag{5.15}$$

Thus, for $h \in \mathcal{D}(\tilde{A})$, by using relations (5.14) and (5.15), we arrive at the equality

$$\|h\|_0^2 = ((I+\tilde{A})h, h) = \frac{1}{2}(y, B_- y) = \frac{1}{2}\|y\|_-^2. \tag{5.16}$$

Therefore, if $y = (I+\tilde{A})h \in \mathcal{H}$ and $x = (I+A)\varphi = (I+\tilde{A})\varphi \in \mathcal{D}(S)$, then, by virtue of (5.16), we obtain

$$\inf_{x \in \mathcal{D}(S)} \|y - x\|_- = \sqrt{2} \inf_{\varphi \in \mathcal{D}(A)} \|h - \varphi\|_0.$$

Since the lineal $\mathcal{D}(A)$ is dense in $\mathcal{D}_0$, the right-hand side of the last equality is equal to zero. Thus, for any $y \in \mathcal{H}$, we have

$$\inf_{x \in \mathcal{D}(S)} \|y - x\|_- = 0,$$

i.e., the lineal $\mathcal{D}(S)$ is B_--dense in $\mathcal{H}$. In view of Theorem 3.5, this yields $\tilde{S}_0 = S_-$. Consequently,

$$A_\mu = (I - S_-)(I + S_-)^{-1} = (I - \tilde{S}_0)(I + \tilde{S}_0)^{-1} = \tilde{A}.$$

■

5.3. Additional Properties of the Operators A_μ *and* A_M. As follows from Theorem 3.4, a self-adjoint operator T belongs to the set $\mathcal{P}_0(S)$ if and only if $S_- \le T \le S_+$, where the operators S_- and S_+ are defined by equalities (3.33).

Consider the operator $R = -S$, which is also an Hermitian contraction. In this case, according to the Theorem 3.4, a self-adjoint operator M belongs to the set $\mathcal{P}_0(R)$ if and only if

$$R_- \le M \le R_+.$$

Let us show that

$$R_- = -S_+ \quad \text{and} \quad R_+ = -S_-. \tag{5.17}$$

In fact, consider the operator $\tilde{R}_0 = -\tilde{S}_0$, which is a norm-preserving self-adjoint extension of the operator R.

By virtue of (3.33), we have

$$R_- = \tilde{R}_0 - \tilde{\tilde{C}}_- = -(\tilde{S}_0 + \tilde{\tilde{C}}_-), \tag{5.18}$$

where $\tilde{\tilde{C}} = \sqrt{\tilde{B}_-}\,\tilde{Q}_-\sqrt{\tilde{B}_-}$. In addition, (3.28) implies that

$$\tilde{B}_- = I + \tilde{R}_0 = I - \tilde{S}_0 = B_+. \tag{5.19}$$

At the same time, $\tilde{Q}_-$ is the orthoprojector onto the subspace $\mathcal{H} \ominus \left(\sqrt{\tilde{B}_-}\,\mathcal{D}(S)\right)$. In view of (5.19), this subspace coincides with the subspace $\mathcal{H} \ominus (\sqrt{B_+}\,\mathcal{D}(S))$. Hence, $\tilde{Q}_- = Q_+$ and, therefore,

$$\tilde{\hat{C}} = \sqrt{B_+}\,Q_+\sqrt{B_+} = \hat{C}_+. \tag{5.20}$$

Inserting (5.20) in (5.18) and taking into account relation (3.33), we obtain the first equality in (5.17). Similarly, it is easy to see that $R_+ = -S_-$. Hence, equalities (5.17) can be rewritten in the form

$$(-S)_- = -S_+ \quad \text{and} \quad (-S)_+ = -S_-.$$

Theorem 5.3. *Assume that A is a closed positive symmetric operator and that the operator A^{-1} is densely defined. Then*

$$(A_\mu)^{-1} = (A^{-1})_M \quad \textit{and} \quad (A_M)^{-1} = (A^{-1})_\mu. \tag{5.21}$$

Proof. It follows from the conditions of the theorem that A^{-1} is a closed positive symmetric operator. According to Theorem 5.1, this means that the operator A^{-1} can be extended to a nonnegative self-adjoint operator.

In the proof of Theorem 5.1, we considered the Hermitian contraction S defined by equality (5.2). Similarly, we can consider the Hermitian contraction

$$R = (I - A^{-1})(I + A^{-1})^{-1}, \qquad \mathcal{D}(R) = (I + A^{-1})\,\mathcal{D}(A^{-1}). \tag{5.22}$$

By comparing (5.22) with (5.2), we conclude that $R = -S$.

Let us show that $1 \notin \sigma_p(S_-)$. Indeed, suppose that $S_- y = y$ for some $y \in \mathcal{H}$. Also let $u \in \mathcal{D}(A)$. In this case, by using (5.2)–(5.4), we get

$$x = u + Au \in \mathcal{D}(S) \quad \text{and} \quad Sx = u - Au.$$

Therefore,

$$2(Au, y) = ((I - S)x, y) = ((I - S_-)x, y) = (x, (I - S_-)y) = 0$$

and, thus, $y \perp \mathcal{R}(A)$. Note that, under the conditions of the theorem, we have

$$\overline{\mathcal{R}(A)} = \overline{\mathcal{D}(A^{-1})} = \mathcal{H}.$$

Consequently, $y = 0$. This enables to conclude that $1 \notin \sigma_p(S_-)$. In exactly the same way, we show that $1 \notin \sigma_p(S_+)$. Hence, by virtue of (5.13), the operators A_μ and A_M are invertible. Moreover, in view of equalities (5.17), we have

$$A_\mu^{-1} = (I+S_-)(I-S_-)^{-1} = (I-R_+)(I+R_+)^{-1} = (A^{-1})_M,$$

$$A_M^{-1} = (I+S_+)(I-S_+)^{-1} = (I-R_-)(I+R_-)^{-1} = (A^{-1})_\mu$$

but this proves relations (5.22).

■

Remarks. A detailed description of Krein's method can be found in Akhiezer and Glazman [1]. Coddington and Snoo [1] generalized Krein's result concerning the existence of two "extreme" nonnegative extensions A_μ and A_M not only to the case of Hermitian (nondensely defined) operators but also to the case of nonnegative linear relations.

6. Squared Symmetric Operators

In the present section, we establish conditions under which the operators B^*B and BB^* are, respectively, the hard and soft extensions of a symmetric operator B^2.

6.1. Operators A^*A and AA^*. It is clear that if A is a bounded operator acting in the space $\mathcal{H}$ ($\mathcal{D}(A) = \mathcal{H}$), then the operators A^*A and AA^* are self-adjoint. Let us show that this property remains true for a broad class of unbounded operators.

Theorem 6.1. *Let A be a closed densely defined operator acting in the space $\mathcal{H}$. Then the operator A^*A is self-adjoint.*

Proof. In the space $\tilde{\mathcal{H}} = \mathcal{H} \oplus \mathcal{H}$, we consider a unitary operator $\tilde{U}$ defined by the equality

$$\tilde{U}\langle x, y\rangle = \langle y, -x\rangle, \tag{6.1}$$

where, as in Subsection 4.4 (Chapter 1), $\langle x, y\rangle$ is an element of the space $\tilde{\mathcal{H}}$. In this notation the equality $(Ax, y) = (x, A^*y)$ can be rewritten in the form

$$\big(\langle x, Ax\rangle, \tilde{U}\langle y, A^*y\rangle\big)_{\tilde{\mathcal{H}}} = 0, \tag{6.2}$$

where $(\cdot,\cdot)_{\tilde{\mathcal{H}}}$ is a scalar product in $\tilde{\mathcal{H}}$ [see equality (4.16) in Chapter 1]. Since $\langle x, Ax\rangle$ and $\langle y, A^*y\rangle$ are elements of the graphs G_A and G_{A^*} of the operators A and A^*, respectively, equality (6.2) means that the subspace $\tilde{U}G_{A^*}$ of the space $\tilde{\mathcal{H}}$ is orthogonal to the subspace G_A. Conversely, if $\langle u, v\rangle \perp G_A$ ($\langle u, v\rangle \in \tilde{\mathcal{H}}$), then $\langle u, v\rangle \in \tilde{U}G_{A^*}$ and $u = -A^*v$. Hence,

$$\tilde{\mathcal{H}} = G_A \oplus \tilde{U}G_{A^*}. \tag{6.3}$$

Let $h \in \mathcal{H}$. Then, by virtue of (6.1) and (6.3), the vector $\langle h, 0\rangle$ can be represented in the form

$$\langle h, 0\rangle = \langle x, Ax\rangle + \langle A^*y, -y\rangle, \tag{6.4}$$

where $x \in \mathcal{D}(A)$ and $y \in \mathcal{D}(A^*)$. It follows from (6.4) that $h = x + A^*y$ and $Ax - y = 0$. Therefore, $h = (I + A^*)x$. Thus, $\mathcal{R}(I + A^*A) = \mathcal{H}$ and, obviously, the operator $I + A^*A$ is Hermitian and invertible.

We now consider the operator $T = (I + A^*A)^{-1}$ and show that T is a self-adjoint contraction in $\mathcal{H}$. Indeed, since $\mathcal{D}(T) = \mathcal{H}$ and $\mathcal{R}(T) = \mathcal{D}(I + A^*A)$, for any x and y from $\mathcal{H}$, we have

$$(Tx, y) = (Tx, (I + A^*A)Ty) = ((I + A^*A)Tx, Ty) = (x, Ty),$$

i.e., $T^* = T$. Moreover, if $y = Tx$, then

$$\|x\|^2 = \|(I + A^*A)y\|^2 = \|y\|^2 + 2\|Ay\|^2 + \|A^*Ay\|^2 \geq \|y\|^2$$

and, hence, $\|Tx\| = \|y\| \leq \|x\|$ ($\forall x \in \mathcal{H}$).

Further, since the operator T is positive (as the inverse operator of a positive operator) and $\mathcal{D}(T) = \mathcal{H}$, the lineal $\mathcal{R}(T) = \mathcal{D}(I + A^*A)$ is dense in $\mathcal{H}$. In fact, if $\varphi \perp \mathcal{R}(T)$, then $(T\varphi, \varphi) = 0$, whence we get $T\varphi = 0$ and, consequently, $\varphi = 0$.

Thus, the operator $I + A^*A$ is defined in a set dense in $\mathcal{H}$. This operator is symmetric and $0 \in \rho(I + A^*A)$. Therefore, the operator $I + A^*A$ and the operator A^*A are self-adjoint.

■

Corollary 6.2. *If A is a closed densely defined operator acting in $\mathcal{H}$, then the operator AA^* is self-adjoint.*

Proof. The proof of Corollary 6.2 follows from Theorem 6.1 with A^* substituted for A.

■

6.2. The Hard and Soft Extensions of a Symmetric Operator B^2

Proposition 6.3. *Let $\hat{A}$ be a self-adjoint extension of a symmetric operator A semibounded from below and let $\mathcal{D}(\hat{A}) \subset \mathcal{D}_0$, where $\mathcal{D}_0$ is the lineal defined in Subsection 2.1. Then the extension $\hat{A}$ coincides with the Friedrichs extension $\tilde{A}$ of the operator A.*

Proof. As shown in Subsection 2.2, the Friedrichs extension $\tilde{A}$ of the symmetric operator A is a restriction of the operator A^* to the lineal $\mathcal{D}(\tilde{A}) = \mathcal{D}(A^*) \cap \mathcal{D}_0$. If the operator $\hat{A}$ satisfies the conditions of the proposition, then it follows from the inclusions

$$\mathcal{D}(\hat{A}) \subset \mathcal{D}_0 \quad \text{and} \quad \mathcal{D}(\hat{A}) \subset \mathcal{D}(A^*)$$

that $\mathcal{D}(\hat{A}) \subset \mathcal{D}(\tilde{A})$. Moreover, the operator $\tilde{A}$ is, at the same time, a restriction of the operator A^*. Thus, the self-adjoint operator $\tilde{A}$ is an extension of the self-adjoint operator $\hat{A}$, which is possible only in the case where $\hat{A} = \tilde{A}$.

■

Theorem 6.4. *If B is a closed symmetric operator such that the operator B^2 is densely defined and $(B^2)^* = B^*B^*$, then the operator B^*B is the Friedrichs (hard) extension of the operator B^2.*

Proof. It is known (see Subsection 2.2) that the Friedrichs extension of the symmetric operator $A = I + B^2$ is defined by equality (2.3), where $\mathcal{D}_0$ is the completion of $\mathcal{D}_0(B^2)$ in the norm

$$\|x\|_0^2 = (Ax, x) = \|x\|^2 + \|Bx\|^2 \quad (x \in \mathcal{D}(B^2)).$$

Since B is a closed operator, the lineal $\mathcal{D}(B)$ with norm $\|\cdot\|_0$ is a Hilbert space. Denote this space by $\mathcal{D}[B]$. It is clear that $\mathcal{D}_0$ is a subspace of $\mathcal{D}[B]$. Let us show that $\mathcal{D}_0 = \mathcal{D}[B]$. Indeed, let $h \in \mathcal{D}[B]$ and $(x, h)_0 = 0$ $(\forall x \in \mathcal{D}_0)$.
Then

$$((I + B^2)x, h) = 0 \quad (\forall x \in \mathcal{D}(B^2)).$$

Hence, $h \in \operatorname{Ker}(I + B^{*2})$. By using the equality $I + B^{*2} = (B^* + iI)(B^* - iI)$, we obtain

$$\operatorname{Ker}(I + B^{*2}) = \operatorname{Ker}(B^* + iI) \dotplus (B^* - iI).$$

Thus, the vector h belongs both to the lineal $\mathcal{D}(B)$ and to the lineal $\mathfrak{N}_{-i} \dotplus \mathfrak{N}_i$. In view of Theorem 4.1 (Chapter 1), this is possible only if $h = 0$.

Hence, $\mathcal{D}_0 = \mathcal{D}[B]$. To complete the proof, it suffices to note that the self-adjoint (according to Theorem 6.1) operator $\hat{A} = I + B^*B$ satisfies the condition $\mathcal{D}(\hat{A}) \subset \mathcal{D}(B) = \mathcal{D}_0$ and apply Proposition 6.3.

■

Theorem 6.5. *Let B be a closed positive symmetric operator in the space $\mathcal{H}$ such that the conditions of Theorem 6.4 are satisfied and the lineal $\mathcal{R}(B^2)$ is dense in $\mathcal{H}$. Then the operator BB^* is the soft extension of the operator B^2.*

Proof. It follows from the conditions of the theorem that the operator B^{-1} exists. Moreover, it is closed and symmetric. Since $\mathcal{R}(B^2)$ is dense in $\mathcal{H}$, the operator

$$A = (B^{-1})^2, \quad \mathcal{D}(A) = \mathcal{R}(B^2),$$

is densely defined. Further, in view of the equality $(B^2)^* = B^*B^*$, it is easy to see that $((B^{-1})^2)^* = ((B^{-1})^*)^2$. By virtue of Theorem 6.4, this means that the hard extension of the operator A has the form $A_\mu = (B^{-1})^* B^{-1}$. Clearly, $A^{-1} = B^2$ is a densely defined symmetric operator in $\mathcal{H}$. It follows from Theorem 5.3 that

$$(B^2)_M = (A^{-1})_M = A_\mu^{-1} = ((B^{-1})^* B^{-1})^{-1} = BB^*.$$

■

Let us illustrate the results established above by a simple example. In the space $L_2(0, 1)$, we consider a symmetric operator

$$B = i\frac{d}{ds}, \quad \mathcal{D}(B) = \mathring{W}_2^1(0, 1),$$

where $W_2^1(0, 1)$ is a Sobolev space. It is clear that

$$B^* = i\frac{d}{ds}, \quad \mathcal{D}(B^*) = W_2^1(0, 1).$$

By virtue of Theorems 6.4 and 6.5, we conclude that the operators

$$A_\mu = B^*B = -\frac{d^2}{ds^2}, \qquad \mathcal{D}(A_\mu) = \left\{ f(s) \in W_2^2(0,1) \,\middle|\, f(0) = f(1) = 0 \right\},$$

and

$$A_M = BB^* = -\frac{d^2}{ds^2}, \qquad \mathcal{D}(A_M) = \left\{ f(s) \in W_2^2(0,1) \,\middle|\, f'(0) = f'(1) = 0 \right\},$$

are the hard and soft extensions of the symmetric operator

$$A = B^2 = -\frac{d^2}{ds^2}, \qquad \mathcal{D}(A) = \mathring{W}_2^2(0,1),$$

respectively.

Remarks. Other proofs of Theorem 6.1 and 6.4 can be found, e.g., in Reed and Simon [1]. Another proof of Proposition 6.3 was given by Krein [2]. Theorem 6.5 is formulated and proved here for the first time.

7. Self-Adjoint Bound-Preserving Extensions of Semibounded Hermitian Operators

In the present section, we consider different conditions for the existence of self-adjoint semibounded bound-preserving extensions of Hermitian operators.

7.1. Preliminary Comments. As in the case of symmetric operators (Subsection 1.1), we introduce semibounded Hermitian operators and the greatest lower bound $m(A)$ of an Hermitian operator A. By repeating of the proof of Proposition 1.1, we establish the following proposition:

Proposition 7.1. *Let A be an Hermitian operator and let $\lambda < m(A)$. Then $\lambda \in \pi(A)$.*

For semibounded Hermitian operators, we can formulate the same von Neumann problem as in the case of symmetric operators:

Is it possible to construct a bound-preserving self-adjoint extension for any semi-bounded Hermitian operator A?

For Hermitian operators, this problem is not always solvable. Indeed, consider the following example:

Example. Let $\{e_1, e_2\}$ be an orthonormal basis in the space $\mathcal{H} = \langle e_1, e_2 \rangle$. In the space $\mathcal{H}$, we consider an Hermitian operator A such that $Ae_1 = e_2$, $\mathcal{D}(A) = \langle e_1 \rangle$. Any extension $\tilde{A}$ of this operator is defined as

$$\tilde{A}e_1 = e_2, \qquad \tilde{A}e_2 = ae_1 + be_2,$$

where a and b are fixed complex numbers. The operator $\tilde{A}$ is self-adjoint provided that $a = 1$ and $\bar{b} = b$.

The eigenvalues of the self-adjoint operator $\tilde{A}$ have the form

$$\lambda_1 = \frac{b - \sqrt{b^2+4}}{2} \quad \text{and} \quad \lambda_2 = \frac{b + \sqrt{b^2+4}}{2}$$

and, hence, $\lambda_1 < 0$ and $\lambda_2 > 0$. Thus, it is impossible to construct a nonnegative self-adjoint extension $\tilde{A}$ of the nonnegative operator A.

Similarly, if the operator A is defined in the space $\mathcal{H} = \langle e_1, e_2 \rangle$ by the formula

$$Ae_1 = ae_1 + be_2 \qquad (\mathcal{D}(A) = \langle e_1 \rangle),$$

where $a \in \mathbb{R}$ and $b \in \mathbb{C}$ are arbitrary numbers, then $m(A) = a$ and any self-adjoint extension $\tilde{A}$ of the operator A has the form

$$\tilde{A}e_1 = Ae_1, \qquad \tilde{A}e_2 = \bar{b}e_1 + ce_2 \qquad (c \in \mathbb{R}).$$

Moreover,

$$m(\tilde{A}) = \frac{a + c - \sqrt{(a-c)^2 + |b|^2}}{2}.$$

It is easy to see that $m(\tilde{A}) < a$ if $b \neq 0$.

Thus, the von Neumann problem is not always solvable for Hermitian operators. In this connection, it is interesting to analyze the conditions under which the von Neumann problem is solvable?

7.2. Sufficient Conditions for the Existence of Bound-Preserving Extensions

Theorem 7.2. *Let A be a closed Hermitian operator such that*

$$(Ax, x) \geq c(x, x) \quad (\forall x \in \mathcal{D}(A)). \tag{7.1}$$

If $c \in \pi(A)$ and $c \notin \sigma_p(QA)$, where Q is the orthoprojector onto the subspace $\overline{\mathcal{D}(A)}$ in the space $\mathcal{H}$, then the operator A admits a bound-preserving extension to a self-adjoint operator $\tilde{A}$.

Proof. Since $c \notin \sigma_p(QA)$, by virtue of Proposition 3.3 (Chapter 1), the domain of definition $\mathcal{D}(A)$ and the defect subspace $\mathfrak{N}_c$ of the operator A are linearly independent. In the lineal $\mathcal{D}(\tilde{A}) = \mathcal{D}(A) \dotplus \mathfrak{N}_c$, we consider an Hermitian operator $\tilde{A}$ defined by the equality

$$\tilde{A}(x_0 + x_c) = Ax_0 + cx_c \quad (x_0 \in \mathcal{D}(A),\ x_0 \in \mathfrak{N}_c).$$

By using the same reasoning as in the proof of Theorem 1.3, we conclude that

$$(\tilde{A}x, x) \geq c(x, x) \quad (\forall x \in \mathcal{D}(\tilde{A}))$$

and, hence, $\tilde{A}$ is an Hermitian bound-preserving extension of the operator A.

Let us show that the operator $\tilde{A}$ is densely defined and its defect numbers are equal to zero. Indeed, let $y \perp \mathcal{D}(\tilde{A})$. Then $y \perp \mathfrak{N}_c$ and, therefore, $y \in \overline{\mathcal{R}(A - cI)}$. Since $c \in \pi(A)$, we see that $\mathcal{R}(A - cI)$ is a subspace of the space $\mathcal{H}$. This implies that $y = (A - cI)x$ where x is an element of $\mathcal{D}(A)$. By applying the operator Q to both sides of this equality, we obtain $0 = (QA - cI)x$. Therefore $x = 0$, because $c \notin \sigma_p(QA)$. Consequently, $y = 0$, i.e., $\overline{\mathcal{D}(\tilde{A})} = \mathcal{H}$.

Further, as in the proof of Theorem 1.3, we show that the defect subspaces $\tilde{\mathfrak{N}}_i$ and $\tilde{\mathfrak{N}}_{-i}$ of the operator $\tilde{A}$ are zero spaces.

This means that, by applying (if necessary) the operation of closure to the operator $\tilde{A}$, we obtain the self-adjoint operator $\overline{\tilde{A}}$ which is a self-adjoint extension of the operator A semibounded from below with the same greatest lower bound c.

■

Let A $(Ae_1 = e_2)$ be the Hermitian operator from the example presented above. Since $(Ae_1, e_1) = 0$, in view of Theorem 7.2, for any $c < 0$, there exists the self-adjoint extension $\tilde{A}$ of the operator A satisfying condition (7.1). At the same time, as shown above, the self-adjoint extension of the operator A satisfying condition (7.1) with $c = 0$ does not exist.

7.3. Necessary and Sufficient Conditions

I. An Hermitian operator A is called *positively closable* if the relations

$$\lim_{n\to\infty}(Ax_n, x_n) = 0 \quad \text{and} \quad \lim_{n\to\infty} Ax_n = h \tag{7.2}$$

imply that $h = 0$.

It is easy to show that any nonnegative symmetric operator is positively closable. Indeed, if A is a nonnegative symmetric operator and relations (7.2) are true, then, by passing to the limit in the inequality

$$|(Ax_n, y)|^2 \le (Ax_n, x_n)(Ay, y) \qquad (\forall\, y \in \mathcal{D}(A)),$$

we obtain $(h, y) = 0$ $(\forall\, y \in \mathcal{D}(A))$ and, therefore, $h = 0$.

Note that the operator A from the example is not positively closable. Indeed, the equality $(Ax, x) = 0$ holds for all $x \in \mathcal{D}(A)$ but $Ax \ne 0$ for $x \ne 0$.

Assume that the operator A satisfies the condition of Theorem 7.2.

Let us show that the nonnegative Hermitian operator

$$B = A - cI \quad (\mathcal{D}(B) = \mathcal{D}(A))$$

is positively closable.

Indeed, let B satisfy condition (7.2), i.e.,

$$(Bx_n, x_n) \to 0 \quad \text{and} \quad Bx_n \to h \quad (n \to \infty).$$

Since $c \in \pi(A)$, we have

$$\| Bx \| = \| (A - cI)x \| \ge k \| x \| \qquad (x \in \mathcal{D}(A))$$

for some $k > 0$. In view of the fact that $Bx_n \to h$, this implies that the sequence $\{x_n\}$ is fundamental and, thus, converges to an element $x \in \mathcal{H}$. Since the operator A is closed, the operator B is also closed and we have $x \in \mathcal{D}(A)$ and $h = Bx$.

Further, in the inequality

$$|(Bx_n, y)|^2 \le (Bx_n, x_n)(By, y) \qquad (y \in \mathcal{D}(A)),$$

we pass to the limit as $n \to \infty$ and conclude that $(h, y) = 0$ for all $y \in \mathcal{D}(A)$, i.e., the vector $h = Bx$ is orthogonal to the lineal $\mathcal{D}(A)$. Hence, $QBx = QAx - cx = 0$ and, clearly, $QAx = cx$. It follows from the relation $c \notin \sigma_p(QA)$ that $x = 0$ and, therefore, $h = Bx = 0$. This proves that the operator B is positively closable.

Proposition 7.3. *If the operator A is positively closable, then the equalities $(Ax, x) = 0$ and $Ax = 0$ are equivalent.*

Proof. Indeed, assume that $(Ax, x) = 0$ and the operator A is positively closable. Consider a sequence $\{x_n\}$ all elements of which are equal to x. In this case, relations (7.2) are true with $Ax = h$. Therefore, $Ax = 0$.

■

Theorem 7.4. *A closed nonnegative Hermitian operator A admits a nonnegative self-adjoint extension if and only if this operator is positively closable.*

Proof. Assume that there exists a nonnegative self-adjoint extension $\tilde{A}$ of a nonnegative Hermitian operator A. Then, for any x and y from $\mathcal{D}(\tilde{A})$, we have

$$|(\tilde{A}x, y)|^2 \leq (\tilde{A}x, x)(\tilde{A}y, y). \tag{7.3}$$

Let $(Ax_n, x_n) \to 0$ and $Ax_n \to h$ as $n \to \infty$. In (7.3), we replace x by x_n and pass in the resulting inequality to the limit. This gives $(h, y) = 0$ for all $y \in \mathcal{D}(\tilde{A})$. Therefore, $h = 0$ and, hence, the operator A is positively closable.

Conversely, suppose that a closed nonnegative operator A acting in a Hilbert space $\mathcal{H}$ is positively closable. Consider an operator B defined by the formula

$$B(Ax) = P_0 x \quad (x \in \mathcal{D}(A)), \tag{7.4}$$

where P_0 is the orthoprojector onto the subspace $\mathcal{H}_0 = \overline{\mathcal{R}(A)}$ in $\mathcal{H}$. The operator B is correctly defined. Indeed, if $Ax = 0$, then, for any $y \in \mathcal{D}(A)$,

$$(Ay, x) = (y, Ax) = 0.$$

Hence, $x \perp \mathcal{H}_0$ but this means that $P_0 x = 0$.

Let us show that B is a positive Hermitian operator. In fact, relation (7.4) implies that

$$(B(Ax), Ay) = (P_0 x, Ay) = (x, Ay) = (Ax, y) = (Ax, P_0 y) = (Ax, B(Ay)) \tag{7.5}$$

for all x and y from $\mathcal{D}(A)$ and, thus, the operator B is Hermitian. Further, it follows from (7.5) that

$$(B(Ax), Ax) = (Ax, x) \geq 0, \tag{7.6}$$

for $y = x$, i.e., B is a nonnegative operator.

Let $(By, y) = 0$, where $y = Ax$. Then, by virtue of (7.6), $(Ax, x) = 0$. Since the operator A is positively closable, in view of Proposition 7.3, we obtain $y = Ax = 0$.

Therefore, $B > 0$ and $\operatorname{Ker} B = \{0\}$. Hence, the operator B is invertible.

We consider two possible cases: $\mathcal{H}_0 = \mathcal{R}(A)$ and $\mathcal{H}_0 \neq \mathcal{R}(A)$.

1. Assume that $\mathcal{H}_0 = \mathcal{R}(A)$ and $\tilde{A}$ is an operator defined by the formula

$$\tilde{A} = B^{-1} P_0. \tag{7.7}$$

It follows from (7.4) and (7.7) that the equality $\tilde{A}x = Ax$ is true for $x \in \mathcal{D}(A)$. Therefore, $\tilde{A}$ is an extension of the operator A. Moreover, since

$$\mathcal{R}(\tilde{A}) = \mathcal{D}(B) = \mathcal{H}_0$$

and the operator B^{-1} is nonnegative, we have

$$(\tilde{A}x, x) = (B^{-1}P_0 x, P_0 x) \geq 0$$

for all $x \in \mathcal{D}(\tilde{A})$.

Note that the fact that the Hermitian operator B acts in $\mathcal{H}_0$ and $\mathcal{D}(B) = \mathcal{H}_0$ implies that B and, hence, B^{-1} are self-adjoint operators acting in the subspace $\mathcal{H}_0$ of the space $\mathcal{H}$. Then the operator $\tilde{A} = B^{-1}P_0 = P_0 B^{-1} P_0$ is a self-adjoint operator acting in $\mathcal{H}$.

Thus, for $\mathcal{R}(A) = \mathcal{H}_0$, the operator $\tilde{A}$ defined by equality (7.7) is a nonnegative self-adjoint extension of the operator A.

2. Let $\mathcal{R}(A) \neq \mathcal{H}_0$. Then, as shown above, B is a positive operator defined in the set $\mathcal{D}(B) = \mathcal{R}(A)$ dense in $\mathcal{H}_0$.

Consider the operator $Q = B + I$. Clearly,

$$(Qx, x) \geq (x, x) \quad (x \in \mathcal{D}(Q) = \mathcal{D}(B)).$$

Denote the Friedrichs extension of the operator Q by $\tilde{Q}$, i.e.,

$$\tilde{Q}x = Q^{*}x \quad (x \in \mathcal{D}(\tilde{Q}) = \mathcal{D}(Q^{*}) \cap \mathcal{D}_0),$$

where $\mathcal{D}_0$ is a subspace of the space $\mathcal{H}_0$ defined in the same way as in Subsection 2.1. Then $\tilde{B} = \tilde{Q} - I$ is a nonnegative self-adjoint extension of the operator B (in the space $\mathcal{H}_0$). Let us show that the operator $\tilde{B}$ is invertible. Indeed, assume that $\tilde{B}y = 0$, i.e., $\tilde{Q}y = y$. Since $y \in \mathcal{D}(\tilde{Q})$, there exists a sequence of elements $\{y_n\}$, $y_n \in \mathcal{D}(Q)$, such that $\| y_n - y \|_0 \to 0$. Then $y_n \to y$ as $n \to \infty$ and

$$\| y_n \|_0^2 = (Qy_n, y_n) = (By_n, y_n) + \| y_n \|^2 \to \| y \|^2,$$

i.e., $(By_n, y_n) \to 0$. Since $y_n = Ax_n$, in view of equality (7.6), we conclude that $(Ax_n, x_n) \to 0$. Moreover, it is obvious that $y_n = Ax_n \to y$. Since the operator A is positively closable, we get $y = 0$. This proves the invertibility of the operator $\tilde{B}$.

Consider the operator

$$\tilde{A} = \tilde{B}^{-1} P_0. \tag{7.8}$$

Since $B \subset \tilde{B}$, by virtue of equalities (7.4) and (7.8), the equality $\tilde{A}x = Ax$ is true for $x \in \mathcal{D}(A)$, i.e., $A \subset \tilde{A}$. Moreover, since $\mathcal{R}(\tilde{A}) = \mathcal{D}(\tilde{B}) \subset \mathcal{H}_0$ and the operator $\tilde{B}^{-1}$ is nonnegative, we have

$$(\tilde{A}x, x) = (\tilde{B}^{-1} P_0 x, P_0 x) \geq 0.$$

for all $x \in \mathcal{D}(\tilde{A})$. In this case, the operator $\tilde{A}$ is self-adjoint because $\tilde{A} = P_0 \tilde{B}^{-1} P_0$ and the operator $\tilde{B}^{-1}$ is self-adjoint.

Thus, for $\mathcal{R}(A) \neq \mathcal{H}_0$, the operator $\tilde{A}$ defined by equality (7.8) is also a nonnegative self-adjoint extension of the nonnegative Hermitian operator A.

■

II. Let A be a nonnegative Hermitian operator acting in a Hilbert space $\mathcal{H}$ with scalar product $(\cdot, \cdot)$. Consider a new "scalar" (pseudoscalar) product $(\cdot, \cdot)_0$ defined by the equality

$$(x, y)_0 = (Ax, y), \quad \{x, y\} \subset \mathcal{D}(A).$$

Let $\mathfrak{N} = \{x \in \mathcal{D}(A) \mid (x, x)_0 = 0\}$. By using the standard procedure, we introduce a pre-Hilbert space (factor space) $\mathcal{D}(A)/\mathfrak{N}$ with the scalar product

$$(\tilde{x}, \tilde{y})_1 = (x, y)_0, \tag{7.9}$$

where $\tilde{x}$ and $\tilde{y}$ are elements of equivalence classes in $\mathcal{D}(A)/\mathfrak{N}$. Completing the space $\mathcal{D}(A)/\mathfrak{N}$ with respect to the scalar product $(\cdot,\cdot)_1$, we obtain the Hilbert space $\tilde{\mathcal{H}}$ and denote the scalar product in it by the same symbol $(\cdot,\cdot)_1$.

It is clear that the zero element $\tilde{0}$ of the space $\tilde{\mathcal{H}}$ coincides with $\mathfrak{N}$.

Assume that there exists a nonnegative self-adjoint extension $\tilde{A}$ of the Hermitian operator A. Since

$$|(Ax,y)|^2 = |(\tilde{A}x,y)|^2 \le (Ax,x)(\tilde{A}y,y) \quad (\forall x \in \mathcal{D}(A),\ y \in \mathcal{D}(\tilde{A})),$$

it is easy to see that

$$\mathfrak{N} = \operatorname{Ker} A. \tag{7.10}$$

Thus, equality (7.10) is a necessary condition for the existence of nonnegative self-adjoint extensions of the Hermitian operator A. In what follows, we always assume that condition (7.10) is satisfied.

Consider an operator V acting from $\tilde{\mathcal{H}}$ into $\mathcal{H}$ as follows:

$$V\tilde{x} = Ax \quad (\forall x \in \mathcal{D}(A)) \tag{7.11}$$

In view of (7.10), the operator V is well defined. Moreover, by virtue of (7.9) and (7.11), we have

$$(Ax,y) = (V\tilde{x},y) = (\tilde{x},\tilde{y})_1 \quad (\{x,y\} \subset \mathcal{D}(A)). \tag{7.12}$$

This means that $V^*y = \tilde{y}$ and, thus, $\mathcal{D}(A) \subset \mathcal{D}(V^*)$.

Consider the set

$$\mathcal{D}_* = \{y \in \mathcal{H} \mid M(y) < \infty\}, \tag{7.13}$$

where

$$M(y) = \sup_{x \in \mathcal{D}(A)} \{|(Ax,y)| \mid (Ax,x) \le 1\}. \tag{7.14}$$

It is easy to see that $\mathcal{D}_*$ is a lineal.

Theorem 7.5. *A nonnegative self-adjoint extension* $\tilde{A}$ *of a nonnegative Hermitian operator* A *exists if and only if the lineal* $\mathcal{D}_*$ *defined by conditions (7.13) and (7.14) is dense in* $\mathcal{H}$.

Proof. Assume that a self-adjoint extension $\tilde{A}$ of the operator A exists. In this case, if $y \in \mathcal{D}(\tilde{A})$, then, for any $x \in \mathcal{D}(A)$ such that $(Ax, x) \leq 1$, we have

$$|(Ax, y)|^2 = |(\tilde{A}x, y)|^2 \leq (Ax, x)(\tilde{A}y, y) \leq (\tilde{A}y, y) < \infty$$

and, hence, $y \in \mathcal{D}_*$. Thus, $\mathcal{D}(\tilde{A}) \subset \mathcal{D}_*$, i.e., $\overline{\mathcal{D}_*} = \mathcal{H}$.

Conversely, assume that the lineal $\mathcal{D}_*$ is dense in $\mathcal{H}$. It follows from (7.13) and (7.14) that

$$|(Ax, y)| \leq M(y) \tag{7.15}$$

for $y \in \mathcal{D}_*$ and $x \in \mathcal{D}(A)$ such that $(Ax, x) \leq 1$. Consider the functional $\varphi(\tilde{x}) = (V\tilde{x}, y)$. By using (7.11) and (7.15), we get

$$|\varphi(\tilde{x})| = |(Ax, y)| \leq M(y) < \infty$$

and, thus, for $y \in \mathcal{D}_*$, the functional φ is bounded in $\mathcal{D}(V)$. Therefore, in view of the Riesz theorem, $y \in \mathcal{D}(V^*)$, i.e., $\mathcal{D}_* \subset \mathcal{D}(V^*)$. This means that the operator V^* is densely defined in $\mathcal{H}$. Thus, the operator V^{**} exists and $V^{**} = \overline{V}$, where $\overline{V}$ is the closure of the operator V. Consequently, for $x \in \mathcal{D}(A)$, we obtain

$$V^{**}V^*x = \overline{V}\tilde{x} = V\tilde{x} = Ax.$$

This means that the nonnegative self-adjoint operator $\tilde{A} = V^{**}V^*$ is an extension of the nonnegative operator A.

■

Note that if the greatest lower bound of the operator A is positive $(m(A) > 0)$, then the nonnegative self-adjoint extension $\tilde{A} = V^{**}V^*$ obtained by applying the outlined procedure is also a nonnegative operator. However, it may happen that $m(\tilde{A}) < m(A)$. To illustrate this observation, we present the following example:

Example. Let $\mathcal{H} = \langle e_1, e_2 \rangle$ be a two-dimensional space with an orthonormal basis $\{e_1, e_2\}$. Assume that the operator A is defined in the lineal $\mathcal{D}(A) = \langle e_1 \rangle$ by the equality

$$Ae_1 = ae_1 + be_2,$$

where $a > 0$ and $b \in \mathbb{C}$ $(b \neq 0)$ are fixed numbers. It is easy to see that A is a nonnegative Hermitian operator and $m(A) = a$.

In the considered case,

$$(\tilde{x}, \tilde{y})_1 = (x, y)_0 = a(x, y), \quad \{x, y\} \subset \mathcal{D}(A). \tag{7.16}$$

In addition, $\mathfrak{N} = \{0\}$ and $\tilde{\mathcal{H}} = \mathcal{D}(A)$. Further, the operator V [equality (7.11)] can be represented in the form

$$Ve_1 = ae_1 + be_2. \tag{7.17}$$

By virtue of (7.16) and (7.17), we can directly show that the operator V^* ($V^*: \mathcal{H} \to \tilde{\mathcal{H}}$) has the form

$$V^* y = \frac{1}{a}(y, Ve_1)e_1 \quad (y \in \mathcal{H}). \tag{7.18}$$

Since the operator V^{**} acts from $\tilde{\mathcal{H}}$ into $\mathcal{H}$, it follows from the equalities

$$(y, Ve_1) = (V^* y, e_1)_1 = (y, V^{**} e_1)$$

that

$$V^{**} e_1 = Ve_1 = Ae_1. \tag{7.19}$$

Equalities (7.18) and (7.19) imply that the operator $\tilde{A} = V^{**} V^*$ is a nonnegative self-adjoint extension of the operator A.

Moreover, since

$$V^* e_2 = \frac{\bar{b}}{a} e_1,$$

we get

$$\tilde{A} e_2 = \frac{\bar{b}}{a} V^{**} e_1 = kAe_1 \quad \left(k = \frac{\bar{b}}{a}\right).$$

Therefore, $\tilde{A}(ke_1 - e_2) = 0$ and, hence, $m(\tilde{A}) = 0 < m(A)$.

Note that Theorem 7.5 can also be formulated for the case $m(A) \neq 0$. For this purpose, it suffices to consider the nonnegative operator $A - m(A)I$ instead the operator A, where $m(A)$ is the greatest lower bound of the operator A.

Remarks. For the first time, necessary and sufficient conditions for the existence of a self-adjoint bound-preserving extension of an Hermitian operator semibounded from below (Theorem 7.2) were established by Ando and Nishio [1].

Later, by using other methods, the problem of self-adjoint extensions of semibounded Hermitian operators preserving the lower bound was studied by Shtraus [3–5]. In particular, as shown in [5], the following theorem is true:

Theorem 7.6. *An Hermitian operator A bounded from below admits a self-adjoint bound-preserving extension if and only if the operator-valued function A_λ defined in the interval $(-\infty, m(A))$ can be majorized by a self-adjoint operator semibounded from below.*

Moreover, in Theorem 7.6, the operator A_λ $(\lambda < m(A))$ is defined in the lineal $\mathcal{D}(A_\lambda) = \mathcal{D}(A) \dotplus \mathfrak{N}_\lambda$, where $\mathfrak{N}_\lambda$ is the defect subspace of the operator A, by the equality

$$A_\lambda(x_0 + x_\lambda) = Ax_0 + \lambda x_\lambda \quad (x_0 \in \mathcal{D}(A),\ x_\lambda \in \mathfrak{N}_\lambda).$$

Recently, Sebestyen and Stochel [1] proposed an original method for the construction of self-adjoint bound-preserving extensions of semibounded Hermitian operators (see Theorem 7.5).

8. Regular $\mathfrak{U}$-Invariant Extensions of Hermitian Operators

In the present section, we propose proofs of the theorems on $\mathfrak{U}$-invariance of the Friedrichs extension of a $\mathfrak{U}$-invariant semibounded operator (Phillips theorem) and on $\mathfrak{U}$-invariance of the self-adjoint Krein extension of a $\mathfrak{U}$-invariant Hermitian contraction. We also present the conditions for the existence of self-adjoint $\mathfrak{U}$-invariant extensions in terms of the von Neumann formulas and in terms of the Shtraus characteristic function.

8.1. Preliminary Concepts and Results. Let A be an Hermitian operator with deficiency index (m, m) $(m \le \infty)$ acting in a Hilbert space $\mathcal{H}$ and let $\mathfrak{U}$ be a set of unitary operators in $\mathcal{H}$ such that the inclusion $U \in \mathfrak{U}$ implies that $U^* \in \mathfrak{U}$.

An operator A is called $\mathfrak{U}$-*invariant*[7] if it commutes with all operators U from the set $\mathfrak{U}$.

As shown by Phillips [2], the Friedrichs extension of a $\mathfrak{U}$-invariant semibounded symmetric operator A is $\mathfrak{U}$-invariant. On the other hand, in the same work, one can find

[7] The term "$\mathfrak{U}$-invariant" was suggested by Kochubei [3].

an example of a 𝔘-invariant symmetric operator with deficiency index (1, 1) without 𝔘-invariant self-adjoint extensions.

Kochubei [3] solved the problem of existence of 𝔘-invariant self-adjoint extensions of a symmetric operator A in terms of the Shtraus characteristic function (see Subsection 8.5).

We also present a proof of the Phillips theorem which differs from the commonly known one and establish conditions for the existence of regular (in particular, self-adjoint) 𝔘-invariant extensions of Hermitian (not only densely defined) 𝔘-invariant operators.

8.2. Some General Properties of 𝔘-invariant Hermitian Operators

Proposition 8.1. *Let A be an Hermitian 𝔘-invariant operator. Then, for any $U \in \mathfrak{U}$ and $\lambda \in \mathbb{C}$,*

$$U: \mathcal{D}(A) \to \mathcal{D}(A) \quad \text{and} \quad U: \mathfrak{N}_\lambda \to \mathfrak{N}_\lambda, \tag{8.1}$$

where $\mathfrak{N}_\lambda$ is a defect subspace of the operator A.

Proof. The first relation in (8.1) follows from the equality $AU = UA$ $(\forall\, U \in \mathfrak{U})$. To prove the second relation in (8.1), we note that, for any $x \in \mathcal{D}(A)$ and $x_\lambda \in \mathfrak{N}_\lambda$,

$$((A - \lambda I)x,\ Ux_\lambda) = (U^*(A - \lambda I)x,\ x_\lambda) = ((A - \lambda I)U^*x, x_\lambda) = 0.$$

Hence, $U: \mathfrak{N}_\lambda \to \mathfrak{N}_\lambda$.

■

Lemma 8.2. *The closure of a 𝔘-invariant operator is also a 𝔘-invariant operator.*

Proof. Assume that A is a 𝔘-invariant operator and $\bar{A}$ is its closure. Clearly, if $x_n \to x$ and $Ax_n \to y$ $(x_n \in \mathcal{D}(A))$, then $x \in \mathcal{D}(A)$ and $y = \bar{A}x$.

Further, let $U \in \mathfrak{U}$. Since $Ux_n \to Ux$ and $Ux_n \in \mathcal{D}(A)$, we have $AUx_n = AUx_n \to Uy$. Therefore, $Ux \in \mathcal{D}(\bar{A})$ and $\bar{A}Ux = Uy = U\bar{A}x$, i.e., $\bar{A}U = U\bar{A}$.

■

Theorem 8.3. *If a closed 𝔘-invariant symmetric operator A has at least one real point of regular type, then the operator A admits a 𝔘-invariant self-adjoint extension*

Proof. Let $\lambda \in \pi(A) \cap \mathbb{R}$. On the lineal $\mathcal{D}(\tilde{A}) = \mathcal{D}(A) \dotplus \mathfrak{N}_\lambda$, we consider a symmetric operator $\tilde{A}$ defined by the equality

$$\tilde{A}(x_0+x_\lambda) = Ax_0 + \lambda x_\lambda \quad (x_0 \in \mathcal{D}(A),\ x_\lambda \in \mathfrak{N}_\lambda).$$

It follows from Proposition 8.1 that the operator $\tilde{A}$ is $\mathfrak{U}$-invariant. As in the proof of Theorem 1.3, we conclude that the defect numbers of the symmetric operator A are equal to zero. The closure of this operator is a self-adjoint operator. By virtue of Lemma 8.2, this operator is $\mathfrak{U}$-invariant.

■

8.3. Semibounded Operators. Phillips Theorem

Theorem 8.4. *The Friedrichs extension of a $\mathfrak{U}$-invariant semibounded symmetric operator A is $\mathfrak{U}$-invariant.*

Proof. Without loss of generality, we assume that the operator A satisfies the inequality

$$(Ax, x) \geq (x, x) \quad (x \in \mathcal{D}(A)).$$

Let $(\cdot,\cdot)_0$ be a scalar product defined on the lineal $\mathcal{D}(A)$ by the equality

$$(x, y)_0 = (Ax, y) \quad (\{x, y\} \subset \mathcal{D}(A)) \tag{8.2}$$

and let $\mathcal{D}_0$ be the completion of $\mathcal{D}(A)$ in the norm $\|\cdot\|_0$ (for details, see Subsection 2.1).

If $U \in \mathfrak{U}$, then, by virtue of (8.1), we get

$$(Ux, Uy)_0 = (AUx, Uy) = (UAx, Uy) = (x, y)_0.$$

Thus, U is a unitary operator in the pre-Hilbert space $\mathcal{D}(A)$.

By extending the operator U to the entire space $\mathcal{D}_0$ by continuity, we arrive at a unitary operator U_0 acting in $\mathcal{D}_0$. Moreover, if $x \in \mathcal{D}_0$ and $\|x_n - x\|_0 \to 0$ $(x_n \in \mathcal{D}(A))$, then

$$\|Ux_n - Ux\| = \|x_n - x\| \leq \|x_n - x\|_0 \to 0 \quad (n \to \infty)$$

and, hence, $Ux_n \to Ux$. This means that $U_0 x = Ux$, i.e., U_0 is the restriction of the operator U to the lineal $\mathcal{D}_0$ $(U_0 = U|_{\mathcal{D}_0})$.

Since $U^* \in \mathfrak{U}$, in exactly the same way as above, we conclude that $U^* : \mathcal{D}_0 \to \mathcal{D}_0$ and $U_0^* = U^* |_{\mathcal{D}_0}$.

Further, by virtue of Theorem 4.1 (Chapter 1) and Proposition 8.1, we have $U^* : \mathcal{D}(A^*) \to \mathcal{D}(A^*)$.

Let $\tilde{A}$ be the Friedrichs extension of the operator A. It follows from equality (2.3) that $\mathcal{D}(\tilde{A}) = \mathcal{D}(A^*) \cap \mathcal{D}_0$. Therefore, $U: \mathcal{D}(\tilde{A}) \to \mathcal{D}(\tilde{A})$. For this reason, in view of the equality

$$(\tilde{A}x, y) = (x, y)_0 \qquad (x \in \mathcal{D}(\tilde{A}),\ y \in \mathcal{D}_0),$$

we obtain

$$(U\tilde{A}x, y) = (\tilde{A}x, U^*y) = (x, U_0^*y)_0 = (U_0x, y)_0 = (\tilde{A}Ux, y)$$

for any $x \in \mathcal{D}(\tilde{A})$ and $y \in \mathcal{D}_0$. Consequently, $U\tilde{A} = \tilde{A}U$.

■

8.4. Condition for the Existence of Regular $\mathfrak{U}$-Invariant Extensions Formulated in Terms of the Von Neumann Formulas. Assume that A is an Hermitian $\mathfrak{U}$-invariant operator and B is its regular extension such that $\lambda \notin \sigma_p(B)$ $(\operatorname{Im}\lambda \neq 0)$. By virtue of Theorem 5.15 (Chapter 1), any vector x from $\mathcal{D}(B)$ admits a representation

$$x = x_0 + x_\lambda + \Phi x_\lambda \qquad (x_0 \in \mathcal{D}(A),\ x_\lambda \in \mathcal{D}(\Phi)), \tag{8.3}$$

where the operator Φ acts from $\mathfrak{N}_\lambda$ into $\mathfrak{N}_{\bar{\lambda}}$. In this case,

$$Bx = Ax_0 + \bar{\lambda}x_\lambda + \lambda\Phi x_\lambda. \tag{8.4}$$

Theorem 8.5. *A regular extension B of a $\mathfrak{U}$-invariant Hermitian operator A defined by equalities (8.3) and (8.4) is $\mathfrak{U}$-invariant if and only if the operator Φ in relations (8.3) and (8.4) is $\mathfrak{U}$-invariant.*

Proof. Assume that the operator B is $\mathfrak{U}$-invariant. Then, for any U from $\mathfrak{U}$ and $x \in \mathcal{D}(B)$, the vector $y = Ux$ belongs to $\mathcal{D}(B)$. Therefore, by using equality (8.3) and Proposition 8.1, we get

$$y = y_0 + y_\lambda + U\Phi U^* y_\lambda, \tag{8.5}$$

where $y_0 = Ux_0 \in \mathcal{D}(A)$ and $y_\lambda = Ux_\lambda \in \mathfrak{N}_\lambda$.

Note that, for fixed λ ($\lambda \notin \sigma_p(B)$), the operator Φ in (8.3) is uniquely defined (see Subsection 5.5, Chapter 1). Therefore, according to (8.3) and (8.5), we have $U\Phi U^* = \Phi$, i.e., $U\Phi = \Phi U$ ($\forall U \in \mathfrak{U}$).

It is clear that the converse assertion is also true: if the operator Φ satisfies the last equality, then the operator B defined by equalities (8.3) and (8.4) is $\mathfrak{U}$-invariant.

■

Theorem 8.6. *Suppose that a $\mathfrak{U}$-invariant symmetric operator A possesses a regular $\mathfrak{U}$-invariant extension B defined by equalities (8.3) and (8.4). If, in this case, the operator Φ in (8.3) and (8.4) is bounded and defined in the entire space $\mathfrak{N}_\lambda$ and $0 \in \rho(\Phi)$, then one can also construct a self-adjoint $\mathfrak{U}$-invariant extension of the operator A.*

Proof. The operator Φ can be represented in the form $\Phi = V|\Phi|$, where $|\Phi| = \sqrt{\Phi^*\Phi}$ and the operator $V\colon \mathfrak{N}_\lambda \to \mathfrak{N}_{\bar{\lambda}}$ is defined by the equality

$$V|\Phi|x = \Phi x. \tag{8.6}$$

Since the operator B is a $\mathfrak{U}$-invariant extension of the operator A, by virtue of Theorem 8.5, any operator $U \in \mathfrak{U}$ commutes with Φ. Hence, the operator $|\Phi|$ acting in the space $\mathfrak{N}_\lambda$ also commutes with U. Taking into account the conditions of the theorem, we prove that the operator $|\Phi|$ maps $\mathfrak{N}_\lambda$ onto $\mathfrak{N}_\lambda$. By virtue of (8.6), we conclude that the operator V is a unitary operator mapping $\mathfrak{N}_\lambda$ onto $\mathfrak{N}_{\bar{\lambda}}$ and, obviously, V commutes with U ($U \in \mathfrak{U}$).

Consider an operator S defined as follows: any vector x from $\mathcal{D}(S)$ admits a representation

$$x = x_0 + x_\lambda + Vx_\lambda \quad (x_0 \in \mathcal{D}(A),\ x_\lambda \in \mathfrak{N}_\lambda)$$

and

$$Sx = Ax_0 + \bar{\lambda}x_\lambda + \lambda Vx_\lambda.$$

The operator S is self-adjoint (see Subsection 4.3, Chapter 1) and commutes with U, i.e., S is a self-adjoint $\mathfrak{U}$-invariant extension of the operator A.

■

8.5. Condition for the Existence of Self-Adjoint $\mathfrak{U}$-Invariant Extensions Formulated in Terms of the Shtraus Characteristic Function. Assume that A is a closed Hermitian operator, λ ($\operatorname{Im}\lambda < 0$) is a fixed number, and $\mathfrak{N}_\lambda$ is a defect subspace of the operator A. As shown in Subsection 6.3 (Chapter 1), the operator A_λ is defined in the lineal $\mathcal{D}(A_\lambda) = \mathcal{D}(A) \dotplus \mathfrak{N}_\lambda$ by the equality

$$A_\lambda(x_0 + x_\lambda) = Ax_0 + \bar{\lambda}x_\lambda, \tag{8.7}$$

where $x_0 \in \mathcal{D}(A)$ and $x_\lambda \in \mathfrak{N}_\lambda$ is a maximal dissipative operator. If, in this case, the operator A is $\mathfrak{U}$-invariant, then, obviously, the operator A_λ is also $\mathfrak{U}$-invariant.

Let μ ($\operatorname{Im}\mu < 0$) be a fixed point. By virtue of Proposition 6.6 and Corollary 6.11 (Chapter 1), we conclude that $A_\lambda \in \mathcal{P}(A)$ and $\mu \in \rho(A_\lambda)$. Therefore, the von Neumann formulas imply that the vector $x = x_0 + x_\lambda$ from $\mathcal{D}(A_\lambda)$ can be represented in the form

$$x = y_0 + y_\mu + \Phi y_\mu, \tag{8.8}$$

where $\mathcal{D}(\Phi) = \mathfrak{N}_\mu$ (since the operator A_λ is maximal) and $\Phi: \mathfrak{N}_\mu \to \mathfrak{N}_{\bar{\mu}}$. It is also clear that, for fixed μ, the operator Φ depends on λ: $\Phi = \Phi(\lambda)$. Further,

$$A_\lambda x = Ay_0 + \bar{\mu}y + \mu\Phi(\lambda)y_\mu. \tag{8.9}$$

In view of (8.8) and (8.9),

$$(A_\lambda - \mu I)x = (A - \mu I)y_0 + (\bar{\mu} - \mu)y_\mu. \tag{8.10}$$

By applying the operator $(A_\lambda - \mu I)^{-1}$ to both sides of equality (8.10) and taking into account the equality $Ay_0 = A_\lambda y_0$, we get

$$x = y_0 + (\bar{\mu} - \mu)(A_\lambda - \mu I)^{-1}y_\mu. \tag{8.11}$$

Equalities (8.8) and (8.11) imply that

$$\Phi(\lambda)y_\mu = -(A_\lambda - \bar{\mu}I)(A_\lambda - \mu I)^{-1}y_\mu. \tag{8.12}$$

In Shtraus [2], the operator

$$C(\lambda) = (A_\lambda - \bar{\mu}I)(A_\lambda - \mu I)^{-1} \tag{8.13}$$

is called the characteristic function of the Hermitian operator A. Thus, the operator $\Phi(\lambda)$ differs from the characteristic function $C(\lambda)$ of the Hermitian operator A only by its sign.

Theorem 8.7. *Assume that, for some fixed* λ $(\operatorname{Im}\lambda < 0)$, *the point* 0 *belongs to the resolvent set of the operator* $C(\lambda)$ $(0 \in \rho(C(\lambda)))$. *Then there exists a self-adjoint* $\mathfrak{U}$*-invariant extension of the* $\mathfrak{U}$*-invariant symmetric operator* A.

Proof. Indeed, if A is a $\mathfrak{U}$-invariant symmetric operator and $U \in \mathfrak{U}$, then, by virtue of equality (8.7), A_λ commutes with U. Hence, it follows from equality (8.13) that the operator $C(\lambda)$ also commutes with the operator U. To complete the proof of Theorem 8.7, it remains to use the fact that $\Phi(\lambda) = -C(\lambda)$ and Theorem 8.6. ■

8.6. Symmetric $\mathfrak{U}$***-Invariant Operator without*** $\mathfrak{U}$***-Invariant Self-Adjoint Extensions.*** In the space $\mathcal{H} = l_2(-\infty, \infty)$, we consider an operator V defined by the formulas[8]

$$\mathcal{D}(V) = \{x \in \mathcal{H} \mid x = (\ldots, x_{-2}, x_{-1}, \boxed{0}, x_1, x_2, \ldots)\}, \tag{8.14}$$

$$Vx = (\ldots, x_{-3}, x_{-2}, \boxed{x_{-1}}, 0, x_1, x_2, \ldots). \tag{8.15}$$

Thus, for any x and y from $\mathcal{D}(V)$, we have $(Vx, Vy) = (x, y)$, i.e., V is an isometric operator. It is also easy to show that $1 \notin \sigma_p(V)$. This enables us to consider an operator A defined by the formulas

$$A = i(V+I)(V-I)^{-1} \qquad (\mathcal{D}(A) = \mathcal{R}(V-I)). \tag{8.16}$$

This operator is Hermitian (as the Cayley transform of an isometric operator).

Let us show that the operator A is densely defined and, hence, symmetric. Indeed, assume that the vector $h = (\ldots, h_{-1}, \boxed{h_0}, h_1, \ldots)$ is orthogonal to the lineal $\mathcal{D}(A)$. Then

$$(Vx, h) = (x, h). \tag{8.17}$$

for all $x \in \mathcal{D}(V)$.

Consider a vector

$$e_m = (\ldots, \delta_{(-2)m}, \delta_{(-1)m}, \boxed{\delta_{0m}}, \delta_{1m}, \delta_{2m}, \ldots),$$

where $m \in \mathbb{Z}$ and δ_{km} is a Kronecker symbol. For any $m \neq 0$, the vector e_m belongs

[8] Frames always denote positions No.0 in the relevant sequences.

to $\mathcal{D}(V)$. Substituting the element $x = e_m$ $(m \neq 0)$ in (8.17), we obtain $h_{m+1} = h_m$ and, hence,

$$h_0 = h_{-1} = h_{-2} = \ldots = a \quad \text{and} \quad h_1 = h_2 = h_3 = \ldots = b,$$

which is possible only for $a = b = 0$. Thus, $h = 0$. Therefore, A is a symmetric operator.

Let $\mathfrak{N}_\lambda$ be a defect subspace of the operator A and let $x_\lambda \in \mathfrak{N}_\lambda$. Then

$$((A - \lambda I)x, x_\lambda) = 0 \quad (\forall\, x \in \mathcal{D}(A)). \tag{8.18}$$

Since $x = (V - I)f$, where $f \in \mathcal{D}(V)$, we conclude that

$$\begin{aligned}(A - \lambda I)x &= (i - \lambda)x + 2i(V - I)^{-1}x \\ &= (i - \lambda)(V - I)f + 2if = (i - \lambda)Vf + (\lambda + i)f.\end{aligned} \tag{8.19}$$

Consequently, by using (8.18) and (8.19), we obtain

$$(i - \lambda)(Vf, x_\lambda) + (i + \lambda)(f, x_\lambda) = 0. \tag{8.20}$$

In particular, for $\lambda = i$, equality (8.20) can be rewritten in the form $(f, x_i) = 0$ $(\forall f \in \mathcal{D}(V))$. Hence,

$$x_i = (\ldots, 0, 0, \boxed{1}, 0, 0, \ldots). \tag{8.21}$$

Thus, $\mathfrak{N}_i = \langle x_i \rangle$, where the vector x_i is defined by equality (8.21). Similarly, for $\lambda = -i$, it follows from equality (8.20) that $(Vf, x_{-i}) = 0$ $(\forall f \in \mathcal{D}(V))$ and, therefore, $\mathfrak{N}_{-i} = \langle x_{-i} \rangle$, where the vector x_{-i} is defined by the equality

$$x_{-i} = (\ldots, 0, 0, \boxed{0}, 1, 0, 0, \ldots). \tag{8.22}$$

Consider a collection $\mathfrak{U}$ of unitary operators U_θ defined as follows: If

$$x = (\ldots, x_{-2}, x_{-1}, \boxed{x_0}, x_1, x_2, \ldots) \in \mathcal{H}, \tag{8.23}$$

then

$$U_\theta x = (\ldots, x_{-2}, x_{-1}, \boxed{x_0}, \theta x_1, \theta x_2, \ldots), \tag{8.24}$$

where $\theta \in \mathbb{C}$, $|\theta| = 1$, and $\theta \neq 1$. By virtue of equalities (8.14)–(8.16), the operator A is $\mathfrak{U}$-invariant. Moreover,

$$U_\theta x_i = x_i \quad \text{and} \quad U_\theta x_{-i} = \theta x_{-i}, \tag{8.25}$$

where x_i and x_{-i} are the vectors from the defect subspaces of the operator A defined by equalities (8.21) and (8.22).

Suppose that S is a $\mathfrak{U}$-invariant self-adjoint extension of the operator A. Hence, if $x \in \mathcal{D}(S)$, then

$$x = x_0 + x_i + \Phi x_i \quad \text{and} \quad Sx = Ax_0 - ix_i + i\Phi x_i,$$

where Φ is a unitary operator mapping $\mathfrak{N}_i$ into $\mathfrak{N}_{-i}$. Thus, since the defect subspaces $\mathfrak{N}_i$ and $\mathfrak{N}_{-i}$ are one-dimensional, we have

$$\Phi x_i = ax_{-i} \quad \text{and} \quad (|a| = 1). \tag{8.26}$$

Furthermore, in view of Theorem 8.5, the operator Φ is also $\mathfrak{U}$-invariant. At the same time, equalities (8.25) and (8.26) imply that

$$\Phi U_\theta x_i = \Phi x_i \quad \text{and} \quad U_\theta \Phi x_i = \theta \Phi x_i.$$

Therefore, the equality $\Phi U_\theta x_i = U_\theta \Phi x_i$ is possible only for $\Phi x_i = 0$ and we arrive at a contradiction.

Consequently, the $\mathfrak{U}$-invariant symmetric operator A under consideration has no $\mathfrak{U}$-invariant self-adjoint extensions. At the same time, regular $\mathfrak{U}$-invariant extensions of the operator A exist. Thus, a dissipative operator A_λ defined in the lineal $\mathcal{D}(A_\lambda) = \mathcal{D}(A) \dotplus \mathfrak{N}_\lambda$ by the equality

$$A_\lambda(x_0 + x_\lambda) = Ax_0 + \bar{\lambda} x_\lambda \quad (x_0 \in \mathcal{D}(A),\ x_\lambda \in \mathfrak{N}_\lambda), \tag{8.27}$$

where $\mathfrak{N}_\lambda$ is a defect subspace of the operator A and λ $(\operatorname{Im}\lambda < 0)$ is a fixed point, is an extension of this sort.

Note that, in the considered case, the operator A_λ has the following property:

Theorem 8.8. *Let A be the symmetric operator defined by equality (8.16). Then the point spectrum of the operator A_λ defined by equality (8.27) covers the entire upper half plane.*

Proof. Let $\lambda \neq i$. Then, by virtue of equality (8.20),

$$(Vf, x_\lambda) = k_\lambda (f, x_\lambda) \qquad \left(k_\lambda = \frac{\lambda + i}{\lambda - i}\right). \tag{8.28}$$

In this case,

$$(|k_\lambda| - 1)\,\mathrm{Im}\,\lambda > 0 \qquad (\mathrm{Im}\,\lambda \neq 0). \tag{8.29}$$

Let

$$x_\lambda = (\ldots, \varphi_{-2}, \varphi_{-1}, \boxed{\varphi_0}, \varphi_1, \varphi_2, \ldots).$$

It directly follows from (8.28) that the equality $\varphi_{m+1} = \bar{k}_\lambda \varphi_m$ is true for all $m \neq 0$ and, hence,,

$$\varphi_n = (\bar{k}_\lambda)^{n-1} \varphi_1 \quad \text{and} \quad \varphi_{-n} = (\bar{k}_\lambda)^{-n} \varphi_0 \qquad (n \in \mathbb{N}). \tag{8.30}$$

Let $\mathrm{Im}\,\lambda > 0$. Then, by using (8.29), we get $|k_\lambda| > 1$. Therefore, in view of (8.30), $|\varphi_n| \geq |\varphi_1|$ $(\forall n \in \mathbb{N})$, which is possible only for $\varphi_1 = 0$. Thus, if $\mathrm{Im}\,\lambda > 0$, then $\varphi_n = 0$ $(n \in \mathbb{N})$. Moreover, since $(\bar{k}_\lambda)^{-1} = k_{\bar{\lambda}}$, we obtain $\varphi_{-n} = k_{\bar{\lambda}}^n \varphi_0$. Consequently, for $\varphi_0 = 1$, we can write

$$x_\lambda = (\ldots, k_{\bar{\lambda}}^2, k_{\bar{\lambda}}, \boxed{1}, 0, 0, \ldots). \tag{8.31}$$

In particular, for $\lambda = i$, we get $k_{\bar{\lambda}} = 0$ and the vector x_λ coincides with the vector x_i defined by equality (8.21).

Similarly, for $\mathrm{Im}\,\lambda < 0$, we obtain $|k_\lambda| < 1$ and, hence, by virtue of (8.30), $\varphi_{-n} = 0$ for all $n \geq 0$. Taking into account equalities (8.30) for $\varphi_1 = 1$, we obtain

$$x_\lambda = (\ldots, 0, 0, \boxed{0}, 1, \bar{k}_\lambda, \bar{k}_\lambda^2, \ldots). \tag{8.32}$$

Let λ_0 $(\mathrm{Im}\,\lambda_0 > 0)$ be a fixed number. Since the operator A_λ defined by equality (8.27) is dissipative, we conclude that $\bar{\lambda}_0 \notin \sigma_p(A_\lambda)$. This implies that, by virtue of the von Neumann formulas, any vector x from $\mathcal{D}(A_\lambda)$ can be represented in the form

$$x = y_0 + y_{\bar{\lambda}_0} + \Phi y_{\bar{\lambda}_0} \qquad (y_0 \in \mathcal{D}(A),\ y_{\bar{\lambda}_0} \in \mathfrak{N}_{\bar{\lambda}_0}),$$

where $y_{\bar{\lambda}_0} = ax_{\bar{\lambda}_0}$ and $\Phi y_{\bar{\lambda}_0} = bx_{\lambda_0}$ (a and b are complex numbers). Finally,

$$x = y_0 + ax_{\bar{\lambda}_0} + bx_{\lambda_0}, \tag{8.33}$$

where $y_0 \in \mathcal{D}(A)$ and, in view of equalities (8.31) and (8.32), the vectors x_{λ_0} and $x_{\bar{\lambda}_0}$ can be represented in the form

$$x_{\lambda_0} = (\ldots, k_{\bar{\lambda}_0}^2, k_{\bar{\lambda}_0}, \boxed{1}, 0, 0, \ldots), \tag{8.34}$$

$$x_{\bar{\lambda}_0} = (\ldots, 0, 0, \boxed{0}, 1, k_{\lambda_0}^{-1}, k_{\lambda_0}^{-2}, \ldots). \tag{8.35}$$

It follows from equalities (8.27), (8.33) and the von Neumann formulas that

$$Ax_0 + \bar{\lambda} x_\lambda = Ay_0 + a\lambda_0 x_{\bar{\lambda}_0} + b\bar{\lambda}_0 x_{\lambda_0}. \tag{8.36}$$

Further,

$$x_0 = (V - I)\varphi \quad \text{and} \quad y_0 = (V - I)\psi \quad (\{\varphi, \psi\} \subset \mathcal{D}(V)) \tag{8.37}$$

and, by using (8.16), we obtain

$$Ax_0 = i(V + I)\varphi \quad \text{and} \quad Ay_0 = i(V + I)\psi.$$

Hence, equality (8.36) can be rewritten in the form

$$i(V + I)\varphi + \bar{\lambda} x_\lambda = i(V + I)\psi + a\lambda_0 x_{\bar{\lambda}_0} + b\bar{\lambda}_0 x_{\lambda_0}. \tag{8.38}$$

Multiplying (8.33) by i and subtracting the expression obtained as a result from (8.38), in view of (8.37), we arrive at the relation

$$2i\varphi + (\bar{\lambda} - i)x_\lambda = 2i\psi + a(\lambda_0 - i)\, x_{\bar{\lambda}_0} + b(\bar{\lambda}_0 - i)\, x_{\lambda_0}. \tag{8.39}$$

Let P be the orthoprojector onto $\mathcal{D}(V)^\perp$ in $\mathcal{H}$. Then $P\varphi = P\psi = 0$ and, in view of equalities (8.32), (8.34), and (8.35), it is easy to see that

$$Px_{\bar{\lambda}_0} = 0, \quad Px_{\lambda_0} = e_0, \quad \text{and} \quad Px_\lambda = 0 \quad (\operatorname{Im}\lambda < 0), \tag{8.40}$$

where

$$e_0 = (\ldots, 0, 0, \boxed{1}, 0, 0, \ldots).$$

By applying the operator P to both sides of equality (8.39) and taking into account equality (8.40), we obtain the equality $b(\bar{\lambda}_0 - i)e_0 = 0$. Consequently, $b = 0$. Thus, for any λ_0 from the upper half plane, we have $\Phi y_{\bar{\lambda}_0} = bx_{\lambda_0} = 0$. Therefore,

$$x = y_0 + y_{\bar{\lambda}_0} \quad \text{and} \quad A_\lambda x = Ay_0 + \lambda_0 y_{\bar{\lambda}_0}.$$

In particular, for $y_0 = 0$, we have $A_\lambda y_{\bar{\lambda}_0} = \lambda_0 y_{\bar{\lambda}_0}$ i.e., $\lambda_0 \in \sigma_p(A_\lambda)$. Thus, the entire upper half plane belongs to the point spectrum of the operator A_λ.

■

8.7. ***Self-Adjoint*** **𝔘*-Invariant Norm-Preserving Extensions of Hermitian Contractions.*** In conclusion, we show that extensions of 𝔘-invariant Hermitian contractions constructed by the Krein method (Theorem 3.1) lead to 𝔘-invariant operators.

Theorem 8.9. *If a strong Hermitian contraction* S *is* 𝔘*-invariant, then its self-adjoint extension* $\tilde{S}$ *constructed by the Krein method is also* 𝔘*-invariant.*

Proof. Let S be a strong 𝔘-invariant Hermitian contraction acting in $\mathcal{H}$. Without loss of generality, we suppose that the lineal $\mathcal{D}(S)$ is a subspace of the space $\mathcal{H}$. If $U \in$ 𝔘, then the operator S commutes both with U and U^*. For this reason,

$$U: \mathcal{D}(S) \to \mathcal{D}(S) \quad \text{and} \quad U^*: \mathcal{D}(S) \to \mathcal{D}(S).$$

Hence, the subspace $\mathcal{D}(S)$ reduces the operator U. This means that the operator U commutes with the orthoprojector P onto $\mathcal{D}(S)$ in $\mathcal{H}$. Therefore, the operators S_1 and S_2 defined by equality (3.2) also commute with U.

By using the fact that the extension $\tilde{S}_1$ of the operator S maps $\mathcal{H}$ into $\mathcal{D}(S)$ and equality (3.6), we conclude that

$$(x, \tilde{S}_1 Uy) = (Sx, Uy) = (SU^*x, y) = (x, U\tilde{S}_1 y) \tag{8.41}$$

for any $x \in \mathcal{D}(S)$ and $y \in \mathcal{H}$. Since the vectors $\tilde{S}_1 Uy$ and $U\tilde{S}_1 y$ belong to the lineal $\mathcal{D}(S)$, it follows from equality (8.41) that these vectors are equal.

Consequently, the operator $\tilde{S}_1$ commutes with U ($U \in$ 𝔘). Moreover, since

$$(Ux, Uy)_0 = (Ux, Uy) - (\tilde{S}_1 Ux, \tilde{S}_1 Uy) = (x, y)_0$$

for all x and y from $\mathcal{H}$, the operator U is unitary with respect to the norm $\|\cdot\|_0$ in the set $\mathcal{H}$ dense in $\hat{\mathcal{H}}$. If we extend the operator U by continuity, then we obtain a unitary operator $\hat{U}$ acting in the space $\hat{\mathcal{H}}$. In this case, the condition $U: \mathcal{D}(S) \to \mathcal{D}(S)$ implies that $\hat{U}: \hat{\mathcal{D}}(S) \to \hat{\mathcal{D}}(S)$, where $\hat{\mathcal{D}}(S)$ is the closure of the lineal $\mathcal{D}(S)$ in the norm $\|\cdot\|_0$.

A similar assertion is also true for the operator U^*. Thus, it is easy to see that $\hat{U}\hat{P} = \hat{P}\hat{U}$, where $\hat{P}$ is the orthoprojector onto the subspace $\hat{\mathcal{D}}(S)$ in $\tilde{\mathcal{H}}$.

Further, since S_2 commutes with U, we have

$$\hat{S}_2\hat{U}x = \hat{U}\hat{S}_2x \qquad (x \in \hat{\mathcal{D}}(S)), \tag{8.42}$$

where $\hat{S}_2$ is an extension of the operator S_2 to the lineal $\hat{\mathcal{D}}(S)$ in the norm $\|\cdot\|_0$. It is clear that equality (8.42) can be rewritten in the form

$$\hat{S}_2\hat{U}\hat{P}x = \hat{U}\hat{S}_2\hat{P}x \qquad (x \in \hat{\mathcal{D}}(S)). \tag{8.43}$$

The fact that the operator $\hat{U}$ commutes with $\hat{P}$ and the operator $\tilde{S}_2$ is the restriction of the operator $\hat{S}_2\hat{P}$ to $\mathcal{H}$ enables us to conclude that

$$\tilde{S}_2Ux = U\tilde{S}_2x \qquad (x \in \mathcal{H}).$$

Therefore, the operators $T = \tilde{S}_1 + \tilde{S}_2$ and T^* and, hence, the operator $\tilde{S} = \frac{1}{2}(T + T^*)$ commute with U.

■

Remark. This section is written on the basis of the works by Phillips [2], Shtraus [3], Kochubei [3] and A. Kuzhel [8].

9. Canonical Dissipative Extensions of Hermitian Operators

In the present section, we introduce the concept of canonical dissipative extensions of Hermitian operators and investigate their properties. Extensions of this kind are essentially used in constructing the spaces of boundary values of Hermitian operators (Chapter 3).

9.1. Auxiliary Statements. The investigation of properties of regular extensions can be significantly simplified by applying the elementary theory of spaces with indefinite metrics. In this subsection, we recall all necessary "indefinite" facts in the form convenient for our subsequent presentation. Their proofs and detailed analysis can be found in Bognar [1] or Azizov and Iokhvidov [1].

Assume that $\mathcal{H}$ is a Hilbert space with scalar product $(\cdot,\cdot)$ and J is a self-adjoint unitary operator acting in $\mathcal{H}$. In $\mathcal{H}$, we introduce an indefinite scalar product (indefinite metric) as

$$[x, y] = (Jx, y) \quad \{x, y\} \subset \mathcal{H}. \tag{9.1}$$

The Hilbert space $\mathcal{H}$ with indefinite metric (9.1) is called a *Krein space* (or a *J-space).*

In the Krein space $\mathcal{H}$, we consider orthoprojectors P_+ and P_- defined by the formulas

$$P_+ = \frac{1}{2}(I+J) \quad \text{and} \quad P_- = \frac{1}{2}(I-J) \tag{9.2}$$

and the corresponding decomposition of the space $\mathcal{H}$

$$\mathcal{H} = \mathcal{H}_+ [\dot{+}] \mathcal{H}_- \quad (\mathcal{H}_+ = P_+\mathcal{H}, \ \mathcal{H}_- = P_-\mathcal{H}), \tag{9.3}$$

where the sign $\dot{+}$ denotes the operation of direct sum and the brackets mean that the subspaces $\mathcal{H}_+$ and $\mathcal{H}_-$ are orthogonal with respect to the indefinite metric (9.1) (i.e., J-orthogonal).

Note that, in equality (9.3), the subspaces $\mathcal{H}_+$ and $\mathcal{H}_-$ also possess the property of orthogonality in a sense of the scalar product $(\cdot,\cdot)$.

Decomposition (9.3) is called the *fundamental decomposition* of the Krein space $\mathcal{H}$.

A lineal $\mathcal{L} \subset \mathcal{H}$ is called *nonnegative (positive)* if

$$[x, x] \geq 0 \quad ([x, x] > 0)$$

for all $x \in \mathcal{L} \setminus \{0\}$.

A positive lineal $\mathcal{L}$ is called *uniformly positive* if there exists a number $\alpha > 0$ such that

$$[x, x] \geq \alpha \|x\|^2 \quad (\forall x \in \mathcal{L}).$$

Nonpositive, negative, and uniformly negative lineals are introduced similarly.

Each of these classes of lineals contains *maximal* lineals, i.e., lineals which do not belong as subsets to any lineal from the same class. It is clear that any maximal uniformly positive (negative) lineal is a (closed) subspace of the Hilbert space $\mathcal{H}$.

Let $\mathcal{L}_+$ be an arbitrary maximal uniformly positive subspace. Its J-orthogonal complement in $\mathcal{H}$, i.e., the space

$$\mathcal{L}_- = \mathcal{L}_+^{[\perp]} = \{y \in \mathcal{H} \mid [x, y] = 0, \ \forall x \in \mathcal{L}_+\},$$

is a maximal uniformly negative subspace. Furthermore, the space $\mathcal{H}$ admits a representation in the form of a J-orthogonal direct sum

$$\mathcal{H} = \mathcal{L}_+ \, [\dotplus] \, \mathcal{L}_-. \tag{9.4}$$

Any decomposition of the form (9.4) is called a *canonical decomposition* of the space $\mathcal{H}$. By virtue of (9.1) and (9.2), we conclude that the fundamental decomposition (9.3) is also canonical.

An operator

$$T_+ = P_-\left(P_+|_{\mathcal{L}_+}\right)^{-1}$$

is called an *angular operator* of the maximal uniformly positive subspace $\mathcal{L}_+$ with respect to $\mathcal{H}_+$. The operator T_+ maps $\mathcal{H}_+$ into $\mathcal{H}_-$. It is a contraction (with $\|T_+\| < 1$) and

$$\mathcal{L}_+ = \{x \mid x = x_+ + T_+x_+, \ x_+ \in \mathcal{H}_+\}. \tag{9.5}$$

Similarly, the angular operator

$$T_- = P_+\left(P_-|_{\mathcal{L}_-}\right)^{-1}$$

of the subspace $\mathcal{L}_-$ with respect to $\mathcal{H}_-$ is a contraction which maps $\mathcal{H}_-$ into $\mathcal{H}_+$. Moreover, $T_- = T_+^*$ and

$$\mathcal{L}_- = \{y \mid y = y_- + T_-y_-, \ y_- \in \mathcal{H}_-\}. \tag{9.6}$$

The self-adjoint operator $T = T_+P_+ + T_-P_-$ acting in the space $\mathcal{H}$ is called an *operator of transition* from the fundamental decomposition (9.3) to the canonical decomposition (9.4).

It follows from $T_- = T_+^*$ that $\|T\| = \|T_+\| = \|T_-\| < 1$.

Proposition 9.1. *Let $P_{\mathcal{L}_+}$ and $P_{\mathcal{L}_-}$ be the projectors onto $\mathcal{L}_+$ and $\mathcal{L}_-$ in $\mathcal{H}$ with respect to decomposition (9.4). Then*

$$P_+ = (I+T)^{-1}(P_{\mathcal{L}_+} + TP_{\mathcal{L}_-}) \quad \text{and} \quad P_- = (I+T)^{-1}(P_{\mathcal{L}_-} + TP_{\mathcal{L}_+})$$

where T is the operator of transition from the fundamental decomposition (9.3) to the canonical decomposition (9.4).

Proof. For any $x \in \mathcal{H}$, in view of equalities (9.2), (9.5), and (9.6), we have

$$(I+T)x = (I+T)(P_+ + P_-)x = l_+ + l_-,$$

where the elements

$$l_+ = (I+T)P_+x \quad \text{and} \quad l_- = (I+T)P_-x, \tag{9.7}$$

belong to $\mathcal{L}_+$ and $\mathcal{L}_-$, respectively. Therefore,

$$(I+T)P_+x = P_{\mathcal{L}_+}(I+T)x.$$

Since $\|T\| < 1$, we conclude

$$P_+ = (I+T)^{-1}(P_{\mathcal{L}_+} + P_{\mathcal{L}_+}T). \tag{9.8}$$

In view of decomposition (9.4), any element $z \in \mathcal{H}$ can be represented in the form $z = l_+ + l_-$, where the elements $l_\pm$ are defined by equalities (9.7). Therefore,

$$P_{\mathcal{L}_+}Tz = P_{\mathcal{L}_+}T(l_+ + l_-) = P_{\mathcal{L}_+}((I+T)TP_+x + (I+T)TP_-x)$$

$$= (I+T)TP_-x = T(I+T)P_-x = TP_{\mathcal{L}_-}z.$$

Thus, $P_{\mathcal{L}_+}T = TP_{\mathcal{L}_-}$ and, hence, equality (9.8) can be rewritten in the form

$$P_+ = (I+T)^{-1}(P_{\mathcal{L}_+} + TP_{\mathcal{L}_-}).$$

At the same time,

$$P_- = I - P_+ = (I+T)^{-1}(I + T - P_{\mathcal{L}_+} - TP_{\mathcal{L}_-}) = (I+T)^{-1}(P_{\mathcal{L}_-} + TP_{\mathcal{L}_+}).$$

∎

Corollary 9.2. *The subspaces $\mathcal{H}_{\pm}$ from decomposition (9.3) admit the following representations:*

$$\mathcal{H}_+ = \{x_+ | x_+ = x - Tx\,,\; x \in \mathcal{L}_+\} \quad \text{and} \quad \mathcal{H}_- = \{y_- | y_- = y - Ty\,,\; y \in \mathcal{L}_-\},$$

where T is the operator of transition from the fundamental decomposition (9.3) to the canonical decomposition (9.4).

Proof. By virtue of Proposition 9.1, one can easily show that

$$P_{\mathcal{L}_+} = (I-T)^{-1}(P_+ - TP_-) \quad \text{and} \quad P_{\mathcal{L}_-} = (I-T)^{-1}(P_- - TP_+).$$

This means that, for any fixed $x_+ \in \mathcal{H}_+$, we have

$$x_+ = P_{\mathcal{L}_+} x_+ + P_{\mathcal{L}_-} x_+ = x + y,$$

where the element

$$x = P_{\mathcal{L}_+} x_+ = (I-T)^{-1} x_+ \qquad (y = P_{\mathcal{L}_-} x_+ = -(I-T)^{-1} T x_+)$$

belongs to the subspace $\mathcal{L}_+$ ($\mathcal{L}_-$). It is clear that $y = -Tx$. Therefore, $x_+ = x - Tx$ ($x \in \mathcal{L}_+$).

Similarly, we conclude that an arbitrary element $y_- \in \mathcal{H}_-$ can be represented in the form $y_- = y - Ty$, where $y = P_{\mathcal{L}_-} y_- \in \mathcal{L}_-$.

■

9.2. Canonical Dissipative Extensions. Let A be a closed Hermitian operator acting in a Hilbert space $\mathcal{H}$. In what follows, we suppose that

$$\mathcal{D}(A) \vee \mathcal{R}(A) = \mathcal{H}. \tag{9.9}$$

This condition is natural because if $\mathcal{H}_0 = \mathcal{D}(A) \vee \mathcal{R}(A) \subset \mathcal{H}$, then the properties of the elements of the subspace $\mathcal{H} \ominus \mathcal{H}_0$ and its dimensionality do not affect the properties of the operator A in the space $\mathcal{H}$.

For this reason, in what follows, without special indications, we consider a closed Hermitian operator A satisfying condition (9.9). Thus, as shown in Chapter 1 (Theorem 3.8), for $\operatorname{Im}\lambda \neq 0$, the defect subspaces $\mathfrak{N}_\lambda$ and $\mathfrak{N}_{\bar\lambda}$ of the operator A are linearly independent.

We now fix a complex number λ ($\operatorname{Im}\lambda \neq 0$). In the lineal

$$\mathcal{N}_\lambda = \mathfrak{N}_\lambda \dotplus \mathfrak{N}_{\bar{\lambda}},$$

we introduce a scalar product by the formula

$$(x, y)_\lambda = 2|\operatorname{Im}\lambda|\big((x_\lambda, y_\lambda) + (x_{\bar{\lambda}}, y_{\bar{\lambda}})\big), \tag{9.10}$$

where

$$x = x_\lambda + x_{\bar{\lambda}} \quad \text{and} \quad y = y_\lambda + y_{\bar{\lambda}} \qquad (\{x_z, y_z\} \subset \mathfrak{N}_z,\ z \in \{\lambda, \bar{\lambda}\}).$$

Obviously, the lineal $\mathcal{N}_\lambda$ is a Hilbert space with respect to the scalar product (9.10). In this space, we consider a self-adjoint unitary operator J defined by the equality

$$J(x_\lambda + x_{\bar{\lambda}}) = \operatorname{sgn}(\operatorname{Im}\lambda)(x_{\bar{\lambda}} - x_\lambda) \tag{9.11}$$

and the indefinite scalar product

$$[x, y]_\lambda = (Jx, y)_\lambda, \qquad \{x, y\} \subset \mathcal{N}_\lambda.$$

By virtue of equalities (9.10) and (9.11), we obtain

$$[x, y]_\lambda = 2\operatorname{Im}\lambda\big((x_{\bar{\lambda}}, y_{\bar{\lambda}}) - (x_\lambda, y_\lambda)\big). \tag{9.12}$$

The space $\mathcal{N}_\lambda$ equipped with the indefinite metric (9.12) is a Krein space. It is easy to see that, for $\operatorname{Im}\lambda < 0$, the subspace $\mathfrak{N}_\lambda$ is maximal uniformly positive and the subspace $\mathfrak{N}_{\bar{\lambda}}$ is maximal uniformly negative in the space $\mathcal{N}_\lambda$. Moreover, by virtue of equalities (9.2) and (9.3), we conclude that the decomposition

$$\mathcal{N}_\lambda = \mathfrak{N}_\lambda [\dotplus] \mathfrak{N}_{\bar{\lambda}} \tag{9.13}$$

is a fundamental decomposition of the Krein space $\mathcal{N}_\lambda$.

Lemma 9.3. *Let A be an Hermitian operator. If $x \in \mathcal{D}(A) \cap \mathcal{N}_\lambda$, then $[x, x]_\lambda = 0$.*

Proof. Since $x \in \mathcal{N}_\lambda$, we have $x = x_\lambda + x_{\bar{\lambda}}$ and

$$((A - zI)y, x_z) = 0, \qquad z \in \{\lambda, \bar{\lambda}\},$$

for all $y \in \mathcal{D}(A)$. Since $x \in \mathcal{D}(A)$, we set $y = x$ in the last equality and obtain

$$(Ax, x_z) = (x, \bar{z}x_z).$$

Therefore,

$$\begin{aligned} 0 = \operatorname{Im}(Ax, x) &= \operatorname{Im}(Ax, x_\lambda + x_{\bar{\lambda}}) \\ &= \operatorname{Im}(x_\lambda + x_{\bar{\lambda}}, \bar{\lambda}x_\lambda + \lambda x_{\bar{\lambda}}) \\ &= 2 \operatorname{Im}\lambda\left(\|x_\lambda\|^2 - \|x_{\bar{\lambda}}\|^2\right) = -[x, x]_\lambda. \end{aligned}$$

■

Let $\mathcal{L}$ be a lineal in $\mathcal{N}_\lambda$ linearly independent of $\mathcal{D}(A)$. We say that an extension F of the Hermitian operator A *is defined by the decomposition*

$$\mathcal{D}(F) = \mathcal{D}(A) \dotplus \mathcal{L} \tag{9.14}$$

if the equality

$$Fx = Ax_0 + \bar{\lambda}x_\lambda + \lambda x_{\bar{\lambda}} \tag{9.15}$$

holds for all

$$x \in \mathcal{D}(F) \quad (x = x_0 + x_\lambda + x_{\bar{\lambda}},\ x_0 \in \mathcal{D}(A),\ x_\lambda + x_{\bar{\lambda}} \in \mathcal{L}).$$

Lemma 9.4. *The operator F defined by decomposition (9.14) is a regular extension of the Hermitian operator A.*

Proof. The proof of Lemma 9.4 follows from Proposition 5.8 (Chapter 1).

■

Lemma 9.5. *The lineal $\mathcal{L}$ in $\mathcal{N}_\lambda$ is uniquely defined by the operator F.*

Proof. Indeed, suppose that the space $\mathcal{N}_\lambda$ contains different lineals $\mathcal{L}_1$ and $\mathcal{L}_2$ such that the decompositions

$$\mathcal{D}(F) = \mathcal{D}(A) \dotplus \mathcal{L}_1 \quad \text{and} \quad \mathcal{D}(F) = \mathcal{D}(A) \dotplus \mathcal{L}_2$$

define the same extension F of the operator A. Then, for any $y = y_\lambda + y_{\bar{\lambda}} \in \mathcal{L}_1 \backslash \mathcal{L}_2$, there exists a nonzero element $x_0 \in \mathcal{D}(A)$ such that $y = x_0 + x_\lambda + x_{\bar{\lambda}}$, where $x_\lambda + x_{\bar{\lambda}} \in \mathcal{L}_2$. Moreover,

$$Fy = \bar{\lambda} y_\lambda + \lambda y_{\bar{\lambda}} = Ax_0 + \bar{\lambda} x_\lambda + \lambda x_{\bar{\lambda}}.$$

Thus,

$$Ax_0 = \bar{\lambda}(y_\lambda - x_\lambda) + \bar{\lambda}(y_{\bar{\lambda}} - x_{\bar{\lambda}}) = \bar{\lambda}\varphi_\lambda + \lambda\varphi_{\bar{\lambda}}, \tag{9.16}$$

where $\varphi_z = y_z - x_z$ $(z \in \{\lambda, \bar{\lambda}\})$. By virtue of (9.16), we obtain

$$0 = ((A - zI)x_0, \varphi_z) = (\bar{z} - z)\|\varphi_z\|^2$$

and, hence, $\varphi_z = 0$, i.e., $x_\lambda = y_\lambda$ and $x_{\bar{\lambda}} = y_{\bar{\lambda}}$, which is possible only for $x_0 = 0$. Thus, $\mathcal{L}_1 \subset \mathcal{L}_2$. Similarly, we get $\mathcal{L}_2 \subset \mathcal{L}_1$. Therefore, $\mathcal{L}_1 = \mathcal{L}_2$.

■

Lemma 9.6. *Assume that extensions F and F' of an Hermitian operator A are defined by the decompositions $\mathcal{D}(F) = \mathcal{D}(A) \dotplus \mathcal{L}$ and $\mathcal{D}(F') = \mathcal{D}(A) \dotplus \mathcal{L}'$, respectively. Then, for any $x \in \mathcal{D}(F)$ and $y \in \mathcal{D}(F')$,*

$$(Fx, y) - (x, F'y) = i[f, g]_\lambda, \tag{9.17}$$

where

$$x = x_0 + f \quad \text{and} \quad y = y_0 + g \quad (\{x_0, y_0\} \subset \mathcal{D}(A),\ f \in \mathcal{L},\ g \in \mathcal{L}').$$

Proof. Under the condition of the lemma, any two vectors $x \in \mathcal{D}(F)$ and $y \in \mathcal{D}(F')$ can be represented in the form

$$x = x_0 + x_\lambda + x_{\bar{\lambda}} \quad \text{and} \quad y = y_0 + y_\lambda + y_{\bar{\lambda}},$$

where $x_\lambda + x_{\bar{\lambda}} = f \in \mathcal{L}$ and $y_\lambda + y_{\bar{\lambda}} = g \in \mathcal{L}'$.

In view of (9.15), it is easy to show that

$$(Fx, y) - (x, F'y) = 2i \operatorname{Im} \lambda \left((x_{\bar{\lambda}}, y_{\bar{\lambda}}) - (x_\lambda, y_\lambda)\right).$$

By virtue of (9.12), this completes the proof.

■

Theorem 9.7. *Any dissipative extension F of a closed Hermitian operator A is defined by decomposition (9.14), where $\mathcal{L}$ is a nonnegative lineal in $\mathcal{N}_\lambda$ and λ is* *a nonreal number. Conversely, any nonnegative lineal* $\mathcal{L}$ *in* $\mathfrak{N}_\lambda$ *linearly independent of* $\mathcal{D}(A)$ *defines a dissipative extension* F *of the operator* A *by decomposition (9.14).*

Proof. Let F be a dissipative extension of the operator A. It follows from Theorem 5.15 and Proposition 6.1 (Chapter 1) that, for any nonreal λ, the extension F is defined by the decomposition $\mathcal{D}(F) = \mathcal{D}(A) \dot{+} \mathcal{L}$, where $\mathcal{L}$ is a lineal in $\mathfrak{N}_\lambda$ which is linearly independent of $\mathcal{D}(A)$. By setting $F' = F$ in (9.17), we obtain

$$2\,\mathrm{Im}\,(Fx, x) = [f,f]_\lambda \geq 0 \qquad (\forall f \in \mathcal{L}). \tag{9.18}$$

Thus, the lineal $\mathcal{L}$ is nonnegative.

On the other hand, equality (9.18) implies that any nonnegative lineal $\mathcal{L}$ in $\mathfrak{N}_\lambda$ linearly independent of $\mathcal{D}(A)$ defines a dissipative extension F of the operator A by decomposition (9.14).

■

A dissipative extension F of a closed Hermitian operator A is called *canonical* if the lineal $\mathcal{L}$ in decomposition (9.14) is a maximal uniformly positive subspace of the Krein space $\mathfrak{N}_\lambda$.

Note that a canonical extension is defined by an arbitrary maximal uniformly positive subspace $\mathcal{L}_+$ of $\mathfrak{N}_\lambda$ because, by virtue of Lemma 9.3, we have $\mathcal{L}_+ \cap \mathcal{D}(A) = \{0\}$. In Corollary 9.10, we will show that the property of an operator F to be a canonical extension does not depend on the choice of a nonreal number λ in the definition of the space $\mathfrak{N}_\lambda$.

Proposition 9.8. *A canonical extension is a densely defined closed maximal dissipative operator.*

Proof. Assume that a dissipative extension F is defined by the decomposition $\mathcal{D}(F) = \mathcal{D}(A) \dot{+} \mathcal{L}_+$ and is canonical. Since, in this decomposition, the lineal $\mathcal{L}_+$ is a maximal uniformly positive subspace of $\mathfrak{N}_\lambda$, we conclude, in view of Theorem 9.7, that F is a maximal dissipative extension.

Let us show that the operator F is densely defined. Without loss of generality, we can assume that $\mathrm{Im}\,\lambda < 0$. Then, by virtue of equality (9.5), the subspace $\mathcal{L}_+$ with respect to the fundamental decomposition (9.13) can be represented in the form

$$\mathcal{L}_+ = \{x \,|\, x = x_\lambda + T_+ x_\lambda,\ x_\lambda \in \mathfrak{N}_\lambda\},$$

where T_+ $(T_+ : \mathfrak{N}_\lambda \to \mathfrak{N}_{\bar{\lambda}})$ is an angular operator of the subspace $\mathcal{L}_+$ with respect to the maximal uniformly positive subspace $\mathfrak{N}_\lambda$. Hence, in view of (9.15), we obtain

$$\mathcal{R}(F-\lambda I) = \mathcal{R}(A-\lambda I) \oplus \mathfrak{N}_\lambda = \mathcal{H}.$$

This equality and Proposition 6.2 (Chapter 1) imply that the operator F is densely defined. At the same time, it follows from Proposition 6.9 (Chapter 1) that the maximal dissipative operator F is closed.

■

In the following theorem, we formulate the characteristic property of canonical dissipative extensions:

Theorem 9.9. *Let F be a densely defined maximal dissipative extension of a closed Hermitian operator A. The extension F is canonical if and only if the following conditions are satisfied:*

$$\mathcal{D}(F) \cap \mathcal{D}(F^*) = \mathcal{D}(A), \tag{9.19}$$

$$\mathcal{D}(F) \cup \mathcal{D}(F^*) = \mathcal{D}(A) + \mathcal{N}_\lambda, \tag{9.20}$$

Proof. It follows from Proposition 6.2 (Chapter 1) that the operator F^* is a regular extension of the operator A. We fix a nonreal number λ. Since the operator F is maximal dissipative, we conclude that at least one number λ or $\bar{\lambda}$ belongs to the resolvent set of the operator F^* (i.e., $\{\lambda, \bar{\lambda}\} \cap \rho(F^*) \neq \emptyset$). Hence, by virtue of the analog of the von Neumann formulas (Theorem 5.15, Chapter 1), the extension F^* is defined by the decomposition

$$\mathcal{D}(F^*) = \mathcal{D}(A) \dotplus \mathcal{L}_-, \tag{9.21}$$

where $\mathcal{L}_- \subset \mathcal{N}_\lambda$.

Assume that the extension F is canonical. Setting $F' = F^*$ in equality (9.17) and taking into account the fact that the lineal $\mathcal{L}_+$ in the decomposition

$$\mathcal{D}(F) = \mathcal{D}(A) \dotplus \mathcal{L}_+ \tag{9.22}$$

is a maximal uniformly positive subspace of $\mathcal{N}_\lambda$, we conclude that $\mathcal{L}_- = \mathcal{L}_+^{[\perp]}$ is a maximal uniformly negative subspace. This assertion and decomposition (9.4) (for $\mathcal{H} = \mathcal{N}_\lambda$) imply equalities (9.19) and (9.20).

Conversely, consider a maximal dissipative extension F satisfying equalities (9.19) and (9.20). In view of Theorem 9.7, the lineal $\mathcal{L}_+$ in decomposition (9.22) is nonnegative. Similarly, by setting $F' = F = F^*$ in (9.17) and using Theorem 6.14 (Chapter 1), we can show that the lineal $\mathcal{L}_-$ in decomposition (9.21) is nonpositive. Further, by virtue of (9.17) (for $F' = F^*$), the lineal $\mathcal{L}_+$ is J-orthogonal to the lineal $\mathcal{L}_-$. Finally, taking into account equalities (9.19) and (9.20), we get the following relations for the lineals $\mathcal{L}_+$ and $\mathcal{L}_-$

$$\mathcal{L}_+ \cap \mathcal{L}_- = \{0\} \quad \text{and} \quad \mathcal{L}_+ [\dot{+}] \mathcal{L}_- = \mathfrak{N}_\lambda. \tag{9.23}$$

As is well known (see Azizov and Iokhvidov [1]), relations (9.23) are possible only if $\mathcal{L}_+$ is a maximal uniformly positive subspace of $\mathfrak{N}_\lambda$. Thus, the operator F is a canonical extension.

■

The following corollary directly follows from Theorem 9.9:

Corollary 9.10. *If F is a canonical dissipative extension of a closed Hermitian operator A, then, for any nonreal λ, the extension F is defined by decomposition (9.22), where the lineal $\mathcal{L}_+$ is a maximal uniformly positive subspace of the Krein space $\mathfrak{N}_\lambda$.*

Remark. In a somewhat different (equivalent) form, the notion of canonical dissipative extensions was introduced and studied by S. Kuzhel in [2].

3. Boundary-value spaces of Hermitian operators

In this chapter, we study regular extensions of Hermitian operators by using boundary-value spaces (BVS). The construction of BVS makes it possible to determine regular extensions in terms of abstract boundary values, which is convenient for applications to the theory of boundary-value problems for differential equations.

We analyze various definitions of BVS of an Hermitian (symmetric) operator and study their general properties. In terms of BVS, we define the characteristic function of an Hermitian operator with arbitrary defect numbers, determine its regular extensions, and describe the nonreal spectra of these extensions.

1. Definition and General Properties of Boundary-Value Spaces

In the present section, we introduce and analyze various definitions of boundary-value spaces for symmetric operators with equal defect numbers, Hermitian operators with equal defect numbers, and Hermitian operators with arbitrary defect numbers.

1.1. Symmetric Operators. Let A be a closed symmetric operator with deficiency index (n, n) $(n \leq \infty)$ acting in a Hilbert space $\mathcal{H}$.

Definition I. *A triple (X, Γ_1, Γ_2), where X is a Hilbert space with scalar product $(\cdot, \cdot)_X$ and Γ_1 and Γ_2 are linear operators mapping $\mathcal{D}(A^*)$ into X, is called a* ***boundary-value space*** *(BVS) of a symmetric operator A if*

(*i*) *for any x and y from $\mathcal{D}(A^*)$,*

$$(A^*x, y) - (x, A^*y) = (\Gamma_1 x, \Gamma_2 y)_X - (\Gamma_2 x, \Gamma_1 y)_X, \tag{1.1}$$

(ii) for any φ and ψ from X, there exists a vector $x \in \mathcal{D}(A^*)$ such that

$$\Gamma_1 x = \varphi \quad \textit{and} \quad \Gamma_2 x = \psi.$$

Note that the use of the term "boundary-value spaces" is motivated by the fact that, in the case where A is a differential operator, the corresponding BVS can be, as a rule, expressed in terms of its boundary values. Thus, the BVS of an operator

$$A = -\frac{d^2}{ds^2}, \qquad \mathcal{D}(A) = \mathring{W}_2^2(0, +\infty),$$

symmetric in the space $L(0, +\infty)$ can be chosen in the form

$$X = \mathbb{C}, \quad \Gamma_1 x = x'(0) \quad \text{and} \quad \Gamma_2 x = x(0) \quad \big(x(s) \in W_2^2(0, +\infty)\big).$$

Theorem 1.1. *For any symmetric operator A with deficiency index (n, n) $(n \leq \infty)$, there exists a BVS (X, Γ_1, Γ_2) with $\dim X = n$.*

Proof. Let V be a unitary operator mapping $\mathfrak{N}_{\bar{\lambda}}$ onto $\mathfrak{N}_\lambda$ $(\operatorname{Im}\lambda \neq 0)$, where $\mathfrak{N}_z$ $(z \in \{\lambda, \bar{\lambda}\})$ are the defect subspaces of the operator A. In the set $\mathcal{D}(A^*)$, we consider linear operators Γ_1 and Γ_2 defined by the equalities

$$\Gamma_1 x = \bar{\lambda} x_\lambda + \lambda V x_{\bar{\lambda}} \quad \text{and} \quad \Gamma_2 x = x_\lambda + V x_{\bar{\lambda}}, \tag{1.2}$$

where

$$x = x_0 + x_\lambda + x_{\bar{\lambda}} \quad (x_0 \in \mathcal{D}(A), \ x_z \in \mathfrak{N}_z, \ z \in \{\lambda, \bar{\lambda}\}).$$

If $x = 0$, then $x_\lambda = x_{\bar{\lambda}} = 0$ and, thus, the operators Γ_1 and Γ_2 are well defined and each of them maps $\mathcal{D}(A^*)$ into $\mathfrak{N}_\lambda$. Let us show that $(\mathfrak{N}_\lambda, \Gamma_1, \Gamma_2)$ is a BVS of the operator A.

Indeed, by virtue of (1.2), for any x and $y = y_0 + y_\lambda + y_{\bar{\lambda}}$ from $\mathcal{D}(A^*)$, we obtain

$$(\Gamma_1 x, \Gamma_2 y) - (\Gamma_2 x, \Gamma_1 y) = (\bar{\lambda} - \lambda)\big((x_\lambda, y_\lambda) - (x_{\bar{\lambda}}, y_{\bar{\lambda}})\big). \tag{1.3}$$

Similarly, by using the von Neumann formulas, we easily show that

$$(A^* x, y) - (x, A^* y) = (\bar{\lambda} - \lambda)\big((x_\lambda, y_\lambda) - (x_{\bar{\lambda}}, y_{\bar{\lambda}})\big). \tag{1.4}$$

Thus, in view of (1.3) and (1.4), the operators A^*, Γ_1 and Γ_2 satisfy equality (1.1).

Assume that φ and ψ are arbitrary vectors from $\mathfrak{N}_\lambda$. Consider a vector $x = x_\lambda + x_{\bar{\lambda}}$, where

$$x_\lambda = \frac{1}{\lambda - \bar{\lambda}}(\lambda\psi - \varphi) \quad \text{and} \quad x_{\bar{\lambda}} = \frac{1}{\lambda - \bar{\lambda}}V^{-1}(\varphi - \bar{\lambda}\psi).$$

These vectors belong to the subspaces $\mathfrak{N}_\lambda$ and $\mathfrak{N}_{\bar{\lambda}}$, respectively, and it is easy to see that

$$\Gamma_1 x = \varphi \quad \text{and} \quad \Gamma_2 x = \psi.$$

Thus, according to Definition I, the triple $(\mathfrak{N}_\lambda, \Gamma_1, \Gamma_2)$ is a boundary-value space of the operator A.

■

Note that the operators Γ_1 and Γ_2 from the BVS (X, Γ_1, Γ_2) and the space X are defined ambiguously. In fact, by choosing different unitary operators V in relations (1.2), we obtain different operators Γ_1 and Γ_2.

In the same way, instead of $\mathfrak{N}_\lambda$, one can take $\mathfrak{N}_{\bar{\lambda}}$ or any space X (with $\dim X = n$) which is not necessarily a subspace of the space $\mathcal{H}$. The indicated ambiguity in the construction of BVS enables us to select the most suitable BVS for each special operator A.

In particular, if A is a symmetric operator semibounded from below, then, in many cases, it is convenient to use a BVS constructed in the following way:

We fix a self-adjoint extension $\tilde{A}$ of the operator A with the greatest lower bound $m(\tilde{A}) = m(A)$ (as shown in Chapter 2, extensions of this sort definitely exist).

Let us show that

$$\mathcal{D}(A^*) = \mathcal{D}(\tilde{A}) \dotplus \operatorname{Ker}(A^* - (m(A) - \alpha)I), \tag{1.5}$$

where α is a fixed positive number.

Indeed, since $m(\tilde{A}) = m(A)$, the operator $R = (\tilde{A} - (m(A) - \alpha)I)^{-1}$ exists and is defined in the entire space $\mathcal{H}$. This means that any element $x \in \mathcal{D}(A^*)$ can be represented in the form

$$x = u + h,$$

where the element $u = R(A^* - (m(A) - \alpha)I)x$ belongs to $\mathcal{D}(\tilde{A})$ and the element $h = x - u$ belongs to $\operatorname{Ker}(A^* - (m(A) - \alpha)I)$. Thus, decomposition (1.5) is true.

We set $X = \operatorname{Ker}(A^* - (m(A) - \alpha)I)$ and consider linear operators $\Gamma_i \colon \mathcal{D}(A^*) \to X$ whose action is described by the formulas

$$\Gamma_1(u+h) = P_X(\tilde{A} - (m(A) - \alpha)I)u \quad \text{and} \quad \Gamma_2(u+h) = h, \tag{1.6}$$

where $u + h \in \mathcal{D}(A^*)$, $u \in \mathcal{D}(\tilde{A})$, $h \in X$, and P_X is the orthoprojector onto X in $\mathcal{H}$

Taking into account decomposition (1.5), we conclude that the operators Γ_i are well defined. We show that these operators satisfy relation (1.1).

Indeed, for any $\{x, y\} \subset \mathcal{D}(A^*)$, where $x = u + h$ and $y = v + z$, we have[9]

$$\begin{aligned}(A^*x, y) - (x, A^*y) &= (\tilde{A}u, z) - (u, (m(A) - \alpha)z) + ((m(A) - \alpha)h, v) - (h, \tilde{A}v)\\ &= (\Gamma_1 x, \Gamma_2 y) - (\Gamma_2 x, \Gamma_1 y).\end{aligned}$$

On the other hand, if φ and ψ are arbitrary elements from X, then it is easy to see that

$$\Gamma_1 x = \varphi \quad \text{and} \quad \Gamma_2 x = \psi,$$

where $x = (\tilde{A} - (m(A) - \alpha)I)^{-1}\varphi + \psi$.

Therefore, the triple (X, Γ_1, Γ_2) is a BVS of the symmetric operator A semi-bounded from below. In what follows, this BVS is called the *canonical positive BVS of the operator A associated with the extension $\tilde{A}$* or, simply, the *canonical positive BVS.*

As stated above, the BVS of a symmetric operator A are defined ambiguously. Among the properties of BVS independent of the form of an operator A, we especially mention the following fact:

Proposition 1.2. *If (X, Γ_1, Γ_2) is a BVS of an operator A, then*

$$\mathcal{D}(A) = \operatorname{Ker}\Gamma_1 \cap \operatorname{Ker}\Gamma_2. \tag{1.7}$$

Proof. Let $y \in \mathcal{D}(A)$. Consider the elements $\varphi = \Gamma_2 y$ and $\psi = -\Gamma_1 y$. By virtue of Definition I, there exists a vector $x \in \mathcal{D}(A^*)$ such that

$$\Gamma_1 x = \Gamma_2 y \quad \text{and} \quad \Gamma_2 x = -\Gamma_1 y.$$

Therefore, equality (1.1) can be rewritten in the form

---

[9] In the case under consideration, the scalar product $(\cdot, \cdot)_X$ in X coincides with $(\cdot, \cdot)$.

$$(A^*x, y) - (x, Ay) = \|\Gamma_1 y\|_X^2 + \|\Gamma_2 y\|_X^2. \tag{1.8}$$

The left-hand side and, hence, the right-hand side of (1.8) are equal to zero and, consequently, $\Gamma_1 y = \Gamma_2 y = 0$.

Thus,

$$\mathcal{D}(A) \subset \operatorname{Ker}\Gamma_1 \cap \operatorname{Ker}\Gamma_2.$$

Conversely, if, for some $y \in \mathcal{D}(A^*)$, the equality $\Gamma_1 y = \Gamma_2 y = 0$ is true, then, in view of (1.1), we have

$$(A^*x, y) = (x, A^*y) \qquad (x \in \mathcal{D}(A^*))$$

and, therefore, $y \in \mathcal{D}(A^*) = \mathcal{D}(A)$.

■

1.2. Hermitian Operators with Equal Defect Numbers. Let A be an Hermitian operator with deficiency index (n, n) $(n \le \infty)$ acting in a Hilbert space $\mathcal{H}$. Recall that, beginning with Subsection 9 (Chapter 2), we assume that the Hermitian operator A satisfies the condition

$$\mathcal{D}(A) \vee \mathcal{R}(A) = \mathcal{H}$$

guaranteeing that its defect subspaces $\mathfrak{N}_\lambda$ and $\mathfrak{N}_{\bar\lambda}$ $(\operatorname{Im}\lambda \ne 0)$ are linearly independent.

In proving Theorem 1.1, we showed that if A is a symmetric operator, then equality (1.4) is true. This implies that, without loss of generality, equality (1.1) in Definition I can be rewritten in the form

$$(\Gamma_1 x, \Gamma_2 y)_X - (\Gamma_2 x, \Gamma_1 y)_X = (\lambda - \bar\lambda)\left[(x_{\bar\lambda}, y_{\bar\lambda}) - (x_\lambda, y_\lambda)\right].$$

Further, by virtue of Proposition 1.2, we have $\mathcal{D}(A) \subset \operatorname{Ker}\Gamma_i$ $(i = 1, 2)$. Hence, instead of the operators Γ_1 and Γ_2, we can consider the restrictions of these operators to the lineal

$$\mathcal{N}_\lambda = \mathfrak{N}_\lambda \dotplus \mathfrak{N}_{\bar\lambda},$$

where λ is a fixed nonreal number.

Thus, we can introduce the following natural definition:

Definition II. *A triple* (X, Γ_1, Γ_2), *where* X *is a Hilbert space with scalar product* $(\cdot, \cdot)_X$ *and* Γ_1 *and* Γ_2 *are linear operators mapping the lineal* $\mathcal{N}_\lambda$ *into* X, *is called a boundary-value space of an Hermitian operator* A *if*

(i) for any $x = x_\lambda + x_{\bar\lambda}$ *and* $y = y_\lambda + y_{\bar\lambda}$ *from* $\mathcal{N}_\lambda$,

$$(\Gamma_1 x, \Gamma_2 y)_X - (\Gamma_2 x, \Gamma_1 y)_X = (\lambda - \bar\lambda)\left[(x_{\bar\lambda}, y_{\bar\lambda}) - (x_\lambda, y_\lambda)\right]; \qquad (1.9)$$

(ii) for any φ *and* ψ *from* X, *there exists a vector* $x \in \mathcal{N}_\lambda$ *such that*

$$\Gamma_1 x = \varphi \quad \textit{and} \quad \Gamma_2 x = \psi.$$

It is clear that if an operator A is symmetric, then, it is easy to see that Definitions I and II are equivalent.

By repeating the proof of Theorem 1.1, we readily obtain the following assertion:

Proposition 1.3. *For any Hermitian operator* A *with deficiency index* (n, n) $(n \leq \infty)$ *such that* $\mathcal{D}(A) \vee \mathcal{R}(A) = \mathcal{H}$, *there exists a BVS* (X, Γ_1, Γ_2) *with* $\dim X = n$.

Proposition 1.4. *If* (X, Γ_1, Γ_2) *is a BVS of an Hermitian operator* A, *then the lineals* Γ_1 *and* Γ_2 *are linearly independent and*

$$\operatorname{Ker}\Gamma_1 \dotplus \operatorname{Ker}\Gamma_2 = \mathfrak{N}_\lambda \dotplus \mathfrak{N}_{\bar\lambda}. \qquad (1.10)$$

Proof. Let $\Gamma_1 x = \Gamma_2 x = 0$, where $x = x_\lambda + x_{\bar\lambda} \in \mathcal{N}_\lambda$. Then, by virtue of (1.9), for any $y = y_\lambda + y_{\bar\lambda} \in \mathcal{N}_\lambda$, we have

$$(x_\lambda, y_\lambda) - (x_{\bar\lambda}, y_{\bar\lambda}) = 0. \qquad (1.11)$$

In particular, for $x_\lambda = y_\lambda$ and $x_{\bar\lambda} = -y_{\bar\lambda}$, equality (1.11) can be rewritten in the form

$$\|x_\lambda\|^2 + \|x_{\bar\lambda}\|^2 = 0$$

and, hence, $x = 0$. Thus, the lineals $\operatorname{Ker}\Gamma_1$ and $\operatorname{Ker}\Gamma_2$ are linearly independent. Further, since $\operatorname{Ker}\Gamma_i \subset \mathcal{N}_\lambda$ $(i = 1, 2)$, we have

$$\operatorname{Ker}\Gamma_1 \dotplus \operatorname{Ker}\Gamma_2 \subset \mathcal{N}_\lambda. \qquad (1.12)$$

Let us show that the inverse inclusion is also true. Indeed, let $y \in \mathcal{N}_\lambda$. In X, we consider the vectors $\varphi = 0$ and $\psi = \Gamma_2 y$. It follows from the definition of BVS that, in $\mathcal{N}_\lambda$, there exists a vector x such that $\Gamma_1 x = 0$ and $\Gamma_2 x = \Gamma_2 y$. Therefore, $x \in \operatorname{Ker}\Gamma_1$ and $y - x \in \operatorname{Ker}\Gamma_2$. Thus, the element $y = x + (y - x)$ belongs to $\operatorname{Ker}\Gamma_1 \dotplus \operatorname{Ker}\Gamma_2$. In view of inclusion (1.12), this proves equality (1.10).

■

Note that if A is an Hermitian operator, then the lineals $\mathcal{D}(A)$ and $\mathcal{N}_\lambda$ may have common nonzero elements. Moreover, the case where $\mathcal{D}(A) \subset \mathcal{N}_\lambda$ is also possible. Thus, if an Hermitian operator A is defined in a two-dimensional space $\mathcal{H} = \langle e_1, e_2 \rangle$ with orthonormal basis $\{e_1, e_2\}$ by the equality

$$Ae_1 = ae_1 + be_2 \quad (\mathcal{D}(A) = \langle e_1 \rangle),$$

where $a \in \mathbb{R}$ and $b \in \mathbb{C}$ $(b \neq 0)$, then

$$\mathfrak{N}_\lambda = \langle \bar{b} e_1 + (\bar{\lambda} - a) e_2 \rangle$$

and, hence, $\mathcal{D}(A) \subset \mathfrak{N}_\lambda \dotplus \mathfrak{N}_{\bar{\lambda}}$.

However, as follows from the linear independence of the lineals $\operatorname{Ker}\Gamma_1$ and $\operatorname{Ker}\Gamma_2$, the operators Γ_1 and Γ_2 cannot simultaneously vanish on nonzero vectors from $\mathcal{D}(A) \cap \mathcal{N}_\lambda$. Thus, Proposition 1.2 is not true for Hermitian operators.

Consider the linear operators

$$\Omega_+ = \frac{1}{\sqrt{2}}(\Gamma_1 + i\Gamma_2) \quad \text{and} \quad \Omega_- = \frac{1}{\sqrt{2}}(\Gamma_1 - i\Gamma_2) \tag{1.13}$$

acting from $\mathcal{N}_\lambda$ into X. Let us show that

$$\operatorname{Ker}\Omega_- \dotplus \operatorname{Ker}\Omega_+ = \mathcal{N}_\lambda. \tag{1.14}$$

For this purpose, we note that the following obvious relation:

$$\operatorname{Ker}\Omega_- \cap \operatorname{Ker}\Omega_+ \subset \operatorname{Ker}\Gamma_1 \cap \operatorname{Ker}\Gamma_2$$

and Proposition 1.4 imply that the lineals $\operatorname{Ker}\Omega_-$ and $\operatorname{Ker}\Omega_+$ are linearly independent. Further, for any $y \in \mathcal{N}_\lambda$, we set $\varphi = (\Gamma_1 + i\Gamma_2)y$.

In view of the definition of BVS, there exists an element $x_- \in \mathcal{N}_\lambda$ such that

$$\Gamma_1 x_- = \frac{1}{2}\varphi \quad \text{and} \quad \Gamma_2 x_- = -\frac{i}{2}\varphi.$$

By using (1.13), it is easy to show that the element x_- in the decomposition $y = x_- + y - x_-$ belongs to the subspace $\operatorname{Ker}\Omega_-$ and the element $y - x_-$ belongs to the subspace Ω_+. Decomposition (1.14) is thus proved.

It follows from equalities (1.9) and (1.13) that

$$(\Omega_+ x, \Omega_+ y)_X - (\Omega_- x, \Omega_- y)_X = [x, y]_\lambda \qquad (\{x, y\} \subset \mathfrak{N}_\lambda), \tag{1.15}$$

where $[\cdot,\cdot]_\lambda$ is an indefinite metric in $\mathfrak{N}_\lambda$ defined by relation (9.12)[10].

In Section 9, we have shown that the lineal $\mathfrak{N}_\lambda$ with indefinite metric (9.12) is a Krein space. By virtue of (1.15), we see that the lineals $\operatorname{Ker}\Omega_-$ and $\operatorname{Ker}\Omega_+$ are mutually J-orthogonal in this space. Moreover, $\operatorname{Ker}\Omega_-$ and $\operatorname{Ker}\Omega_+$ are the positive and negative lineals, respectively. Further, in view of (1.14), we have the following decomposition:

$$\operatorname{Ker}\Omega_- \; [\dot{+}] \; \operatorname{Ker}\Omega_+ = \mathfrak{N}_\lambda. \tag{1.16}$$

By analogy with the proof of Theorem 9.9, we can assert that decomposition (1.16) is canonical for the Krein space $\mathfrak{N}_\lambda$ and the lineal $\operatorname{Ker}\Omega_-$ $(\operatorname{Ker}\Omega_+)$ is a maximal uniformly positive (negative) subspace of $\mathfrak{N}_\lambda$.

Proposition 1.5. *If* (X, Γ_1, Γ_2) *is a BVS of an Hermitian operator* A, *then*

$$\dim X = \dim \mathfrak{N}_\lambda = \dim \mathfrak{N}_{\bar\lambda} \qquad (\operatorname{Im}\lambda \neq 0).$$

Proof. By virtue of the definition of BVS and decomposition (1.16), relations (1.13) imply that the operator Ω_+ bijectively maps $\operatorname{Ker}\Omega_-$ onto X.

Therefore,

$$\dim \operatorname{Ker}\Omega_- = \dim X. \tag{1.17}$$

Without loss of generality, we can assume that $\operatorname{Im}\lambda < 0$. Then, as shown in Section 9, the subspace $\mathfrak{N}_\lambda$, just as the subspace $\operatorname{Ker}\Omega_-$ is maximal and uniformly positive in $\mathfrak{N}_\lambda$. It is well known that the dimensions of these subspaces coincide (see Azizov and Iokhvidov [1]). In view of (1.17), this means that $\dim \mathfrak{N}_\lambda = \dim X$. In exactly the same way as in the case of the maximal uniformly negative subspace $\operatorname{Ker}\Omega_+$, we conclude that $\dim \mathfrak{N}_{\bar\lambda} = \dim X$, which completes the proof.

■

[10] Here and in what follows, in all references to Section 9 (Chapter 2), we simply write Section 9 without indicating the chapter because this does not lead to any ambiguities.

1.3. Hermitian Operators with Arbitrary Defect Numbers. Let A be a closed Hermitian operator with arbitrary defect numbers (n, m) $(n, m \le \infty)$ acting in a Hilbert space $\mathcal{H}$.

It follows from Proposition 1.5 that, for the Hermitian operator A with unequal defect numbers, the BVS (X, Γ_1, Γ_2) does not exist in the form of Definition II. An approach to the definition of BVS proposed below enables us to solve this problem.

Let F be a canonical dissipative extension of the operator A.

Definition III. A quadruple (X_+, X_-, Y_+, Y_-), *where* $X_\pm$ *are Hilbert spaces with scalar products* $(\cdot,\cdot)_\pm$, *and* Y_+ *and* Y_- *are linear operators mapping* $\mathcal{D}(F)$ *and* $\mathcal{D}(F^*)$ *onto* X_+ *and* X_-, *respectively, is called a boundary-value space of the Hermitian operator* A *if*

$$\text{(i)} \qquad (Fx, x) - (x, Fx) = i\|Y_+x\|_+^2 \quad (x \in \mathcal{D}(F)), \tag{1.18}$$

$$\text{(ii)} \qquad (F^*y, y) - (y, F^*y) = -i\|Y_-x\|_-^2 \quad (y \in \mathcal{D}(F^*)). \tag{1.19}$$

It follows from Definition III that a BVS of the operator A is, in fact, constructed according to its canonical dissipative extension F. In order to emphasize this connection, we sometimes say that (X_+, X_-, Y_+, Y_-) is a BVS *associated* with the extension F.

Theorem 1.6. *For any canonical dissipative extension* F *of a closed Hermitian operator* A *there exists a BVS* (X_+, X_-, Y_+, Y_-) *associated with this extension.*

Proof. Recall that (see Section 9) any canonical extension F of the operator A is defined by a decomposition

$$\mathcal{D}(F) = \mathcal{D}(A) \dotplus \mathcal{L}_+, \tag{1.20}$$

where $\mathcal{L}_+$ is a maximal uniformly positive subspace of $\mathcal{N}_\lambda$. In view of the definition of uniform positivity, one can easily see that $\mathcal{L}_+$ is a Hilbert space with respect to the scalar product $(\cdot,\cdot)_+ = [\cdot,\cdot]_\lambda$. We denote this Hilbert space by X_+.

For any element $x \in \mathcal{D}(F)$ $(x = x_0 + x_+,\ x_0 \in \mathcal{D}(A),\ x_+ \in \mathcal{L}_+)$, we set

$$Y_+x = x_+.$$

It is clear that Y_+ is a linear operator mapping $\mathcal{D}(F)$ onto X_+. In view of the definitions of the space X_+ and the operator Y_+, equality (9.17) (with $F' = F$) implies equality (1.18).

In view of Proposition 9.8, the canonical extension F is densely defined in $\mathcal{H}$. Consequently, the operator F^* exists. In Theorem 9.9, it was proved that the extension F^* is defined by the decomposition

$$\mathcal{D}(F^*) = \mathcal{D}(A) \dotplus \mathcal{L}_-, \tag{1.21}$$

where $\mathcal{L}_- = \mathcal{L}_+^{[\perp]}$ is a maximal uniformly negative subspace of $\mathcal{N}_\lambda$. Similarly, we obtain that $\mathcal{L}_-$ is a Hilbert space with respect to the scalar product $(\cdot,\cdot)_- = -[\cdot,\cdot]_\lambda$. We denote this space by X_-. Further, if we introduce the operator Y_- ($Y_-: \mathcal{D}(F^*) \to X_-$) defined by the equality

$$Y_- y = y_- \quad (y \in \mathcal{D}(F^*)),$$

where $y = y_0 + y_-$, $y_0 \in \mathcal{D}(A)$, and $y_- \in \mathcal{L}_-$, and take into account equality (9.17) (with $F' = F = F^*$), then we arrive at equality (1.19).

Thus, the quadruple (X_+, X_-, Y_+, Y_-) is a BVS of the Hermitian operator A in the sense of Definition III. This BVS is associated with the canonical dissipative extension F of the operator A.

■

Corollary 1.7. *If (X_+, X_-, Y_+, Y_-) is a BVS of an operator A, then*

$$\operatorname{Ker} Y_+ = \operatorname{Ker} Y_- = \mathcal{D}(A).$$

Proof. Assume that F is a canonical dissipative extension of the Hermitian operator A associated with the BVS (X_+, X_-, Y_+, Y_-). By virtue of Lemma 9.6 and decomposition (1.20) and (1.21), equalities (1.18) and (1.19) can be rewritten in the form

$$[x_+, x_+]_\lambda = \| Y_+ x_+ \|_+^2 \quad \text{and} \quad -[y_-, y_-]_\lambda = \| Y_- y_- \|_-^2, \tag{1.22}$$

where $x_+ \in \mathcal{L}_+$ and $y_- \in \mathcal{L}_-$. Since the subspace $\mathcal{L}_+$ ($\mathcal{L}_-$) is uniformly positive (negative), Corollary 1.7 is proved.

■

Note that the deficiency index of the operator A uniquely specifies the dimensionalities of the spaces $X_\pm$ in a BVS (X_+, X_-, Y_+, Y_-).

Corollary 1.8. *If (X_+, X_-, Y_+, Y_-) is a BVS of a closed Hermitian operator A, then the dimensionality of the space X_+ (X_-) is equal to the defect number of the operator A in the lower (upper) half plane.*

Proof. Let F be the canonical dissipative extension associated with the BVS (X_+, X_-, Y_+, Y_-). It follows from equalities (1.22) that the operator Y_+ (Y_-) establishes a one-to-one correspondence between the subspaces $\mathcal{L}_+$ $(\mathcal{L}_-)$ and X_+ (X_-). Hence,

$$\dim \mathcal{L}_+ = \dim X_+ \quad \text{and} \quad \dim \mathcal{L}_- = \dim X_- .$$

As in the proof of Proposition 1.5, we conclude that

$$\dim X_+ = \dim \mathfrak{N}_\lambda \quad \text{and} \quad \dim X_- = \dim \mathfrak{N}_{\bar{\lambda}} \qquad (\operatorname{Im}\lambda < 0).$$

■

Theorem 1.9. *If* A *is a closed Hermitian operator with deficiency index* (n, n), *then the definition of the BVS* (X_+, X_-, Y_+, Y_-) *of the operator* A *is equivalent to the definition of the BVS* (X, Γ_1, Γ_2) *from Subsection 1.2.*

Proof. Assume that an operator A satisfies the conditions of the theorem and (X_+, X_-, Y_+, Y_-) is its BVS. By virtue of Corollary 1.8, we have

$$\dim X_+ = \dim X_- = n.$$

This means that, without loss of generality, we can assume that the spaces $X_\pm$ coincide, i.e.,

$$X = X_+ = X_- .$$

Let F be a canonical dissipative extension of the operator A associated with the BVS (X, X, Y_+, Y_-). This extension is defined by decomposition (1.20), where $\mathcal{L}_+$ is a maximal uniformly positive subspace of $\mathcal{N}_\lambda$. Thus, in view of (9.4), the Krein space $\mathcal{N}_\lambda$ can be represented in the form

$$\mathcal{N}_\lambda = \mathcal{L}_+ [\dot{+}] \mathcal{L}_- \qquad (\mathcal{L}_- = \mathcal{L}_+^{[\perp]}). \tag{1.23}$$

Consider linear operators Γ_i $(\Gamma_i\colon \mathcal{N}_\lambda \to X,\ i = 1, 2)$ defined by the formulas

$$\Gamma_1 x = \frac{1}{\sqrt{2}}(Y_+x_+ + Y_-x_-) \quad \text{and} \quad \Gamma_2 x = \frac{i}{\sqrt{2}}(-Y_+x_+ + Y_-x_-), \tag{1.24}$$

where $x = x_+ + x_- \in \mathcal{N}_\lambda$, $x_+ \in \mathcal{L}_+$, and $x_- \in \mathcal{L}_-$.

By virtue of (1.23), we conclude that the operators Γ_1 and Γ_2 are correctly defined in $\mathcal{N}_\lambda$.

Let us show that the triple (X, Γ_1, Γ_2) is a BVS of the Hermitian operator A in the sense of Definition II. Indeed, by using (1.24), one can easily show that

$$\alpha = (\Gamma_1 x, \Gamma_2 y)_X - (\Gamma_2 x, \Gamma_1 y)_X = i(Y_+ x_+, Y_+ y_+)_X - i(Y_- x_-, Y_- y_-)_X, \tag{1.25}$$

where $\{x, y\} \subset \mathcal{N}_\lambda$, $x = x_+ + x_-$, and $y = y_+ + y_-$. By virtue of equalities (1.22) and (1.23), this yields

$$\alpha = i[x_+, y_+]_\lambda + i[x_-, y_-]_\lambda = i[x, y]_\lambda. \tag{1.26}$$

In view of the definition of an indefinite metric [equality (9.12)], relations (1.25) and (1.26) imply equality (1.9) in Definition II.

Further, we show that, for all vectors φ and ψ from X, there exist a vector $x \in \mathcal{N}_\lambda$ such that

$$\Gamma_1 x = \varphi \quad \text{and} \quad \Gamma_2 x = \psi.$$

To prove this fact, it suffices to note that the operator Y_+ (Y_-) establishes a one-to-one correspondence between the subspaces $\mathcal{L}_+$ $(\mathcal{L}_-)$ and X (X) and to consider the vector $x = x_+ + x_-$ whose elements $x_\pm$ are unique solutions of the following equations:

$$Y_+ x_+ = \frac{1}{\sqrt{2}}(\varphi + i\psi) \quad \text{and} \quad Y_- x_- = \frac{1}{\sqrt{2}}(\varphi - i\psi),$$

respectively. Thus, the triple (X, Γ_1, Γ_2) is a BVS of the Hermitian operator A.

Conversely, consider a BVS (X, Γ_1, Γ_2) of the Hermitian operator A in the sense of Definition II.

In terms of this BVS, we construct the canonical decomposition of the space $\mathcal{N}_\lambda$ of the form (1.16), where the operators $\Omega_\pm$ are defined by equalities (1.13).

Consider the canonical dissipative extension F of the operator A defined by decomposition (1.20) with the maximal uniformly positive subspace $\mathcal{L}_+ = \operatorname{Ker} \Omega_-$.

For any element $x \in \mathcal{D}(F)$ $(x = x_0 + x_+,\ x_0 \in \mathcal{D}(A),\ x_+ \in \mathcal{L}_+)$, we set

$$Y_+ x = \Omega_+ x_+.$$

By virtue of the definition of BVS (X, Γ_1, Γ_2), it is easy to see that $\mathcal{R}(\Omega_+) = X$. Therefore, the operator Y_+ maps $\mathcal{D}(F)$ onto X.

It follows from equality (1.15) that

$$\| Y_+ x \|_X^2 = \| \Omega_+ x_+ \|_X^2 = [x_+, x_+]_\lambda \quad (x \in \mathcal{D}(F)).$$

In view relation (9.17) (with $F' = F$), this implies equality (1.18) in Definition III.

Similarly, the extension F^* is defined by decomposition (1.21), where $\mathcal{L}_- = \operatorname{Ker}\Omega_+$ is a maximal uniformly negative subspace. The operator Y_- is defined by the equality

$$Y_- y = \Omega_- y_-,$$

where $y = y_0 + y_-$, $y_0 \in \mathcal{D}(A)$, and $y_- \in \mathcal{L}_-$. This operator maps $\mathcal{D}(F^*)$ onto X. Taking into account relations (9.17) and (1.15), we see that relation (1.19) in the Definition III is true.

Thus, the quadruple (X, X, Y_+, Y_-) is a BVS of the Hermitian operator A.

The proof of Theorem 1.9 is completed.

■

1.4. Examples of BVS

1. In the space $\mathcal{H} = L_2(0, 1)$, we consider a symmetric operator A defined by the formulas[11]

$$A = -\frac{d^2}{ds^2}, \qquad \mathcal{D}(A) = \mathring{W}_2^2(0,1).$$

The deficiency index of this operator is equal to $(2, 2)$. Its adjoint operator A^* is given by the formula

$$A^* = -\frac{d^2}{ds^2}, \qquad \mathcal{D}(A^*) = W_2^2(0,1).$$

It is easy to show that a BVS (X, Γ_1, Γ_2) of the operator A in the sense of Definition I can be chosen as follows:

$$X = \mathbb{C}^2, \quad \Gamma_1 x = \langle -x(0), x(1) \rangle \quad \text{and} \quad \Gamma_2 x = \langle x'(0), x'(1) \rangle \quad (x(s) \in \mathcal{D}(A^*)),$$

It is clear that, for the considered operator A, we can also construct BVS both in the sense of Definition II and in the sense of Definition III.

In particular, the BVS (X_+, X_-, Y_+, Y_-) associated with the canonical dissipative extension

$$F = -\frac{d^2}{ds^2}, \quad \mathcal{D}(F) = \{x(s) \in W_2^2(0,1) \mid x(0) = -ix'(0), \ x(1) = ix'(1)\},$$

---

[11] $W_\beta^\alpha(a, b)$ is a Sobolev space.

of the operator A can be chosen in the form

$$X_+ = X_- = \mathbb{C}^2, \quad Y_+ = \sqrt{2}\Gamma_1|_{\mathcal{D}(F)} \text{ and } Y_- = \sqrt{2}\Gamma_1|_{\mathcal{D}(F^*)}.$$

2. In the space $\mathcal{H} = L_2(0, \infty)$, we consider the symmetric operator

$$A = -i\frac{d}{ds}, \quad \mathcal{D}(A) = \overset{\circ}{W}{}_2^1(0, \infty).$$

Since the deficiency index of the operator A is equal to $(0, 1)$ the construction of its BVS is possible only in the sense of Definition III. It is easy to see that the operator

$$A^* = -i\frac{d}{ds}, \quad \mathcal{D}(A^*) = W_2^1(0, \infty),$$

adjoint to the operator A is a canonical dissipative extension of the operator A. This enables us to construct the BVS (X_+, X_-, Y_+, Y_-) of the operator A in the following form:

$$X_+ = \mathbb{C}, \quad Y_+x = x(0) \quad (x(s) \in \mathcal{D}(A^*)),$$

$$X_- = \{0\}, \quad Y_-y = 0 \quad (y(s) \in \mathcal{D}(A)).$$

3. Let $\mathfrak{H}$ be a Hilbert space with scalar product $(\cdot, \cdot)_{\mathfrak{H}}$. We consider a Hilbert space $\mathcal{H} = L_2(\mathfrak{H}, (0, \infty))$ of measurable vector functions $x(s)$ $(s \in (0, \infty))$ with values in $\mathfrak{H}$ and such that

$$(x, x)_{\mathcal{H}} = \int_0^\infty (x(s), x(s))_{\mathfrak{H}}\, ds < \infty. \tag{1.27}$$

Note that the space $\mathcal{H}$ is the tensor product of the spaces $L_2(0, \infty)$ and $\mathfrak{H}$, i.e., $\mathcal{H} = L_2(0, \infty) \otimes \mathfrak{H}$.

Let J be a self-adjoint unitary operator in $\mathfrak{H}$. This operator defines an indefinite metric in the space $\mathfrak{H}$ by the formula

$$[x, y] = (Jx, y)_{\mathfrak{H}} \quad \{x, y\} \subset \mathfrak{H}. \tag{1.28}$$

The space $\mathfrak{H}$ is a Krein space with respect to the indefinite metric (1.28).

In the space $\mathcal{H}$, consider the symmetric operator

$$A = -iJ\frac{d}{ds}, \qquad \mathcal{D}(A) = \overset{\circ}{W}{}_2^1(\mathfrak{H}, (0, \infty)) = \overset{\circ}{W}{}_2^1(0, \infty) \otimes \mathfrak{H}.$$

Its adjoint operator has the form

$$A^* = -iJ\frac{d}{ds}, \qquad \mathcal{D}(A^*) = W_2^1(\mathfrak{H}, (0, \infty)).$$

In view of (1.27) and (1.28), we can directly show that

$$(A^*x, y)_{\mathcal{H}} - (x, A^*y)_{\mathcal{H}} = i\,[x(0), y(0)]. \tag{1.29}$$

Consider the fundamental decomposition of the Krein space $\mathfrak{H}$

$$\mathfrak{H} = \mathfrak{H}_+ \,[\dot{+}]\, \mathfrak{H}_- \quad (\mathfrak{H}_+ = \mathcal{R}(I+J),\ \mathfrak{H}_- = \mathcal{R}(I-J)). \tag{1.30}$$

In view of (1.28), it is easy to see that the restriction of the indefinite metric $[\cdot,\cdot]$ to the subspaces $\mathfrak{H}_\pm$ coincides (to within the sign) with the original scalar product $(\cdot,\cdot)_{\mathfrak{H}}$, i.e.,

$$[x, x] = (x, x)_{\mathfrak{H}} \quad (x \in \mathfrak{H}_+) \quad \text{and} \quad [y, y] = -(y, y)_{\mathfrak{H}} \quad (y \in \mathfrak{H}_-).$$

Equality (1.29) and Theorem 9.9 imply that the operator

$$F = -iJ\frac{d}{ds}, \qquad \mathcal{D}(F) = \{x(s) \in \mathcal{D}(A^*) \,|\, x(0) \in \mathfrak{H}_+\} \tag{1.31}$$

is a canonical dissipative extension of the operator A. In this case,

$$F^* = -iJ\frac{d}{ds}, \qquad \mathcal{D}(F^*) = \{y(s) \in \mathcal{D}(A^*) \,|\, y(0) \in \mathfrak{H}_-\}.$$

The BVS (X_+, X_-, Y_+, Y_-) associated with the canonical extension F can be chosen in the form

$$X_+ = \mathfrak{H}_+, \quad Y_+x = x(0) \quad (x(s) \in \mathcal{D}(F)),$$

$$X_- = \mathfrak{H}_-, \quad Y_-y = y(0) \quad (y(s) \in \mathcal{D}(F^*)).$$

Note that, by setting $\mathfrak{H} = \mathbb{C}$ and $J = I$, we obtain the BVS from Example 2.

4. In the space $\mathcal{H} = L_2(\mathbb{R}^3)$, we consider a symmetric operator A defined as the restriction of the Laplacian $-\Delta$ to the set of all functions from $W_2^2(\mathbb{R}^3)$ vanishing at the point $x = 0$, namely,

$$A = -\Delta, \qquad \mathcal{D}(A) = \left\{u(x) \in W_2^2(\mathbb{R}^3) \,\middle|\, u(0) = 0\right\}.$$

The operator A is nonnegative $(m(A) = 0)$ and its deficiency index is equal to $(1, 1)$. By virtue of decomposition (1.5) (with $\alpha = 1$), we obtain

$$\mathcal{D}(A^*) = \mathcal{D}(\tilde{A}) \dotplus X,$$

where $\tilde{A} = -\Delta$, $\mathcal{D}(\tilde{A}) = W_2^2(\mathbb{R}^3)$, is a self-adjoint operator in $L_2(\mathbb{R}^3)$ and

$$X = \operatorname{Ker}(A^* + I) = \left\langle \frac{e^{-|x|}}{|x|} \right\rangle \qquad \left(x \in \mathbb{R}^3,\ |x| = \sqrt{x_1^2 + x_2^2 + x_3^2}\ \right).$$

In view of equalities (1.6), we can directly show that the triple (X, Γ_1, Γ_2), where $\Gamma_i \colon \mathcal{D}(A^*) \to X$,

$$\Gamma_1 f = 2(\alpha_1(f) + \alpha_2(f))\frac{e^{-|x|}}{|x|}, \qquad \Gamma_2 f = \alpha_2(f)\frac{e^{-|x|}}{|x|},$$

$$\alpha_1(f) = \lim_{|x| \to 0} \frac{d(|x| f(x))}{d|x|}, \qquad \text{and} \qquad \alpha_2(f) = \lim_{|x| \to 0} |x| f(x),$$

is a canonical positive BVS of the operator A constructed according to the extension $\tilde{A}$.

In the general case where an operator A symmetric in $L_2(\mathbb{R}^n)$ is a restriction of a formal self-adjoint elliptic expression $\mathcal{A}$ of order $m > n/2$ to a set of functions $u(x) \in W_2^2(\mathbb{R}^n)$ vanishing in a certain subset of $\mathbb{R}^n$ of measure zero, examples of BVS of the same type were constructed by Kochubei in [6].

5. In the space $\mathcal{H} = L_2(0, 1)$, we consider the Hermitian operator

$$A = -i\frac{d}{ds}, \qquad \mathcal{D}(A) = \left\{x(s) \in \mathring{W}_2^1(0, 1) \,\middle|\, (x, \tau) = 0\right\},$$

where $\tau(s)$ is a fixed function from $W_2^1(0, 1)$. The operator A can be regarded as a restriction of the symmetric operator

$$A' = -i\frac{d}{ds}, \qquad \mathcal{D}(A') = \overset{\circ}{W}{}_2^1(0,1).$$

In Subsection 2.2, we analyze this problem in a more general case. In particular, we show that if $\tau(s)$ is not a function of the form $\alpha e^{\beta s}$, where $\{\alpha, \beta\} \subset \mathbb{C}$, then the Hermitian operator A satisfies the condition $\mathcal{D}(A) \vee \mathcal{R}(A) = \mathcal{H}$. Moreover, its defect subspaces have the form

$$\mathfrak{N}_\lambda = \left\langle e^{i\bar{\lambda}s}, P_\lambda(\hat{A} - \bar{\lambda}I)^{-1}\tau \right\rangle \qquad (\operatorname{Im}\lambda \neq 0),$$

where $\hat{A}$ is an arbitrary self-adjoint extension of the symmetric operator A' and P_λ is the orthoprojector onto $\mathcal{R}(A' - \lambda I)$ in $L_2(0, 1)$.

It is easy to see that the triple $(\mathfrak{N}_i, \Gamma_1, \Gamma_2)$, where Γ_i $(i = 1, 2)$ are linear operators defined by the formulas

$$\Gamma_1(x_i + x_{-i}) = x_i + \Phi x_{-i} \quad \text{and} \quad \Gamma_2(x_i + x_{-i}) = ix_i - i\Phi x_{-i}$$

$(x_\lambda \in \mathfrak{N}_\lambda,\ \lambda \in \{i, -i\}$, and Φ is an isometric operator from $\mathfrak{N}_{-i}$ onto $\mathfrak{N}_i)$, is a BVS of the Hermitian operator A in the sense of Definition II.

***Remarks.* I.** The problem of description of self-adjoint extensions of symmetric operators in terms of "abstract boundary conditions" (leading in the case of differential operators directly to boundary-value problems) has a long history (see M. Gorbachuk, V. Gorbachuk, and Kochubei [1]). For the first time it was posed by Calkin [1].

Talyush [1] was the first who introduced BVS in the form of Definition I and used this concept for the investigation of dissipative extensions of symmetric operators with equal and finite defect numbers. At the same time, it should be noted that he considered the operators Γ_i under an additional assumption that $\Gamma_1 x = \Gamma_2 x = 0$ $(x \in \mathcal{D}(A))$ and did not prove the existence of these operators.

Kochubei [1] proved the existence of operators Γ_i satisfying the conditions of Definition I for any symmetric operator A with equal but not necessarily finite defect numbers (Theorem 1.1). In the same work, the triple (X, Γ_1, Γ_2) was called the boundary-value space of a symmetric operator A.

Later, Bruk [1] showed that the equality $\Gamma_1 x = \Gamma_2 x = 0$ $(x \in \mathcal{D}(A))$ used as a condition in the definition of BVS suggested by Kochubei is not independent and follows from the other conditions of this definition (see Proposition 1.2).

In the case of symmetric operators semibounded from below, Kochubei [7] defined positive BVS which form an important subclass of BVS and studied their properties. In many cases, the application of BVS of this type enables one to give a natural description of different classes of extensions of symmetric operators. In our book, we use (Chapter 4) canonical positive BVS [equalities (1.5) and (1.6)]. The idea of their construction appeared under the influence of the work of Vishik [1].

Derkach [1] generalized the concept of BVS to the case of symmetric operators acting in a Krein space.

Storozh [1] analyzed the case of symmetric operators with unequal defect numbers.

Various examples of BVS and their applications to the investigation of different classes of extensions of symmetric operators can be found in the monograph by V. Gorbachuk and M. Gorbachuk [1] (see also Kochubei [4] and Mikhailets [1]).

II. In the case of Hermitian (nondensely defined) operators, the direct application of BVS in the form of Definition I is impossible because the operator A^* does not exist. In different works, this problem was solved by using different methods.

For the first time, the concept of BVS of an Hermitian operator with equal defect numbers was introduced and studied by A. Kuzhel and Karpenko [1, 2]. These works were based on the application of an analog of the von Neumann formulas (see also S. Kuzhel [1]).

Later, by using linear relations, the concept of BVS of an Hermitian operator was introduced by Malamud [2]. In [2], Derkach generalized this definition to the case of Hermitian operators acting in a Krein space.

Finally, the third approach to the definition of BVS is based on the application, instead of the operator A^*, of a dissipative extension of an Hermitian operator A with "good" properties (canonical dissipative extension). On the basis of this idea, S. Kuzhel [2] formulated the definition of BVS for closed Hermitian operators with arbitrary defect numbers (Subsection 1.3).

2. Description of Regular Extensions of Hermitian Operators in Terms of Boundary-Value Spaces

In the present section, by using BVS in the form of Definition III we describe regular extensions of Hermitian operators. We present a detailed analysis of an important special case where the original operator is a restriction of a symmetric operator to a subspace of finite codimensionality.

2.1. General Case. Let A be a closed Hermitian operator with arbitrary defect numbers acting in a Hilbert space $\mathcal{H}$. As earlier, we assume that the operator A satisfies condition (9.9).

Let (X_+, X_-, Y_+, Y_-) be a BVS of the operator A in the form of Definition III and let K be a linear operator acting from X_+ into X_-.

The operator K is called *admissible for the BVS* (X_+, X_-, Y_+, Y_-) (or simply *admissible*) if for all elements $x \in \mathcal{D}(F)$ and $y \in \mathcal{D}(F^*)$ such that

$$KY_+x = Y_-y, \tag{2.1}$$

the inclusion $x + y \in \mathcal{D}(A)$ implies that

$$Y_+x = Y_-y = 0.$$

Theorem 2.1. *Suppose that* $\tilde{A}$ *is a regular extension of a closed Hermitian operator* A *and* μ *is a nonreal number* ($\operatorname{Im}\mu < 0$) *which does not belong to the point spectrum* $\sigma_p(\tilde{A})$. *Then there exists a BVS* (X_+, X_-, Y_+, Y_-) *of the operator* A *and an operator* K *admissible for this BVS such that*

$$\mathcal{D}(\tilde{A}) = \{x+y \mid KY_+x = Y_-y,\ x \in \mathcal{D}(F),\ y \in \mathcal{D}(F^*)\} \tag{2.2}$$

and

$$\tilde{A}(x+y) = Fx + F^*y, \tag{2.3}$$

where F *is a canonical dissipative extension of the operator* A *associated with the BVS* (X_+, X_-, Y_+, Y_-).

Conversely, any admissible operator K *defines a regular extension* $\tilde{A}$ *of the operator* A *by formulas (2.2) and (2.3).*

Proof. Since there exists μ ($\operatorname{Im}\mu < 0$) such that $\mu \notin \sigma_p(\tilde{A})$, by virtue of the von Neumann formulas (Subsection 5.5, Chapter 1), the regular extension $\tilde{A}$ is defined by the decomposition

$$\mathcal{D}(\tilde{A}) = \mathcal{D}(A) \dotplus \{x_\mu + \Phi x_\mu \mid x_\mu \in \mathcal{D}(\Phi) \subset \mathfrak{N}_\mu\}, \tag{2.4}$$

where Φ is a linear operator acting from $\mathfrak{N}_\mu$ into $\mathfrak{N}_{\bar{\mu}}$. Recall that the subspace $\mathfrak{N}_\mu$ ($\mathfrak{N}_{\bar{\mu}}$) is a maximal uniformly positive (negative) subspace of $\mathcal{N}_\mu$ (Subsection 9.2).

Assume that F is a canonical dissipative extension of the operator A defined by decomposition (1.20) for $\mathcal{L}_+ = \mathfrak{N}_\mu$. As in the proof of Theorem 1.6, we construct a BVS (X_+, X_-, Y_+, Y_-) of the Hermitian operator A associated with the extension F. Taking into account the form of the operators $Y_\pm$ in this BVS and setting

$$K = \Phi, \quad \mathcal{D}(K) = \mathcal{D}(\Phi),$$

we transform relation (2.4) into (2.2).

Let us show that the operator K is admissible. In fact, in view of relations (1.20), (1.21), and (2.4), it is easy to see that any two elements $x \in \mathcal{D}(F)$ and $y \in \mathcal{D}(F^*)$ satisfy relation (2.1) if and only if

$$x = x_0 + x_\mu \quad \text{and} \quad y = y_0 + \Phi x_\mu \quad \{x_0, y_0\} \subset \mathcal{D}(A). \tag{2.5}$$

This means that the sum $x + y$ belongs to $\mathcal{D}(A)$ if and only if $x_\mu + \Phi x_\mu \in \mathcal{D}(A)$. It follows from (2.4) that this relation is possible only under the condition $x_\mu = 0$. Then $Y_+ x = Y_- y = 0$. Thus, the operator K is admissible.

Further, by using (2.5) and the fact that the operators $\tilde{A}$, F and F^* are defined by decompositions (2.4), (1.20), and (1.21), respectively, we obtain

$$\begin{aligned} \tilde{A}(x+y) &= A(x_0 + y_0) + \bar{\mu} x_\mu + \mu \Phi x_\mu \\ &= (Ax_0 + \bar{\mu} x_\mu) + (Ay_0 + \mu \Phi x_\mu) = Fx + F^* y, \end{aligned} \tag{2.6}$$

which proves equality (2.3).

Conversely, let K be an admissible operator with respect to a BVS (X_+, X_-, Y_+, Y_-) of the operator A. We now show that relations (2.2) and (2.3) define a regular extension of the operator A.

Indeed, the canonical dissipative extension F associated with the BVS (X_+, X_-, Y_+, Y_-) is defined by decomposition (1.20), where $\mathcal{L}_+$ is a maximal uniformly positive subspace of $\mathcal{N}_\lambda$ (λ is an arbitrary nonreal number). In view of Corollary 1.7 and the fact that the operator K is admissible, we can rewrite equality (2.2) in the form

$$\mathcal{D}(\tilde{A}) = \mathcal{D}(A) \dotplus \mathcal{L}, \tag{2.7}$$

where

$$\mathcal{L} = \{x_+ + y_- \mid KY_+ x_+ = Y_- y_-,\ x_+ \in \mathcal{L}_+,\ y_- \in \mathcal{L}_- = \mathcal{L}_+^{[\perp]}\} \tag{2.8}$$

is a lineal in $\mathcal{N}_\lambda$.

Let us show that the operator $\tilde{A}$ is defined by decomposition (2.7). Without loss of generality, we assume that $\operatorname{Im} \lambda < 0$. Then relations (9.5) and (9.6) (with $\mathcal{H}_+ = \mathfrak{N}_\lambda$ and $\mathcal{H}_- = \mathfrak{N}_{\bar{\lambda}}$) imply that the elements $x_+ \in \mathcal{L}_+$ and $y_- \in \mathcal{L}_-$ can be represented in the form

$$x_+ = x_\lambda + T_+ x_\lambda \quad \text{and} \quad y_- = y_{\bar{\lambda}} + T_- y_{\bar{\lambda}} \quad (\operatorname{Im} \lambda < 0), \tag{2.9}$$

where T_+ (T_-) is an angular operator of the subspace $\mathcal{L}_+$ ($\mathcal{L}_-$) with respect to the subspace $\mathfrak{N}_\lambda$ ($\mathfrak{N}_{\bar{\lambda}}$). In this case,

$$Fx_+ = \bar{\lambda}x_\lambda + \lambda T_+ x_\lambda \quad \text{and} \quad F^* y_- = \lambda y_{\bar{\lambda}} + \bar{\lambda} T_- y_{\bar{\lambda}}. \tag{2.10}$$

It follows from relations (2.3) and (2.7)–(2.10) that the operator $\tilde{A}$ is defined by decomposition (2.7) and, hence, it is a regular extension of the operator A (see Lemma 9.4).

■

Remark. It follows from Theorem 2.1 [relations (2.7) and (2.8)] and Lemma 9.5 that the admissible operator K in formula (2.2) is uniquely defined by the extension $\tilde{A}$.

Corollary 2.2. *If a regular extension $\tilde{A}$ of a closed Hermitian operator A in a BVS (X_+, X_-, Y_+, Y_-) is defined by formulas (2.2) and (2.3), then the operator $\tilde{A}$ is closed if and only if the admissible operator K is closed.*

Proof. In the lineal $\mathcal{D}(\tilde{A})$, we define a scalar product

$$(x, y)_{\tilde{A}} = (x, y) + (\tilde{A}x, \tilde{A}y) \qquad \{x, y\} \subset \mathcal{D}(\tilde{A}). \tag{2.11}$$

It is obvious that the lineal $\mathcal{D}(\tilde{A})$ with the scalar product (2.11) is a Hilbert space if and only if the operator $\tilde{A}$ is closed. As in Theorem 2.1, in view of equalities (2.2) and (2.3), we conclude that the regular extension $\tilde{A}$ is defined by decomposition (2.7), where $\mathcal{L} \subset \mathcal{N}_\lambda$, and λ is an arbitrary nonreal number. Without loss of generality, we set $\lambda = -i$. By virtue of (9.10), (9.15), and (2.11), we have

$$(x, y)_{\tilde{A}} = (x_0 + f,\ y_0 + g) + (Ax_0 + if_{-i} - if_i,\ Ay_0 + ig_{-i} - ig_i)$$

$$= (x_0, y_0)_{\tilde{A}} + (f, g)_{-i}, \tag{2.12}$$

where $x = x_0 + f$ and $y = y_0 + g$ ($\{x_0, y_0\} \subset \mathcal{D}(A)$, $\{f, g\} \subset \mathcal{L}$). By using (2.12), we can rewrite decomposition (2.7) in the form

$$\mathcal{D}(\tilde{A}) = \mathcal{D}(A) \oplus_{\tilde{A}} \mathcal{L}, \tag{2.13}$$

where the lineal $\mathcal{D}(A)$ is a closed subspace of the Hilbert space $\left(\mathcal{D}(\tilde{A}), \|\cdot\|_{\tilde{A}}\right)$ (since the operator A is closed). Therefore, since the scalar products $(\cdot,\cdot)_{\tilde{A}}$ and $(\cdot,\cdot)_{-i}$ coincide in $\mathcal{L} \subset \mathcal{N}_{-i}$, decomposition (2.13) implies that the operator $\tilde{A}$ is closed if and only if the lineal $\mathcal{L}$ is closed in $\mathcal{N}_{-i}$.

In the space $\mathcal{N}_{-i}$, we define a new scalar product

$$(x, y)'_{-i} = [x_+, y_+]_{-i} - [x_-, y_-]_{-i}, \tag{2.14}$$

where $x = x_+ + x_-$ and $y = y_+ + y_-$ ($\{x_+, y_+\} \subset \mathcal{L}_+$, $\{x_-, y_-\} \subset \mathcal{L}_-$). Since the lineals $\mathcal{L}_\pm$ are maximal uniformly definite subspaces of $\mathcal{N}_{-i}$ and their direct sum is a canonical decomposition of the space $\mathcal{N}_{-i}$, we conclude that the norm $\|\cdot\|'_{-i} = \sqrt{(\cdot,\cdot)'_{-i}}$ is equivalent to the norm $\|\cdot\|_{-i}$. Further, by virtue of equalities (1.22), (2.8), and (2.14), we obtain

$$\|x\|'^2_{-i} = \|Y_+x_+\|^2_+ + \|KY_+x_+\|^2_-$$

for any $x = x_+ + x_- \in \mathcal{L}$. It follows from this equality that the lineal $\mathcal{L}$ is closed in $\mathcal{N}_{-i}$ if and only if the operator K is closed.

■

2.2. Special Cases

I. By considering different subsets of the set $\mathcal{P}(A)$ of regular extensions of an Hermitian operator A, we make the statement of Theorem 2.1 more accurate. In the following theorem, we show that the condition of dissipativeness of an extension $\tilde{A}$ enables us to describe this extension in terms of *an arbitrary* BVS of the Hermitian operator A.

Theorem 2.3. *Let $\tilde{A}$ be a dissipative extension of a closed Hermitian operator A. Then, for any BVS (X_+, X_-, Y_+, Y_-) of the operator A, there exists an admissible contracting operator K such that the extension $\tilde{A}$ is defined by relations (2.2) and (2.3).*

Conversely, any admissible contraction K from X_+ into X_- determines a dissipative extension $\tilde{A}$ of the operator A by relations (2.2) and (2.3).

Proof. According to Theorem 9.7, any dissipative extension $\tilde{A}$ of the operator A is defined by the decomposition

$$\mathcal{D}(\tilde{A}) = \mathcal{D}(A) \dotplus \mathcal{L}, \tag{2.15}$$

where $\mathcal{L}$ is a nonnegative lineal in $\mathcal{N}_\lambda$.

Let (X_+, X_-, Y_+, Y_-) be a BVS of the operator A. The extension F associated with this BVS is defined by decomposition (1.20), where $\mathcal{L}_+$ is a maximal uniformly positive subspace of $\mathcal{N}_\lambda$. It is easy to see that any nonnegative lineal $\mathcal{L}$ in $\mathcal{N}_\lambda$ can be represented in the form

$$\mathcal{L} = \{x_+ + \Phi x_+ \mid x_+ \in \mathcal{D}(\Phi) \subset \mathcal{L}_+\}, \tag{2.16}$$

where Φ is a linear operator acting from $\mathcal{L}_+$ into $\mathcal{L}_- = \mathcal{L}_+^{[\perp]}$ such that

$$-[\Phi x_+, \Phi x_+]_\lambda \le [x_+, x_+]_\lambda \quad (x_+ \in \mathcal{D}(\Phi)). \tag{2.17}$$

Since the operator Y_+ determines one-to-one correspondence between $\mathcal{L}_+$ and X_+, there exists the inverse operator Y_+^{-1} which maps X_+ onto $\mathcal{L}_+$. Hence, the linear operator

$$K = Y_- \Phi Y_+^{-1}, \qquad \mathcal{D}(K) = \mathcal{R}\left(Y_+ |_{\mathcal{D}(\Phi)}\right) \tag{2.18}$$

is correctly defined and acts from the the space X_+ into the space X_-. By using equalities (1.22) and (2.17), we can write

$$\| K Y_+ x_+ \|_-^2 = \| Y_- \Phi x_+ \|_-^2 \le \| Y_+ x_+ \|_+^2 \quad (x_+ \in \mathcal{D}(\Phi)).$$

Hence, the operator K is contracting. In view of the definition of the operator K and equality (2.16), we transform decomposition (2.15) into equality (2.2).

We prove that the operator K is admissible for the BVS (X_+, X_-, Y_+, Y_-) by analogy with the proof of Theorem 2.1 (by substituting $x_+ \in \mathcal{L}_+$ for $x_\mu \in \mathfrak{N}_\mu$).

Let us show that the law of action of the operator $\tilde{A}$ is determined by equality (2.3).

Indeed, by virtue of (2.9), any element $x_+ \in \mathcal{D}(\Phi)$ admits the representation

$$x_+ = x_\lambda + T_+ x_\lambda, \qquad \Phi x_+ = y_{\bar{\lambda}} + T_- y_{\bar{\lambda}} \quad (\operatorname{Im}\lambda < 0), \tag{2.19}$$

where $x_\lambda \in \mathfrak{N}_\lambda$, $y_{\bar{\lambda}} \in \mathfrak{N}_{\bar{\lambda}}$, and T_+ (T_-) is an angular operator of the subspace $\mathcal{L}_+$ $(\mathcal{L}_-)$.

By using the same reasoning as in the proof of equality (2.6) (with $\mu = \lambda$) and (2.19), we conclude that equality (2.3) is true.

Conversely, let us show that any admissible contraction K $(K\colon X_+ \to X_-)$ defines a dissipative extension $\tilde{A}$.

Indeed, it follows from Theorem 2.1 that any admissible operator K determines a regular extension $\tilde{A}$ of the operator A by formulas (2.2) and (2.3). This extension is defined by decomposition (2.7), where the lineal $\mathcal{L}$ is defined by equality (2.8). Taking into account (1.22), for any $f = x_+ + y_- \in \mathcal{L}$, we obtain

$$[f, f]_\lambda = [x_+, x_+]_\lambda + [y_-, y_-]_\lambda = \|Y_+ x_+\|_+^2 - \|K Y_+ x_+\|_-^2 \tag{2.20}$$

In the considered case, the operator K is contracting. By virtue of (2.20), this means that the lineal $\mathcal{L}$ in decomposition (2.7) is a nonnegative lineal in $\mathcal{N}_\lambda$. Therefore, by virtue of Theorem 9.7, we conclude that the extension $\tilde{A}$ is dissipative.

■

Corollary 2.4. *An extension $\tilde{A}$ of a closed Hermitian operator A is a closed maximal dissipative extension if and only if the admissible contraction K in formula (2.2) is defined in the entire space X_+.*

Proof. Assume that $\tilde{A}$ is a closed maximal dissipative extension of the operator A. Let us show that the admissible contraction K in equality (2.2) corresponding to this extension is defined in the entire space X_+.

Indeed, by virtue of Corollary 2.2, the contraction K is a closed operator. Hence, the lineal $\mathcal{D}(K)$ is closed in X_+. Suppose that $\mathcal{D}(K) \subset X_+$ and consider the operator

$$K' = KP,$$

where P is the orthoprojector onto the subspace $\mathcal{D}(K)$ of the space X_+. It is clear that K' is a contraction. We prove that it is admissible for the BVS.

By using decompositions (1.20) and (1.21) and Corollary 1.7, one can readily show that the definition of an admissible operator can be rewritten in the following (equivalent) form:

An operator K is admissible if the lineal

$$\mathcal{L}_K = \{x_+ + y_- \mid K Y_+ x_+ = Y_- y_-,\ x_+ \in \mathcal{L}_+,\ y_- \in \mathcal{L}_-\} \tag{2.21}$$

is linearly independent of $\mathcal{D}(A)$.

The definition of the operator K' and the first equality in (1.22) yield

$$\mathcal{L}_{K'} = \mathcal{L}_K [\dot{+}] \{x_+ \in \mathcal{L}_+ \mid P Y_+ x_+ = 0\}. \tag{2.22}$$

Assume that there exists a nonzero element $\psi \in \mathcal{L}'_K \cap \mathcal{D}(A)$. Since K is an admissible contraction, we have $\mathcal{L}_K \cap \mathcal{D}(A) = \{0\}$. Hence, in view of decomposition (2.22), the element ψ can be represented in the form

$$\psi = f + x_+ \quad (f \in \mathcal{L}_K,\ x_+ \in \mathcal{L}_+,\ PY_+x_+ = 0),$$

where $x_+ \neq 0$. Moreover,

$$[\psi, \psi]_\lambda = [f, f]_\lambda + [x_+, x_+]_\lambda. \tag{2.23}$$

It follows from equality (2.20) that $\mathcal{L}_K$ is a nonnegative lineal. Hence, $[f, f]_\lambda \geq 0$. At the same time, since $\mathcal{L}_+$ is a uniformly positive subspace, we obtain

$$[x_+, x_+]_\lambda > 0.$$

Thus, taking into account relation (2.23), we conclude that $[\psi, \psi]_\lambda > 0$.

On the other hand, by virtue of Lemma 9.3, we have $[\psi, \psi]_\lambda = 0$. Contradiction. This means that $\psi = 0$ and, therefore, the contraction K' is admissible.

By construction, we have $K' \supset K$. Theorem 2.3 easily implies that this inclusion is equivalent to the inclusion $\tilde{A}' \supset \tilde{A}$, i.e.,

$$K' \supset K \Leftrightarrow \tilde{A}' \supset \tilde{A}, \tag{2.24}$$

where $\tilde{A}'$ is a dissipative extension corresponding to the admissible contraction K'. By using the fact that $\tilde{A}$ is a maximal dissipative extension, we conclude that $\tilde{A}' = \tilde{A}$. Then $K' = K$ and, hence, $\mathcal{D}(K) = X_+$.

The converse assertion that if $\mathcal{D}(K) = X$, then the dissipative extension $\tilde{A}$ is maximal and closed immediately follows from Theorem 2.3 and relation (2.24).

■

It is obvious that, in the case where A is a symmetric operator $\left(\overline{\mathcal{D}(A)} = \mathcal{H}\right)$, it follows from equality (2.21) that any operator K acting from X_+ into X_- is admissible. By using the results presented above, we arrive at the following statement:

Corollary 2.5. *For any BVS of a closed symmetric operator A, relations (2.2) and (2.3) establish a one-to-one correspondence between the set of dissipative extensions of the operator A and the set of contracting operators acting from X_+ into X_-.*

Furthermore, an extension $\tilde{A}$ ismaximal dissipative, maximal symmetric, or self-adjoint if and only if the operator K defined in the entire space X_+ is a contraction, an isometric operator $(\mathcal{R}(K) \subset X_-)$, or a unitary operator $(\mathcal{R}(K) = X_-)$, respectively.

II. Usually, in applications, an Hermitian operator A is defined as the restriction of a symmetric (densely defined) operator A' to a subspace of finite codimensionality. In what follows, we analyze this case in detail.

Let A' be a closed symmetric operator with arbitrary defect numbers in a Hilbert space $\mathcal{H}$. Assume that $\mathcal{H}_0$ is a finite-dimensional subspace of $\mathcal{H}$. The operator A is introduced by the equalities

$$A = A'\big|_{\mathcal{D}(A)} \quad \text{and} \quad \mathcal{D}(A) = \mathcal{D}(A') \cap (\mathcal{H} \ominus \mathcal{H}_0). \tag{2.25}$$

It is clear that A is a closed Hermitian operator acting in the Hilbert space $\mathcal{H}$.

Proposition 2.6. *The domain of definition of the operator* A *is dense in* $\mathcal{H} \ominus \mathcal{H}_0$ *and*

$$\mathcal{D}(A') = \mathcal{D}(A) \dotplus \mathcal{L},$$

where $\mathcal{L}$ *is a lineal in* $\mathcal{D}(A')$ *such that* $\dim \mathcal{L} = \dim \mathcal{H}_0$.

Proof. Let $\{e_i\}_{i=1}^n$ be an orthonormal basis in $\mathcal{H}_0$ $(\dim \mathcal{H}_0 = n)$. We define an operator A_1 as the restriction of the operator A' to the lineal

$$\mathcal{D}(A_1) = \mathcal{D}(A') \cap (\mathcal{H} \ominus \langle e_1 \rangle).$$

In $\mathcal{D}(A')$, we consider a scalar product

$$(x, y)_{A'} = (x, y) + (A'x, A'y) \quad (\{x, y\} \subset \mathcal{D}(A')), \tag{2.26}$$

Since the operator A' is closed, we find that $\mathcal{D}(A')$ is a Hilbert space with respect to the scalar product (2.26).

Consider a linear functional $F(x) = (x, e_1)$ in $\mathcal{D}(A')$. By virtue of (2.26), the functional F is continuous in the Hilbert space $(\mathcal{D}(A'), \|\cdot\|_{A'})$ and, hence, there exists an element $e_1' \in \mathcal{D}(A')$ such that

$$F(x) = (x, e_1')_{A'} \quad (x \in \mathcal{D}(A')).$$

In view of the equality $\operatorname{Ker} F = \mathcal{D}(A_1)$, this yields

$$\mathcal{D}(A') = \mathcal{D}(A_1) \dotplus \langle e_1' \rangle. \tag{2.27}$$

Let $z \perp \mathcal{D}(A_1)$. The linear functional $\Phi(x) = (x, z)$ $(x \in \mathcal{D}(A'))$ is continuous in $(\mathcal{D}(A'), \|\cdot\|_{A'})$. Moreover,

$$\operatorname{Ker} F = \operatorname{Ker} \Phi.$$

Consequently, $F(x) = \alpha\Phi(x)$ $(\alpha \in \mathbb{C})$ or, which is the same, $(x, e_1) = \alpha(x, z)$ $(x \in \mathcal{D}(A'))$. Hence, by using the equality $\overline{\mathcal{D}(A')} = \mathcal{H}$, we obtain $e_1 = \bar{\alpha} z$. The equality

$$\overline{\mathcal{D}(A_1)} = \mathcal{H} \ominus \langle e_1 \rangle \tag{2.28}$$

is thus proved.

Hermitian operators A_i $(i \geq 2)$ are defined as restrictions of the symmetric operators A_{i-1} acting in $\mathcal{H}_{i-1} = \mathcal{H} \ominus \langle e_1, \ldots, e_{i-1} \rangle$ to the lineal

$$\mathcal{D}(A_i) = \mathcal{D}(A_{i-1}) \cap (\mathcal{H}_{i-1} \ominus \langle e_i \rangle).$$

Acting as above and using equalities (2.27) and (2.28), after finitely many steps $(i = 2, \ldots, n)$, we arrive at the assertion of Proposition 2.6.

■

Recall that a BVS of an Hermitian operator A is defined only under the condition that $\mathcal{D}(A) \vee \mathcal{R}(A) = \mathcal{H}$ (see Section 1). The following simple condition under which this equality is true immediately follows from Proposition 2.6.

Corollary 2.7. *Assume that an Hermitian operator A is defined by equality (2.25), where $\mathcal{H}_0 \subset \mathcal{D}(A'^*)$. Then the following equalities are equivalent:*

1. $\mathcal{D}(A) \vee \mathcal{R}(A) = \mathcal{H}$.

2. $\mathcal{R}\left(A'^*|_{\mathcal{H}_0}\right) \cap \mathcal{H}_0 = \{0\}$.

Let $\hat{A}$ be a fixed self-adjoint extension of the symmetric operator A'. Note that, in view of Theorem 4.6 (Chapter 1), this extension exists and acts either in the space $\hat{\mathcal{H}} = \mathcal{H}$ or in a broader space $\hat{\mathcal{H}} \supset \mathcal{H}$ (the case of an operator A with unequal defect numbers).

We set

$$T_\lambda = P_\lambda (\hat{A} - \bar{\lambda} I)^{-1} \qquad (\operatorname{Im} \lambda \neq 0), \tag{2.29}$$

where P_λ is the orthoprojector onto $\mathcal{R}(A' - \lambda I)$ in $\hat{\mathcal{H}}$.

Proposition 2.8. *For any nonreal* λ, *the defect subspace* $\mathfrak{N}_\lambda$ *of the Hermitian operator* A *has the form*

$$\mathfrak{N}_\lambda = \mathfrak{N}'_\lambda \oplus \mathcal{R}\left(T_\lambda|_{\mathcal{H}_0}\right),$$

where $\mathfrak{N}'_\lambda$ *is the defect subspace of the symmetric operator* A'..

Proof. By virtue of the inclusion $A' \supset A$, it is easy to see that

$$\mathfrak{N}_\lambda = \mathfrak{N}'_\lambda \oplus \mathcal{M},$$

where $\mathcal{M} = \mathcal{R}(A' - \lambda I) \ominus \mathcal{R}(A - \lambda I)$.

In view of relation (2.29), we obtain

$$((A' - \lambda I)x,\ T_\lambda y) = (x, y) \tag{2.30}$$

for all $x \in \mathcal{D}(A')$ and $y \in \mathcal{H}$. According to the definition of the Hermitian operator A [equality (2.25)], this yields $\mathcal{R}\left(T_\lambda|_{\mathcal{H}_0}\right) \subseteq \mathcal{M}$.

It follows from equality (2.30) that

$$\operatorname{Ker}\left(T_\lambda|_{\mathcal{H}_0}\right) = \{0\}.$$

Consequently,

$$\dim \mathcal{R}\left(T_\lambda|_{\mathcal{H}_0}\right) = \dim \mathcal{H}_0 = n.$$

At the same time, by virtue of Proposition 2.6, we have $\dim \mathcal{M} = n$. Therefore, $\mathcal{M} = \mathcal{R}\left(T_\lambda|_{\mathcal{H}_0}\right)$.

■

It is clear that, by virtue of Proposition 2.8, in the set of canonical dissipative extensions of the Hermitian operator A, it is reasonable to select the extension F defined by the decomposition

$$\mathcal{D}(F) = \mathcal{D}(A) \dotplus \mathfrak{N}_\lambda \qquad (\operatorname{Im}\lambda < 0). \tag{2.31}$$

As in the proof of Theorem 1.6, we construct the BVS (X_+, X_-, Y_+, Y_-) associated with the extension F. By using the results of Subsection 2.1, we can describe dissipative extensions of the operator A in terms of this BVS.

We now illustrate the results presented above by a simple example.

In the space $\mathcal{H} = L_2(0, 1)$, we consider a symmetric operator

$$A' = -i\frac{d}{ds}, \quad \mathcal{D}(A') = \mathring{W}_2^1(0, 1).$$

We fix $\tau(s) \in W_2^1(0, 1)$ such that $\tau(s) \notin \{\alpha e^{\beta s} \mid \{\alpha, \beta\} \subset \mathbb{C}\}$ and consider the following Hermitian operator:

$$A = -i\frac{d}{ds}, \quad \mathcal{D}(A) = \{x(s) \in \mathcal{D}(A') \mid (x, \tau) = 0\}.$$

By virtue of Corollary 2.7, the defect subspaces $\mathfrak{N}_\lambda$ and $\mathfrak{N}_{\bar{\lambda}}$ of the Hermitian operator A are linearly independent. Moreover, in view of Proposition 2.8, we have

$$\mathfrak{N}_\lambda = \langle e^{i\bar{\lambda}s} \rangle \oplus \langle P_\lambda (\hat{A} - \bar{\lambda} I)^{-1} \tau \rangle, \tag{2.32}$$

where

$$\hat{A} = -i\frac{d}{ds}, \quad \mathcal{D}(\hat{A}) = \{x(s) \in W_2^2(0, 1) \mid x(0) = x(1)\},$$

is a self-adjoint operator in $L_2(0, 1)$. It is easy to see that, in the considered case, the orthoprojector P_λ and the resolvent $(\hat{A} - \bar{\lambda} I)^{-1}$ are defined by the formulas

$$(P_\lambda x)(s) = x(s) - \frac{1}{\alpha}\left(x(s), e^{i\bar{\lambda}s}\right) e^{i\bar{\lambda}s},$$

$$((\hat{A} - \bar{\lambda} I)^{-1} x)(s) = c e^{i\bar{\lambda}s} + e^{i\bar{\lambda}s} \int_0^s x(\xi) e^{-i\bar{\lambda}\xi} d\xi,$$

where $\alpha = \left(e^{i\bar{\lambda}s}, e^{i\bar{\lambda}s}\right)$ and c is a constant determined from the condition

$$(\hat{A} - \bar{\lambda} I)^{-1} x \in \mathcal{D}(\hat{A}).$$

This means that equality (2.32) can be rewritten in the form

$$\mathfrak{N}_\lambda = \left\langle e^{i\bar{\lambda}s}, \psi_\lambda(s) \right\rangle,$$

where

$$\psi_\lambda(s) = e^{i\bar{\lambda}s}\int_0^s \tau(\xi)e^{-i\bar{\lambda}\xi}d\xi.$$

Let F be a canonical dissipative extension of the Hermitian operator A defined by decomposition (2.31). The BVS (X_+, X_-, Y_+, Y_-) associated with the extension F has the form

$$X_+ = \mathfrak{N}_\lambda \quad \text{and} \quad X_- = \mathfrak{N}_{\bar{\lambda}} \quad (\operatorname{Im}\lambda < 0)$$

and the linear operators $Y_+\colon \mathcal{D}(F) \to X_+$ and $Y_-\colon \mathcal{D}(F^*) \to X_-$ are defined by the formulas

$$Y_+ x = \sqrt{2|\operatorname{Im}\lambda|}\, x_+ \quad \text{and} \quad Y_- y = \sqrt{2|\operatorname{Im}\lambda|}\, y_-,$$

where

$$x = x_0 + x_+ \in \mathcal{D}(F) \quad \text{and} \quad y = y_0 + y_- \in \mathcal{D}(F^*),$$

$$\{x_0, y_0\} \subset \mathcal{D}(A), \quad x_+ \in \mathfrak{N}_\lambda, \quad y_- \in \mathfrak{N}_{\bar{\lambda}}.$$

By using Theorem 2.3, we obtain a description of all dissipative extensions of the Hermitian operator A in terms of this BVS.

Remarks. This section is an extended and revised version of the work by S. Kuzhel [1].

To describe various classes of proper extensions of symmetric (densely defined) operators with equal defect numbers, it is convenient to use BVS in the form of Definition I. A presentation of the corresponding results and detailed bibliography can be found in the monograph by M. Gorbachuk and V. Gorbachuk [1] and survey by M. Gorbachuk, V. Gorbachuk and Kochubei [1].

The case of the Hermitian operator A with equal defect numbers defined by relations (2.25) was studied by other methods by Kochubei [2].

Under the additional assumption that its defect numbers are finite, the self-adjoint extensions of this operator were considered by Coddington [1]. In the case of linear relations, a similar problem was solved by Coddington and Dijksma [1] and Kochubei [4].

3. Characteristic Functions of Hermitian Operators

In this section, on the basis of Definition III of BVS, we introduce the characteristic function of an Hermitian operator with arbitrary defect numbers and investigate its properties.

3.1. Definition of Characteristic Function. Let A be a closed Hermitian operator acting in a Hilbert space $\mathcal{H}$ and let (X_+, X_-, Y_+, Y_-) be its BVS.

For any nonreal number λ $(\operatorname{Im}\lambda < 0)$, we consider a canonical dissipative extension A_λ of the Hermitian operator A defined by the decomposition

$$\mathcal{D}(A_\lambda) = \mathcal{D}(A) \dot{+} \mathfrak{N}_\lambda \quad (\operatorname{Im}\lambda < 0). \tag{3.1}$$

By virtue of Theorem 2.3 and Corollary 2.4, the extension A_λ is determined as follows:

$$\mathcal{D}(A_\lambda) = \{x + y \mid C(\lambda) Y_+ x = Y_- y, \ x \in \mathcal{D}(F), \ y \in \mathcal{D}(F^*)\}, \tag{3.2}$$

$$A_\lambda(x + y) = Fx + F^* y, \tag{3.3}$$

where the contracting operator $C(\lambda)$ $(C(\lambda): X_+ \to X_-)$ is defined in the entire space X_+.

The operator function $C(\lambda)$ $(\operatorname{Im}\lambda < 0)$ is called the *characteristic function* of the Hermitian operator A.

It is clear that the form of the characteristic function $C(\lambda)$ depends on the choice of BVS.

For example, let $(X_+^0, X_-^0, Y_+^0, Y_-^0)$ be the BVS of an Hermitian operator A constructed in the proof of Theorem 1.6 (where $\mathcal{L}_+ = \mathfrak{N}_\mu$ and $\lambda = \mu$ $(\operatorname{Im}\mu < 0)$ is a fixed point).

In view of Subsection 8.5 (Chapter 1), it is easy to see that, in this BVS, the characteristic function $C_0(\lambda)$ has the following form:

$$C_0(\lambda) = -(A_\lambda - \bar{\mu}I)(A_\lambda - \mu I)^{-1}. \tag{3.4}$$

Let us clarify the relationship between the characteristic functions of an Hermitian operator A that correspond to different BVS. For this purpose, we use a known result from the theory of interspherical transformations, presented below in a form convenient for our discussion (the proof of this result can be found in Krein and Shmulyan [1] or Azizov and Iokhvidov [1]):

Let X_1 and X_2 be Hilbert spaces. In the Hilbert space $\hat{X} = X_1 \oplus X_2$, every bounded operator T $(\mathcal{D}(T) = \hat{X})$ can be represented in the form of an operator-valued matrix

$$T = \begin{pmatrix} T_{11} & T_{12} \\ T_{21} & T_{22} \end{pmatrix}, \qquad T_{ij} \in \mathcal{B}(X_j, X_i). \tag{3.5}$$

By using this matrix, we construct the following linear fractional transformation:

$$\Phi_T(Z) = (T_{21} + T_{22} Z)(T_{11} + T_{12} Z)^{-1}. \tag{3.6}$$

We consider this transformation in its natural domain of definition which consists of all elements $Z \in \mathcal{B}(X_1, X_2)$ such that $0 \in \rho(T_{11} + T_{12} Z)$. Obviously, for any $Z \in \mathcal{D}(\Phi_T)$, the linear operator $\Phi_T(Z)$ acts from X_1 into X_2.

The linear fractional transformation (3.5) is called *interspherical* if

(a) $\{Z \mid Z \in \mathcal{B}(X_1, X_2),\ \|Z\| \le 1\} \subset \mathcal{D}(\Phi_T)$;

(b) the condition $\|Z\| < 1$ $(\|Z\| \le 1)$ implies that

$$\|\Phi_T(Z)\| < 1 \quad (\|\Phi_T(Z)\| \le 1).$$

Lemma 3.1. *The transformation Φ_T is interspherical if $\mathcal{R}(T) = \hat{X}$ and, for any $x \in \mathcal{D}(T) = \hat{X}$, the following equality is true:*

$$\|P_1 T x\|^2_{X_1} - \|P_2 T x\|^2_{X_2} = \|P_1 x\|^2_{X_1} - \|P_2 x\|^2_{X_2},$$

where P_i $(i = 1, 2)$ is the orthoprojector onto the subspace X_i in $\hat{X}$.

Let $C(\lambda)$ and $C'(\lambda)$ be the characteristic functions of the Hermitian operator A in the BVS (X_+, X_-, Y_+, Y_-) and (X'_+, X'_-, Y'_+, Y'_-), respectively. By virtue of Corollary 1.8, we have

$$\dim X_+ = \dim X'_+ \quad \text{and} \quad \dim X_- = \dim X'_-.$$

Hence, there exist isometric mappings X'_+ onto X_+ and X'_- onto X_-; we denote these mappings by V_+ and V_-, respectively.

Theorem 3.2. *The following relation is true:*

$$C'(\lambda) = V_-^{-1}\Phi_T(C(\lambda))V_+,$$

where $\Phi_T(\cdot)$ *is an interspherical transformation.*

Proof. Parallel with the BVS (X'_+, X'_-, Y'_+, Y'_-), we consider the BVS $(X_+, X_-, \tilde{Y}_+, \tilde{Y}_-)$, where $\tilde{Y}_+ = V_+Y'_+$ and $\tilde{Y}_- = V_-Y'_-$. Clearly, if $\tilde{C}(\lambda)$ is a characteristic function of the operator A in the BVS $(X_+, X_-, \tilde{Y}_+, \tilde{Y}_-)$, then

$$C'(\lambda) = V_-^{-1}(\tilde{C}(\lambda))V_+. \tag{3.7}$$

Assume that F and $\tilde{F}$ are the canonical dissipative extensions of the operator A associated with the BVS (X_+, X_-, Y_+, Y_-) and $(X_+, X_-, \tilde{Y}_+, \tilde{Y}_-)$, respectively. These extensions are defined by the choice of maximal uniformly positive subspaces $\mathcal{L}_+$ and $\tilde{\mathcal{L}}_+$ of the Krein space $\mathcal{N}_\mu$ (μ is a fixed nonreal number) in decomposition (1.20). In this case,

$$\mathcal{N}_\mu = \mathcal{L}_+ [\dot{+}] \mathcal{L}_- = \tilde{\mathcal{L}}_+ [\dot{+}] \tilde{\mathcal{L}}_-, \tag{3.8}$$

where $\mathcal{L}_- = \mathcal{L}_+^{[\perp]}$ and $\tilde{\mathcal{L}}_- = \tilde{\mathcal{L}}_+^{[\perp]}$.

In view of (3.8), there exists a unique representation of an arbitrary element $f \in \mathcal{N}_\mu$ in the form

$$f = x_+ + y_- = \tilde{x}_+ + \tilde{y}_-, \tag{3.9}$$

where $x_+ \in \mathcal{L}_+$, $y_- \in \mathcal{L}_-$, $\tilde{x}_+ \in \tilde{\mathcal{L}}_+$, and $\tilde{y}_- \in \tilde{\mathcal{L}}_-$. By using decomposition (3.9), we define a linear operator T in the Hilbert space $\hat{X} = X_+ \oplus X_-$ according to the formula

$$T\langle Y_+x_+, Y_-y_-\rangle = \langle \tilde{Y}_+\tilde{x}_+, \tilde{Y}_-\tilde{y}_-\rangle. \tag{3.10}$$

Taking into account the definition of BVS and Corollary 1.7, we establish that the operator T is well defined in the entire space $\hat{X}$ and $\mathcal{R}(T) = \hat{X}$. Furthermore, by virtue of equalities (1.22), (3.9), and (3.10), we get

$$\|\tilde{Y}_+\tilde{x}_+\|_{X_+}^2 - \|\tilde{Y}_-\tilde{y}_-\|_{X_-}^2 = [f,f]_\mu = \|Y_+x_+\|_{X_+}^2 - \|Y_-y_-\|_{X_-}^2.$$

Consequently, Lemma 3.1 implies that the transformation Φ_T, where the operator T is defined by (3.10), is interspherical.

In view of Corollary 9.10 and Corollary 1.7, relations (3.2) and (3.3) imply that the canonical extension A_λ is defined by the decomposition

$$\mathcal{D}(A_\lambda) = \mathcal{D}(A) \dotplus \mathcal{L},$$

where

$$\mathcal{L} = \{ f = x_+ + y_- \mid C(\lambda) Y_+ x_+ = Y_- y_-, \ x_+ \in \mathcal{L}_+, \ y_- \in \mathcal{L}_- \}$$

is a maximal uniformly positive subspace in $\mathcal{N}_\mu$. It follows from the definition of characteristic function that the subspace $\mathcal{L}$ can also be represented in the form

$$\mathcal{L} = \{ f = \tilde{x}_+ + \tilde{y}_- \mid \tilde{C}(\lambda) \tilde{Y}_+ \tilde{x}_+ = \tilde{Y}_- \tilde{y}_-, \ \tilde{x}_+ \in \tilde{\mathcal{L}}_+, \ \tilde{y}_- \in \tilde{\mathcal{L}}_- \}.$$

Therefore, considering equality (3.10) only for $f \in \mathcal{L}$ and representing the operator T in the form (3.5), where $X_+ = X_1$ and $X_- = X_2$, we obtain

$$\tilde{Y}_+ \tilde{x}_+ = (T_{11} + T_{12} C(\lambda)) Y_+ x_+ \quad \text{and} \quad \tilde{C}(\lambda) \tilde{Y}_+ \tilde{x}_+ = (T_{21} + T_{22} C(\lambda)) Y_+ x_+. \qquad (3.11)$$

Since the transformation Φ_T is interspherical, we have $0 \in \rho(T_{11} + T_{12} C(\lambda))$. Hence, it follows from (3.11) that $\tilde{C}(\lambda) = \Phi_T(C(\lambda))$, which, by virtue of equality (3.7), proves the theorem.

■

Corollary 3.3. *If $C(\lambda)$ is a characteristic function of an Hermitian operator A, then $\| C(\lambda) \| < 1$ for any λ $(\operatorname{Im} \lambda < 0)$.*

Proof. By virtue of Corollary 9.10, the canonical extension A_λ is defined by the decomposition

$$\mathcal{D}(A_\lambda) = \mathcal{D}(A) \dotplus \mathcal{L},$$

where $\mathcal{L}$ is a maximal uniformly positive subspace in $\mathcal{N}_\mu$. In Subsection 8.5 (Chapter 2), it was shown that a characteristic function of the form (3.4) coincides with the angular operator of the subspace $\mathcal{L}$ with respect to the maximal uniformly positive subspace $\mathfrak{N}_\mu$. Hence (see Section 9), this characteristic function satisfies the inequality

$$\| C_0(\lambda) \| < 1 \qquad (\operatorname{Im} \lambda < 0). \tag{3.12}$$

Since the transformation Φ_T is interspherical, inequality (3.12) holds for any characteristic function $C(\lambda)$ of the Hermitian operator A.

■

3.2. Condition of Unitary Equivalence of Simple Hermitian Operators. Recall that an Hermitian operator A acting in a Hilbert space $\mathcal{H}$ is called *simple* if it does not induce a self-adjoint operator in any nonzero subspace of the space $\mathcal{H}$.

Theorem 3.4. *Closed simple Hermitian operators A_1 and A_2 acting in a Hilbert space $\mathcal{H}$ are unitary equivalent if and only if some of their characteristic functions coincide.*

Proof. Let the operators A_1 and A_2 be unitary equivalent in $\mathcal{H}$. Then there exists a unitary operator V in $\mathcal{H}$ such that

$$A_2 = VA_1V^{-1}. \tag{3.13}$$

One can easily verify that, in this case, the defect subspaces $\mathfrak{N}_\lambda(A_1)$ and $\mathfrak{N}_\lambda(A_2)$ of the operators A_1 and A_2 satisfy the equality

$$V\mathfrak{N}_\lambda(A_1) = \mathfrak{N}_\lambda(A_2). \tag{3.14}$$

Denote by $A_{1\lambda}$ and $A_{2\lambda}$ the canonical dissipative extensions of the operators A_1 and A_2 defined by decomposition (3.1). In view of equalities (3.13) and (3.14), we get

$$A_{2\lambda} = VA_{1\lambda}V^{-1}. \tag{3.15}$$

Assume that (X_+, X_-, Y_+, Y_-) is a BVS of the operator A_1 associated with the canonical extension F_1. It follows from Theorem 9.9 and equalities (3.13) and (3.14) that the operator

$$F_2 = VF_1V^{-1} \tag{3.16}$$

is a canonical dissipative extension of the Hermitian operator A_2. Moreover, the quadruple $(X_+, X_-, Y_+V^{-1}, Y_-V^{-1})$ is the BVS of the operator A_2 associated with the canonical extension F_2.

Let $C^1(\lambda)$ and $C^2(\lambda)$ be the characteristic functions of the Hermitian operators A_1 and A_2 in the BVS (X_+, X_-, Y_+, Y_-) and $(X_+, X_-, Y_+V^{-1}, Y_-V^{-1})$, respectively.

By virtue of equalities (3.15) and (3.16), an element $f = x + y$ $(x \in \mathcal{D}(F_1), y \in \mathcal{D}(F_1^*))$ belongs to $\mathcal{D}(A_{1\lambda})$ if and only if the element $Vf = Vx + Vy$ $(Vx \in \mathcal{D}(F_2), Vy \in \mathcal{D}(F_2^*))$ belongs to $\mathcal{D}(A_{2\lambda})$. In view of (3.2), this means that the characteristic functions in the BVS considered above coincide (i.e., $C^1(\lambda) = C^2(\lambda)$).

Conversely, let certain characteristic functions $C^1(\lambda)$ and $C^2(\lambda)$ of the Hermitian operators A_1 and A_2, respectively, coincide.

This is possible only in the case where the BVS $\left(X_+^i, X_-^i, Y_+^i, Y_-^i\right)$ $(i = 1, 2)$ corresponding to these functions have the same spaces

$$X_+^1 = X_+^2 = X_+ \quad \text{and} \quad X_-^1 = X_-^2 = X_-.$$

Thus, the characteristic function $C^i(\lambda)$ is defined in the BVS $\left(X_+, X_+, Y_+^i, Y_-^i\right)$ $(i = 1, 2)$. This BVS is associated with some canonical dissipative extension F_i of the Hermitian operator A_i. The extension F_i is defined by the decomposition

$$\mathcal{D}(F_i) = \mathcal{D}(A_i) \dotplus \mathcal{L}_+^i,$$

where $\mathcal{L}_+^i$ is a maximal uniformly positive subspace in the Krein space

$$\mathcal{N}_\mu^i = \mathfrak{N}_\mu(A_i) \dotplus \mathfrak{N}_{\bar{\mu}}(A_i)$$

(μ is a fixed point from the lower half plane). Therefore,

$$\mathcal{N}_\mu^i = \mathcal{L}_+^i \,[+]\, \mathcal{L}_-^i \quad (i = 1, 2), \tag{3.17}$$

where $\mathcal{L}_-^i$ is a J-orthogonal complement of the subspace $\mathcal{L}_+^i$ in $\mathcal{N}_\mu^i$.

Since the operator Y_+^i $\left(Y_-^i\right)$ determines the one-to-one correspondence of the subspaces $\mathcal{L}_+^i$ $(\mathcal{L}_-^i)$ and X_+ (X_-), it follows from equality (3.17) that, for any element $f_1 = x_+^1 + y_-^1 \in \mathcal{N}_\mu^1$, there exists a unique element $f_2 = x_+^2 + y_-^2$ $\left(x_+^i \in \mathcal{L}_+^i,\ y_-^i \in \mathcal{L}_-^i\right)$ such that

$$Y_+^1 x_+^1 = Y_+^2 x_+^2 \quad \text{and} \quad Y_-^1 y_-^1 = Y_-^2 y_-^2.$$

Consequently, the operator $\Omega f_1 = f_2$ is well defined in $\mathcal{N}_\mu^1$ and maps $\mathcal{N}_\mu^1$ onto $\mathcal{N}_\mu^2$. Moreover, in view of (1.22), we have

$$[\Omega f_1, \Omega f_1]_\mu^{(2)} = [f_1, f_1]_\mu^{(1)}, \tag{3.18}$$

where $[\cdot,\cdot]^{(i)}$ is the indefinite metric in the Krein space $\mathcal{N}_\mu^i$ defined by (9.12).

By virtue of Corollary 9.10, the canonical extensions $A_{i\lambda}$ ($i = 1, 2$ and λ is an arbitrary point from the lower half plane) are defined by the decomposition

$$\mathcal{D}(A_{i\lambda}) = \mathcal{D}(A_i) \dotplus \mathcal{L}^i(\lambda),$$

where $\mathcal{L}^i(\lambda)$ is a maximal uniformly positive subspace in $\mathcal{N}_\mu^i$.

Taking into account the condition $C^1(\lambda) = C^2(\lambda)$, equality (3.2), and the definition of the operator Ω, we obtain

$$\Omega\, \mathcal{L}^1(\lambda) = \mathcal{L}^2(\lambda). \tag{3.19}$$

In particular, for $\lambda = \mu$, relations (3.1) and (3.19) imply that

$$\Omega\, \mathfrak{N}_\mu(A_1) = \mathfrak{N}_\mu(A_2). \tag{3.20}$$

By virtue of decomposition (9.13) (with $\lambda = \mu$) and relation (3.18), equality (3.20) yields

$$\Omega\, \mathfrak{N}_{\bar\mu}(A_1) = \mathfrak{N}_{\bar\mu}(A_2). \tag{3.21}$$

In view of the definition of scalar product in $\mathcal{N}_\mu^i$ (see (9.10)), we conclude that the operator Ω maps the subspace $\mathfrak{N}_z(A_1)$ onto the subspace $\mathfrak{N}_z(A_2)$ $\quad z \in \{\mu, \bar\mu\}$) isometrically (in the norm of the space $\mathcal{H}$).

Let $C_0^i(\lambda)$ be a characteristic function of the operator A_i of the form (3.4). By analogy with the proof of Corollary 3.3, we establish that $C_0^i(\lambda)$ coincides with the angular operator of the subspace $\mathcal{L}^i(\lambda)$ with respect to the subspace $\mathfrak{N}_\mu(A_i)$.

Thus,

$$\mathcal{L}^i(\lambda) = \left\{(I + C_0^i(\lambda))x_\mu^i \;\middle|\; x_\mu^i \in \mathfrak{N}_\mu(A_i)\right\}.$$

Consequently, by virtue of equalities (3.19)–(3.21), we have

$$C_0^2(\lambda) = \Omega C_0^1(\lambda)\Omega^{-1}.$$

Note that, in this equality, the operator Ω does not depend on λ. The unitary equivalence of the operators A_1 and A_2 now follows from the results of Shtraus [1, Theorem 4]. Theorem 3.4 is proved.

■

Remarks. In terms of BVS (in the form of Definition I), the characteristic function of a symmetric operator with equal defect numbers was first defined by Kochubei [5]. In this case, the obtained characteristic function depends on the choice of BVS and, in a particular case (equality (3.4)), coincides with the Shtraus characteristic function [2].

Later, Derkach and Malamud [1] used BVS of this type for the definition of the Weyl function of a symmetric operator. Note that, for fixed BVS, the characteristic function $C(\lambda)$ and the Weyl function $M(\lambda)$ of a symmetric operator are related by the simple linear fractional transformation

$$M(\lambda) = -i(I+C(\lambda))(I-C(\lambda))^{-1},$$

which enables us to investigate only one of these functions. For the generalization of the concept of Weyl function to the case of Hermitian operators with equal defect numbers and different applications of this theory, see Derkach and Malamud [2, 3].

In this section, unlike works mentioned above, we have used BVS in the form of Definition III. This enables us to generalize the results of Kochubei [5] to the case of Hermitian operators with arbitrary defect numbers.

4. Spectral Properties of Regular Extensions

In this section, we describe the nonreal spectra of regular extensions of Hermitian operators with arbitrary defect numbers in terms of BVS.

4.1. Auxiliary Statements. Let $\tilde{A}$ be a regular extension of a closed Hermitian operator A. Assume that the point spectrum of the operator $\tilde{A}$ does not cover the entire lower half plane. By virtue of Theorem 2.1, this means that there exists a BVS (X_+, X_-, Y_+, Y_-) of the Hermitian operator A such that the operator $\tilde{A}$ is defined by the relations

$$\mathcal{D}(\tilde{A}) = \{x+y \mid KY_+x = Y_-y,\ x\in \mathcal{D}(F),\ y\in \mathcal{D}(F^*)\}, \tag{4.1}$$

$$\tilde{A}\,(x+y) = Fx+F^*y, \tag{4.2}$$

where F is the canonical dissipative extension of the operator A associated with the BVS (X_+, X_-, Y_+, Y_-).

Note that the admissible operator K in formula (4.1) is uniquely defined by the extension $\tilde{A}$ (see the remark to Theorem 2.1). For this reason, in what follows, we denote the extension $\tilde{A}$ by A_K.

Thus, the extension A_K ($\equiv \tilde{A}$) is defined by relations (4.1) and (4.2). By analogy with the proof of Theorem 2.1, we conclude that the regular extension A_K is defined by the decomposition

$$\mathcal{D}(A_K) = \mathcal{D}(A) \dotplus \mathcal{L}, \tag{4.3}$$

where the lineal $\mathcal{L}$ is contained in $\mathcal{N}_\mu$ (μ is an arbitrary nonreal number) and is defined by equality (2.8). Note that the subspace $\mathcal{L}_+$ in (2.8) is a maximal uniformly positive subspace in $\mathcal{N}_\mu$ and, hence, the decomposition

$$\mathcal{N}_\mu = \mathcal{L}_+ [\dotplus] \mathcal{L}_-, \tag{4.4}$$

is a canonical decomposition of the Krein space $\mathcal{N}_\mu$.

Recall that the decomposition

$$\mathcal{N}_\mu = \mathfrak{N}_\mu [\dotplus] \mathfrak{N}_{\bar{\mu}}, \tag{4.5}$$

where $\mathfrak{N}_z$ $(z \in \{\mu, \bar{\mu}\})$ are the defect subspaces of the operator A, is the fundamental decomposition of the space $\mathcal{N}_\mu$ (see Subsection 9.2).

Let P_μ and $P_{\bar{\mu}}$ be the orthoprojectors onto $\mathfrak{N}_\mu$ and $\mathfrak{N}_{\bar{\mu}}$ in $\mathcal{N}_\mu$ with respect to decomposition (4.5). Since the extension A_K is defined by decomposition (4.3), by virtue of (9.15), we conclude that, for any $x \in \mathcal{D}(A_K)$ $(x = x_0 + f,\ x_0 \in \mathcal{D}(A)\ f \in \mathcal{L})$, the following equality is true:

$$(A_K - \mu I)x = (A - \mu I)x_0 + (\bar{\mu} - \mu)P_\mu f.$$

By virtue of Theorem 3.1 (Chapter 1), this equality yields

$$\mu \notin \sigma_p(A_K) \Leftrightarrow \mathcal{L} \cap \mathfrak{N}_{\bar{\mu}} = \{0\}, \tag{4.6}$$

where the symbol $\Leftrightarrow$ denotes the equivalence of the corresponding statements.

Let $\mu \notin \sigma_p(A_K)$. In view of (4.6), the lineal $\mathcal{L}$ can be represented in the form

$$\mathcal{L} = \{f | f = P_\mu f + \Phi P_\mu f,\ P_\mu f \in \mathcal{D}(\Phi) \subset \mathfrak{N}_\mu\},\tag{4.7}$$

where the linear operator Φ $(\Phi : \mathfrak{N}_\mu \to \mathfrak{N}_{\bar{\mu}})$ is defined by the equality

$$\Phi P_\mu f = P_{\bar{\mu}} f \quad (f \in \mathcal{L}).\tag{4.8}$$

In this case, the operator $(A_K - \mu I)^{-1}$ in its domain of definition

$$\mathcal{D}((A_K - \mu I)^{-1}) = \mathcal{R}(A_K - \mu I) = \mathcal{R}(A - \mu I) \oplus \mathcal{D}(\Phi)\tag{4.9}$$

acts according to the following law:

$$(A_K - \mu I)^{-1}(\psi + x_\mu) = (A - \mu I)^{-1}\psi + \frac{1}{\bar{\mu} - \mu}(I + \Phi)x_\mu,\tag{4.10}$$

where $\psi \in \mathcal{R}(A - \mu I)$ and $x_\mu \in \mathcal{D}(\Phi)$.

Lemma 4.1. *Assume that the extension A_K is defined by decomposition (4.3), a nonreal number μ does not belong to the point spectrum of the operator A_K, and the operator Φ is defined by equality (4.8). Then*

(i) $\mu \in \sigma_r(A_K) \Leftrightarrow \overline{\mathcal{R}\left(P_\mu |_{\mathcal{L}}\right)} \neq \mathfrak{N}_\mu$;

(ii) $\mu \in \sigma_c(A_K) \Leftrightarrow \overline{\mathcal{R}\left(P_\mu |_{\mathcal{L}}\right)} = \mathfrak{N}_\mu$ *and Φ is an unbounded operator.*

Proof. The proof follows from relations (4.9) and (4.10). ■

4.2. Description of Nonreal Spectra of Regular Extensions

Theorem 4.2. *Assume that the regular extension A_K of a closed Hermitian operator A is defined in a BVS (X_+, X_-, Y_+, Y_-) by formulas (4.1) and (4.2) and let $C(\cdot)$ be a characteristic function of the operator A in this BVS. Then, for any μ from the upper half plane, the following assertions are true:*

(i) $\mu \in \sigma_p(A_K) \Leftrightarrow \operatorname{Ker}(K - C(\bar{\mu})) \neq 0$;

(ii) $\mu \in \sigma_\alpha(A_K) \Leftrightarrow 0 \in \sigma_\alpha(K - C(\bar{\mu}))$, $\alpha \in \{r, c\}$.

Proof. We fix an arbitrary point μ in the upper half plane and set $\lambda = \bar{\mu}$. It is clear that $\operatorname{Im}\lambda < 0$. In Subsection 4.1, we have shown that the extension A_K is defined by decomposition (4.3) with a lineal $\mathcal{L} \subset \mathcal{N}_\lambda$ of the form (2.8).

Note that the operator Y_+ (Y_-) in equality (2.8) defines a one-to-one correspondence between the subspace $\mathcal{L}_+$ ($\mathcal{L}_-$) and the space X_+ (X_-) and, therefore, the operator Y_+ (Y_-) is invertible. Somewhat arbitrarily, we denote this operator by Y_+^{-1} (Y_-^{-1}).

Thus, in what follows, the operator Y_+^{-1} (Y_-^{-1}) maps the space X_+ (X_-) onto the subspace $\mathcal{L}_+$ ($\mathcal{L}_-$). With this notation, we rewrite equality (2.8) in the form

$$\mathcal{L} = \left\{ f \,\middle|\, f = x_+ + Y_-^{-1} K Y_+ x_+, \; x_+ \in \mathcal{R}\left(Y_+^{-1}\big|_{\mathcal{D}(K)}\right) \subset \mathcal{L}_+ \right\}, \tag{4.11}$$

where the operator $Y_-^{-1} K Y_+$ acts from $\mathcal{L}_+$ into $\mathcal{L}_-$.

Recall (see Subsection 3.1) that the characteristic function $C(\lambda)$ in the BVS (X_+, X_-, Y_+, Y_-) determines the canonical dissipative extension A_λ of the operator A by formulas (3.2) and (3.3).

As in the case of the operator A_K considered above, we show that the extension A_λ is defined by the decomposition

$$\mathcal{D}(A_\lambda) = \mathcal{D}(A) \dotplus \mathcal{L}',$$

where the lineal $\mathcal{L}' \subset \mathcal{N}_\lambda$ is determined by equality (4.11) with $K = C(\lambda)$. Moreover, by virtue of equality (3.1) and Lemma 9.5, we conclude that $\mathcal{L}' = \mathfrak{N}_\lambda$, where $\mathfrak{N}_\lambda$ is the defect subspace of the operator A.

Thus,

$$\mathfrak{N}_\lambda = \left\{ x_\lambda \,\middle|\, x_\lambda = x_+ + Y_-^{-1} C(\lambda) Y_+ x_+, \; x_+ \in \mathcal{L}_+ \right\}, \tag{4.12}$$

where the operator $Y_-^{-1} C(\lambda) Y_+$ is defined in the entire subspace $\mathcal{L}_+$ and acts from $\mathcal{L}_+$ into $\mathcal{L}_-$.

By using decomposition (4.5) and equalities (1.22), we easily show that

$$\mathfrak{N}_{\bar{\lambda}} = \mathcal{N}_\lambda [-] \mathfrak{N}_\lambda = \left\{ x_{\bar{\lambda}} \,\middle|\, x_{\bar{\lambda}} = y_- + Y_+^{-1} C^*(\lambda) Y_- y_-, \; y_- \in \mathcal{L}_- \right\}, \tag{4.13}$$

where the operator $Y_+^{-1} C^*(\lambda) Y_-$ is defined in the entire subspace $\mathcal{L}_-$ and acts from $\mathcal{L}_-$ into $\mathcal{L}_+$.

In view of Corollary 9.2 (with $\mathcal{H}_+ = \mathfrak{N}_\lambda$ and $\mathcal{H}_- = \mathfrak{N}_{\bar{\lambda}}$) and equalities (4.12) and (4.13), we conclude that the operator

$$T = -Y_-^{-1} C(\lambda) Y_+ P_{\mathcal{L}_+} - Y_+^{-1} C^*(\lambda) Y_- P_{\mathcal{L}_-} \tag{4.14}$$

[where $P_{\mathcal{L}_\pm}$ are the projectors onto $\mathcal{L}_\pm$ in $\mathcal{N}_\lambda$ with respect to decomposition (4.4)] is an operator of transition from the fundamental decomposition (4.5) to the canonical decomposition (4.4).

If the operator of transition T is known, then, by virtue of Proposition 9.1, we can express the orthoprojector P_λ ($P_{\bar\lambda}$) onto $\mathfrak{N}_\lambda$ ($\mathfrak{N}_{\bar\lambda}$) in the space $\mathcal{N}_\lambda$ in terms of the projectors $P_{\mathcal{L}_\pm}$.

In particular, by using the fact, that any element $f \in \mathcal{L}$ has the form (4.11), we obtain

$$P_\lambda f = (I+T)^{-1} Y_+^{-1} (I - C^*(\lambda) K) Y_+ x_+ \tag{4.15}$$

and

$$P_{\bar\lambda} f = (I+T)^{-1} Y_-^{-1} (K - C(\lambda)) Y_+ x_+. \tag{4.16}$$

By using these equalities and Lemma 9.1, we can now complete the proof of the theorem.

Indeed, assertion (4.6) implies that the point $\mu = \bar\lambda$ belongs to the point spectrum of the operator A_K if and only if there exists a nonzero element $f \in \mathcal{L}$ such that $P_{\bar\lambda} f = 0$.

By virtue of (4.16), this equality is equivalent to the equality

$$(K - C(\lambda)) Y_+ x_+ = 0,$$

where, in view of (4.11), $Y_+ x_+ \neq 0$. Thus,

$$\mu \in \sigma_p(A_K) \Leftrightarrow \operatorname{Ker}(K - C(\bar\mu)) \neq \{0\}. \tag{4.17}$$

Further, relation (i) in Lemma 4.1 implies that the condition $\mu = \bar\lambda \in \sigma_c(A_K)$ is equivalent to the condition

$$\overline{\mathcal{R}\left(P_{\bar\lambda}\big|_{\mathcal{L}}\right)} \neq \mathfrak{N}_{\bar\lambda}. \tag{4.18}$$

It follows from (4.16) and representations (2.8) and (4.11) of the lineal $\mathcal{L}$ that

$$\mathcal{R}\left(P_{\bar\lambda}\big|_{\mathcal{L}}\right) = \mathcal{R}\left((I+T)^{-1} Y_-^{-1} (K - C(\lambda))\right). \tag{4.19}$$

Note that the definition of an operator of translation (Subsection 9.1) and equality (9.6) (with $\mathcal{H}_- = \mathfrak{N}_{\bar\lambda}$) imply that the operator $(I+T)^{-1}$ is a continuous one-to-one mapping of the subspace $\mathcal{L}_-$ onto the subspace $\mathfrak{N}_{\bar{\lambda}}$. On the other hand, it follows from equality (1.22) that the operator Y_-^{-1} determines a continuous one-to-one correspondence between the space X_- and the subspace $\mathcal{L}_-$. Hence, it follows from equality (4.19) that condition (4.18) is equivalent to the condition

$$\overline{\mathcal{R}(K - C(\lambda))} \neq X_-.$$

In view of (4.17), this means that the assertions $\mu = \bar{\lambda} \in \sigma_r(A_K)$ and $0 \in \sigma_r(K - C(\bar{\mu}))$ are equivalent.

Thus,

$$\mu \in \sigma_r(A_K) \Leftrightarrow 0 \in \sigma_r(K - C(\bar{\mu})). \tag{4.20}$$

Let $\mu = \bar{\lambda} \in \sigma_c(A_K)$. By virtue of (4.17) and (4.20), we obtain

$$0 \in \sigma_c(K - C(\lambda)) \cup \rho(K - C(\lambda)). \tag{4.21}$$

According to assertions (i) and (ii) in Lemma 4.1, the condition $\mu \in \sigma_c(A_K)$ is equivalent to the property of the operator Φ defined by equality (4.8) to be unbounded. Substituting the expressions for the operators P_λ and $P_{\bar{\lambda}}$ from equalities (4.15) and (4.16) into (4.8) and taking into account (4.21), we get

$$\Phi = (I+T)^{-1} Y_+^{-1}\big(I - C^*(\lambda)K\big)(K - C(\lambda))^{-1} Y_-(I+T). \tag{4.22}$$

It follows from the properties of the operators Y_-^{-1} and $(I+T)^{-1}$ considered above that the operator $Y_-(I+T)$ determines a continuous one-to-one correspondence between the subspace $\mathfrak{N}_{\bar{\lambda}}$ and the space X_-. It is easy to see that the mapping $(I+T)^{-1} Y_+^{-1}$ of the space X_+ onto the subspace $\mathfrak{N}_\lambda$ possesses similar properties. Therefore, representation (4.22) implies that the operator Φ is unbounded if and only if the following operator is unbounded:

$$S = \big(I - C^*(\lambda)K\big)(K - C(\lambda))^{-1} \quad (S\colon X_- \to X_+). \tag{4.23}$$

We now rewrite the operator S in the form

$$S = \big(I - C^*(\lambda)C(\lambda)\big)(K - C(\lambda))^{-1} - C^*(\lambda). \tag{4.24}$$

Note that

$$0 \in \rho\big(I - C^*(\lambda)C(\lambda)\big)$$

because

$$\| C(\lambda) \| = \| C^*(\lambda) \| < 1$$

(see Corollary 3.3). Hence, by virtue of (4.21), equality (4.24) implies that the operator S is unbounded if and only if the inclusion $0 \in \sigma_c(K - C(\lambda))$ is true. Thus,

$$\mu \in \sigma_c(A_K) \Leftrightarrow 0 \in \sigma_c(K - C(\bar{\mu})).$$

■

Theorem 4.3. *Assume that a regular extension* A_K *is defined by the conditions of Theorem 4.2. Then, for any nonreal* μ *from the lower half plane* ($\operatorname{Im}\mu < 0$), *the following assertion are true:*

(i) $\mu \in \sigma_\alpha(A_K) \Leftrightarrow 0 \in \sigma_\alpha\big(I - C^*(\mu)K\big)$, $\alpha \in \{p, r\}$;

(ii) $\mu \in \text{ç}(A_K) \Leftrightarrow$ *the "operator* $K\big(I - C^*(\mu)K\big)^{-1}$ *acting from* X_+ *into* X_- *is densely defined and unbounded".*

Proof. The proof of this theorem similar to the proof of Theorem 4.2. For this reason, we present only a sketch of the proof.

We set $\mu = \lambda$. As earlier, the operator A_K is defined by decomposition (4.3), where $\mathcal{L} \subset \mathfrak{N}_\lambda$ is a lineal of the form (4.11). For any $f \in \mathcal{L}$, relations (4.15) and (4.16) are true.

By virtue of (4.6), $\mu \in \sigma_p(A_K)$ if and only if there exists a nonzero element $f \in \mathcal{L}$ such that $P_\lambda f = 0$. It follows from (4.15) that this equality is equivalent to the equality

$$\big(I - C^*(\lambda)K\big)Y_+ x_+ = 0,$$

where $Y_+ x_+ \neq 0$. Therefore,

$$\mu \in \sigma_p(A_K) \Leftrightarrow 0 \in \sigma_p\big(I - C^*(\mu)K\big).$$

For the residual spectrum of the operator A_K, the reasoning is exactly the same as in the proof of Theorem 4.2 [by substituting λ for $\bar{\lambda}$ and replacing equality (4.16) with equality (4.15)]. As a result, we obtain

$$\mu \in \sigma_r(A_K) \Leftrightarrow 0 \in \sigma_r\big(I - C^*(\mu)K\big).$$

Further, the principal difference between the continuous spectrum of the operator A_K and its analog in Theorem 4.2 is that the representation of the operator Φ takes the form

$$\Phi = (I+T)^{-1} Y_-^{-1}(K - C(\lambda))\big(I - C^*(\lambda)K\big)^{-1} Y_+(I+T).$$

Hence, the condition $\mu = \lambda \in \sigma_c(A_K)$ is equivalent to the fact that the operator

$$S = (K - C(\lambda))\big(I - C^*(\lambda)K\big)^{-1} \qquad (S\colon X_+ \to X_-)$$

is unbounded.

Rewriting the operator S in the form

$$S = \big(I - C(\lambda)C^*(\lambda)\big)K\big(I - C^*(\lambda)K\big)^{-1} - C(\lambda),$$

we see that it is unbounded if and only if the operator $K\big(I - C^*(\lambda)K\big)^{-1}$ densely defined in the space X_+ is unbounded.

■

Corollary 4.4. *If the operator K which determines the regular extension A_K in the BVS (X_+, X_-, Y_+, Y_-) is bounded, then condition (ii) in Theorem 4.3 can be rewritten in the form*

$$\mu \in \sigma_c(A_K) \Leftrightarrow 0 \in \sigma_c\big(I - C^*(\mu)K\big) \qquad (\operatorname{Im}\mu < 0).$$

Proof. If $\mu \in \sigma_c(A_K)$, then, by virtue of statement (i) in Theorem 4.3, the operator $\big(I - C^*(\mu)K\big)^{-1}$ exists and is densely defined in X_+. Taking into account the equality

$$C^*(\mu)K\big(I - C^*(\mu)K\big)^{-1} = \big(I - C^*(\mu)K\big)^{-1} - I$$

and the fact that the operator K is bounded, we show that the operator $K\big(I - C^*(\mu)K\big)^{-1}$ densely defined in X_+ is unbounded if and only if $0 \in \sigma_c\big(I - C^*(\mu)K\big)$.

Consequently,

$$\mu \in \sigma(A_K) \Leftrightarrow 0 \in \sigma_c\big(I - C^*(\mu)K\big).$$

■

Let us now illustrate the results obtained above by the following simple example:

In the Hilbert space $\mathcal{H} = L_2(\mathfrak{H}, (0, \infty))$, we consider a symmetric operator

$$A = -iJ\frac{d}{ds}, \qquad \mathcal{D}(A) = \mathring{W}_2^1(\mathfrak{H}, (0, \infty)),$$

where J is a self-adjoint unitary operator acting in the auxiliary Hilbert space $\mathfrak{H}$.

Let (X_+, X_-, Y_+, Y_-) be the BVS of the operator A constructed in item 3 of Subsection 1.4. In this BVS, relations (4.1) and (4.2) determining the regular extension $\tilde{A} = A_K$ of the symmetric operator A, take the form

$$\mathcal{D}(A_K) = \{ x(s) \in W_2^1(\mathfrak{H}, (0, \infty)) \mid KP_+x(0) = P_-x(0) \}, \tag{4.25}$$

$$A_K x(s) = -iJ\frac{d}{ds}x(s) \quad (x(s) \in \mathcal{D}(A_K)), \tag{4.26}$$

where P_+ and P_- are, respectively, the orthoprojectors onto the subspaces $\mathfrak{H}_+$ and $\mathfrak{H}_-$ in the space $\mathfrak{H}$ (here, $\mathfrak{H}_+$ and $\mathfrak{H}_-$ are the subspaces from the fundamental decomposition (1.30) of the Krein space $\mathfrak{H}$) and K $(\mathcal{D}(K) \subseteq \mathfrak{H}_+)$ is a linear operator from $\mathfrak{H}_+$ into $\mathfrak{H}_-$.

If K is an arbitrary contracting operator from $\mathfrak{H}_+$ into $\mathfrak{H}_+$, then relations (4.25) and (4.26) determine an arbitrary dissipative extension A_K of the operator A.

It is easy to show that, for any nonreal number λ from the lower half plane, the defect subspace $\mathfrak{N}_\lambda$ of the operator A has the form

$$\mathfrak{N}_\lambda = \left\{ e^{i\bar{\lambda}s}x(0) \,\middle|\, x(0) \in \mathfrak{H}_+ \right\} \quad (\operatorname{Im}\lambda < 0).$$

This equality implies that the canonical dissipative extension A_λ of the operator A defined by decomposition (3.1) coincides with the canonical dissipative extension F defined by formulas (1.31) for any λ $(\operatorname{Im}\lambda < 0)$, i.e.,

$$A_\lambda = F \quad (\forall \lambda, \ \operatorname{Im}\lambda < 0). \tag{4.27}$$

Recall that the BVS (X_+, X_-, Y_+, Y_-) of the operator A from item 3 in Subsection 1.4 is associated with the operator F. Therefore, equality (4.27) and the definition of the characteristic function $C(\lambda)$ (see Subsection 3.1) imply that, in this BVS,

$$C(\lambda) = 0 \quad (\forall \lambda, \ \operatorname{Im}\lambda < 0).$$

By using this equality in Theorems 4.2 and 4.3, we obtain the description of the nonreal spectrum of the regular extension A_K. In particular, for points μ from the upper half plane $\mathbb{C}^+ = \{\mu \mid \operatorname{Im}\mu > 0\}$, we can write

$$\mathbb{C}^+ \subset \sigma_p(A_K) \Leftrightarrow \mathrm{Ke}K \neq \{0\}.$$

If the operator K is invertible, then

$$\mathbb{C}^+ \subset \sigma_\alpha(A_K) \Leftrightarrow 0 \in \sigma_\alpha(A_K), \quad \alpha \in \{r, c\}.$$

Similarly, for the lower half plane $\mathbb{C}^- = \{\mu \mid \operatorname{Im}\mu < 0\}$ we conclude that

$$\sigma_p(A_K) \cap \mathbb{C}^- = \emptyset,$$

$$\mathbb{C}^- \subset \sigma_r(A_K) \Leftrightarrow \overline{\mathcal{D}(K)} \subset \mathfrak{H}_+,$$

and

$$\mathbb{C}^- \subset \sigma_c(A_K) \Leftrightarrow \text{the "operator } K \text{ is densely defined and unbounded".}$$

Remarks. The description of the nonreal spectra of dissipative extensions of symmetric operators in the terms of BVS was constructed by Kochubei [5]. In this work, he used BVS defined in the sense of Definition I.

In the case of almost solvable extensions of symmetric operators, similar results were obtained by Derkach and Malamud [1].

4. Examples and applications

1. Quasidifferential Operators

In this section, we discuss some general properties of quasidifferential operators.

1.1. Quasiderivatives. In the Hilbert space $L_2(a, b)$, where (a, b) is a finite interval, we consider the set of all real functions $\{p_k(s)\}_{k=0}^n$ such that

$$\int_a^b \frac{ds}{|p_0(s)|} < \infty \quad \text{and} \quad \int_a^b |p_k(s)|ds < \infty \qquad (k = 1, \dots, n). \tag{1.1}$$

For some functions f from $L_2(a, b)$, the expressions $f^{[k]}$ $(k = 0, \dots, 2n)$ defined as

$$\begin{gathered} f^{[k]} = D^k f \quad (k = 0, \dots, n-1), \\ f^{[n]} = p_0 D^n f, \\ f^{[n+k]} = p_k f^{[n-k]} - D f^{[n+k-1]} \qquad (k = 1, \dots, n), \end{gathered} \tag{1.2}$$

where

$$(Df)(s) = \frac{df(s)}{ds},$$

are meaningful.

The functions $f^{[k]}$ are called *quasiderivatives* of the function f.

Thus,

$$f^{[2n]} = p_n f - D[p_{n-1}D - D[p_{n-2}D^2 - \dots - D[p_1 D^{n-1} - D[p_0 D^n f]\dots].$$

If, for any k $(k = 0, \dots, n)$, the function $p_k(s)$ is differentiable up to the $(n-k)$th order, inclusively, then, for the function f differentiable up to the $(2n)$th order, inclusively, the quasiderivative $f^{[2n]}$ can be written in the form

$$f^{[2n]} = \sum_{k=0}^{n} (-1)^k D^k (p_{n-k} D^k f). \tag{1.3}$$

Consider the lineal $\mathcal{D}$ of functions f from $L_2(a, b)$ whose quasiderivatives $f^{[k]}$ $(k = 0, \dots, 2n-1)$ exist and are absolutely continuous and the quasiderivative $f^{[2n]}$ belongs to $L_2(a, b)$.

Note that the lineal $\mathcal{D}$ depends on the choice of the set of functions $\{p_k(s)\}_{k=0}^{n}$.

In particular, for $p_0(s) \equiv 1$ and $p_k(s) \equiv 0$ $(k = 1, \dots, n)$, the lineal $\mathcal{D}$ coincides with the sets of functions from $W_2^{2n}(a, b)$.

Let $f \in \mathcal{D}$. If, in the last equality in (1.2), we substitute $n-k+1$ for k, then we get

$$D f^{[2n-k]} = p_{n-k+1} f^{[k-1]} - f^{[2n+1-k]} \qquad (k = 1, \dots, n).$$

By using this equality and (1.2), we prove that the following relation is true for all f and g from $\mathcal{D}$

$$\sum_{k=0}^{n-1} p_{n-k} f^{[k]} g^{[k]} - D \sum_{k=1}^{n} f^{[2n-k]} g^{[k-1]} = f^{[2n]} g - \frac{1}{p_0} f^{[n]} g^{[n]}. \tag{1.4}$$

In view of (1.4), we arrive at the well-known Lagrange identity

$$f^{[2n]} \bar{g} - f \bar{g}^{[2n]} = \frac{d}{ds} [f, g]_s, \tag{1.5}$$

where

$$[f, g]_s = \sum_{k=1}^{n} \left(f^{[k-1]}(s) \bar{g}^{[2n-k]}(s) - f^{[2n-k]}(s) \bar{g}^{[k-1]}(s) \right). \tag{1.6}$$

By integrating relation (1.5) from a and s $(a \le s \le b)$, we establish the following important Lagrange identity in the integral form:

$$\int_a^b \left(f^{[2n]}\bar{g} - f\bar{g}^{[2n]}\right)ds = [f,g]_s - [f,g]_a. \tag{1.7}$$

In particular, for $s = b$, we have

$$\left(f^{[2n]}, g\right) - \left(f, g^{[2n]}\right) = [f,g]_b - [f,g]_a. \tag{1.8}$$

1.2. Quasidifferential Systems

Theorem 1.1. *Assume that* c_k $(k = 0, \ldots, 2n-1)$ *are arbitrary complex numbers and* s_0 *is a fixed point from* $[a, b]$. *Then, for any function* h *from* $L_2(a, b)$, *the quasidifferential system*

$$f^{[2n]} = h, \quad f^{[k]}(s_0) = c_k \quad (k = 0, \ldots, 2n-1)$$

possesses a unique solution from $\mathcal{D}$.

The proof of this theorem can be found in Naimark [4, 5].

Remark. Theorem 1.1 remains true if, instead of the equation $f^{[2n]} = h$, we consider the equation

$$f^{[2n]} - \lambda f = h, \tag{1.9}$$

where λ is an arbitrary complex number.

Consider the lineal

$$\mathfrak{N}_0 = \{\varphi \in \mathcal{D} \mid \varphi^{[2n]} = 0\}. \tag{1.10}$$

By virtue of Theorem 1.1, we have $\dim \mathfrak{N}_0 = 2n$. In particular, the real solutions φ_m $(m = 1, \ldots, 2n)$ of the equation $\varphi^{[2n]} = 0$ with the initial conditions

$$\varphi^{[k-1]}(a) = \delta_{mk} \quad (k = 1, \ldots, 2n), \tag{1.11}$$

where δ_{mk} is the Kronecker symbol, form a basis of the lineal $\mathfrak{N}_0$.

***1.3. Quasidifferential Operator** A.* In the space $L_2(a, b)$, we consider a linear operator A defined in the lineal

$$\mathcal{D}(A) = \{f \in \mathcal{D} \,|\, f^{[k]}(a) = f^{[k]}(b) = 0 \ (k = 0, \dots, 2n-1)\} \tag{1.12}$$

by the equality

$$Af = f^{[2n]} \quad (f \in \mathcal{D}(A)). \tag{1.13}$$

By virtue of (1.6) and (1.12), the equalities

$$[f, g]_a = [f, g]_b = 0$$

hold for all $f \in \mathcal{D}(A)$ and $g \in \mathcal{D}$. This means that, in view of equalities (1.8) and (1.13), the following relation is true:

$$(Af, g) = (f, g^{[2n]}) \quad (f \in \mathcal{D}(A),\ g \in \mathcal{D}) \tag{1.14}$$

In particular, if $g \in \mathcal{D}(A)$, then it follows from equalities (1.13) and (1.14) that the operator A is Hermitian.

Proposition 1.2. *The space $L_2(a, b)$ can be represented in the form*

$$L_2(a, b) = \mathcal{R}(A) \oplus \mathfrak{N}_0, \tag{1.15}$$

where the lineal $\mathfrak{N}_0$ is defined by equality (1.10).

Proof. Equalities (1.10) and (1.14) imply that the lineals $\mathcal{R}(A)$ and $\mathfrak{N}_0$ are orthogonal in $L_2(a, b)$.

Let $h \in L_2(a, b) \ominus (\mathcal{R}(A) \oplus \mathfrak{N}_0)$. Theorem 1.1 proves that there exists an element $f \in \mathcal{D}$ such that

$$f^{[2n]} = h \quad \text{and} \quad f^{[k]}(b) = 0 \quad (k = 0, \dots, 2n-1). \tag{1.16}$$

By setting in equality (1.8) $g = \varphi_m$ $(m = 1, \dots, 2n)$, where the elements $\varphi_1, \varphi_2, \dots,$ φ_{2n} form a basis in the space $\mathfrak{N}_0$ and satisfy conditions (1.11), we obtain

$$\left(f^{[2n]}, \varphi_m\right) - \left(f, \varphi_m^{[2n]}\right) = [f, \varphi_m]_b - [f, \varphi_m]_a. \tag{1.17}$$

By virtue of (1.10) and the fact that $f^{[2n]} = h \perp \mathfrak{N}_0$, we conclude that the left-hand side of (1.17) is equal to zero. Thus,

$$[f, \varphi_m]_b - [f, \varphi_m]_a = 0 \quad (m = 1, \dots, 2n). \tag{1.18}$$

In view of relations (1.6) and (1.16), we find that $[f, \varphi_m]_b = 0$. Therefore, by using (1.6), (1.11) and (1.18), we get

$$0 = [f, \varphi_m]_a = \begin{cases} -f^{[2n-m]}(a) & (1 \le m \le n), \\ f^{[2n-m]}(a) & (m > n). \end{cases}$$

Hence, $f^{[k]}(a) = 0$ for $k = 0, \dots, 2n-1$. This assertion and equalities (1.16) imply that $f \in \mathcal{D}(A)$ and $h = Af \in \mathcal{R}(A)$. By assumption, $h \perp \mathcal{R}(A)$.

Thus, $h = 0$ and, therefore, decomposition (1.15) is true.

■

Corollary 1.3. *The operator A is symmetric in $L_2(a, b)$.*

Proof. As indicated above, A is an Hermitian operator. Therefore, to complete the proof, it remains to show that the operator A is densely defined.

Indeed, let $h \perp \mathcal{D}(A)$. It follows from Theorem 1.1 that there exists an element $g \in \mathcal{D}$ such that $h = g^{[2n]}$. In view of equality (1.14), this means that

$$(Af, g) = (f, g^{[2n]}) = (\,, h) = 0 \quad (\forall f \in \mathcal{D}(A)).$$

Hence, by virtue of Proposition 1.2, we obtain $g \in \mathfrak{N}_0$. By using (1.10), we conclude, that $h = g^{[2n]} = 0$. Thus, $\overline{\mathcal{D}(A)} = L_2(a, b)$.

■

Theorem 1.4. *If $p_0(s) > 0$ for $s \in [a, b]$, then the operator A is a semi-bounded from below.*

Proof. Assume that the element f in equality (1.4) belongs to $\mathcal{D}(A)$ and $g = \bar{f}$. Integrating this equality on the segment $[a, b]$, in view conditions (1.12), we get (changing the order of integration and taking into account the fact that $f^{[n-k]} = D^{n-k}f$ for $k \ge 1$)

$$(Af, f) = \int_a^b \left| f^{[n]}(s) \right|^2 \frac{1}{p_0(s)} ds + \sum_{k=1}^{n} \int_a^b p_k(s) \left| D^{n-k} f(s) \right|^2 ds. \tag{1.19}$$

In what follows, it is convenient to denote the mth derivative of the function f by $f^{(m)}$ instead of $D^m f$. Hence, $f^{(m)} = D^m f$ $(m \in \mathbb{N})$.

Let us show that, for $f \in \mathcal{D}(A)$ and $k = 1, \dots, n-1$,

$$f^{(n-k)}(s) = \int_a^s \frac{(s-\xi)^{k-1}}{(k-1)!} f^{(n)}(\xi)\,d\xi. \tag{1.20}$$

Indeed, for $k = 1$, identity (1.20) is obvious. Assume that this identity is true for $k = m-1$ $(< n-1)$. Then, by differentiating the difference

$$f^{(n-m)}(s) - \int_a^s \frac{(s-\xi)^{m-1}}{(m-1)!} f^{(n)}(\xi)\,d\xi \tag{1.21}$$

with respect to s, we prove that

$$f^{(n-(m-1))}(s) - \int_a^s \frac{(s-\xi)^{(m-1)-1}}{((m-1)-1)!} f^{(n)}(\xi)\,d\xi = 0.$$

Consequently, difference (1.21) is equal to a certain constant c. Since $f^{[n-m]}(a) = f^{(n-m)}(a) = 0$, we have $c = 0$. This proves equality (1.20).

Let

$$v_k(s,\xi) = \begin{cases} \dfrac{(s-\xi)^{k-1}}{(k-1)!} & (\xi \le s), \\ 0 & (\xi > s). \end{cases}$$

Then relation (1.20) can be rewritten in the form

$$f^{(n-k)}(s) = \int_a^b v_k(s,\xi) f^{(n)}(\xi)\,d\xi, \tag{1.22}$$

where $f \in \mathcal{D}(A)$ and $k = 1, \dots, n-1$.

We set

$$\mathcal{K}(\xi, \eta) = -\sum_{k=1}^{n} \int_a^b p_k(s) v_k(s,\xi) v_k(s,\eta)\,ds, \tag{1.23}$$

where $\{\xi, \eta\} \subset [a, b]$. By virtue of relations (1.22) and (1.23), one can easily show that

$$\int_a^b \int_a^b \mathcal{K}(\xi, \eta) f^{(n)}(\eta) \overline{f^{(n)}(\xi)}\, d\eta\, d\xi = -\sum_{k=1}^{n} \int_a^b p_k(s) \left| f^{(n-k)}(s) \right|^2 ds. \tag{1.24}$$

Hence, in view of (1.19) and (1.24), we obtain

$$(Af, f) = \int_a^b \left| f^{[n]}(s) \right|^2 \frac{1}{p_0(s)} ds - \int_a^b \int_a^b \mathcal{K}(\xi, \eta) f^{(n)}(\eta) \overline{f^{(n)}(\xi)}\, d\eta\, d\xi. \tag{1.25}$$

Let $\mathcal{H}_0$ be the Hilbert space of all measurable functions such that[12]

$$\int_a^b |f(s)|^2 \frac{1}{p_0(s)} ds < \infty.$$

A scalar product in $\mathcal{H}_0$ is defined by the equality

$$(f, g)_0 = \int_a^b f(s) \overline{g(s)} \frac{1}{p_0(s)} ds. \tag{1.26}$$

In the space $\mathcal{H}_0$, we consider an integral operator K whose kernel $\mathcal{K}(\xi, \eta)$ is symmetric and continuous

$$(Kf)(\xi) = \int_a^b \mathcal{K}(\xi, \eta) f(\eta) \frac{1}{p_0(\eta)} d\eta. \tag{1.27}$$

By virtue of (1.2), we can write $f^{[n]} = p_0 f^{(n)}$. Hence, it follows from (1.26) and (1.27) that

$$(Kf^{[n]}, f^{[n]})_0 = \int_a^b \int_a^b \mathcal{K}(\xi, \eta) f^{(n)}(\eta) f^{(n)}(\xi)\, d\eta\, d\xi. \tag{1.28}$$

In view of (1.26) and (1.28), equality (1.25) can be rewritten in the form

$$(Af, f) = (f^{[n]}, f^{[n]})_0 - (Kf^{[n]}, f^{[n]})_0. \tag{1.29}$$

Let $\varphi_1, \varphi_2, \varphi_3, \ldots$ be the orthogonal system of eigenfunctions of the operator K and let $\lambda_1, \lambda_2, \lambda_3, \ldots$ be the corresponding system of characteristic numbers. Without

---

12 It is clear that the lineal $\mathcal{D}$ is contained in $\mathcal{H}_0$.

loss of generality, we can assume that $|\lambda_1| \le |\lambda_2| \le |\lambda_3| \le \dots$. The Hilbert–Schmidt theorem implies that

$$(Kf^{[n]}, f^{[n]})_0 = \sum_{m=1}^{\infty} \frac{1}{\lambda_m} |(f^{[n]}, \varphi_m)_0|^2.$$

Since $1/\lambda_m \to 0$ as $m \to \infty$, the inequality $\frac{1}{|\lambda_m|} \le 1$ holds beginning with a certain number $m = m_0$. If, in this case,

$$(f^{[n]}, \varphi_m)_0 = 0 \quad (m = 1, \dots, m_0 - 1),$$

then

$$(Kf^{[n]}, f^{[n]})_0 = \sum_{m=m_0}^{\infty} \frac{1}{\lambda_m} |(f^{[n]}, \varphi_m)_0|^2 \le \sum_{m=m_0}^{\infty} |(f^{[n]}, \varphi_m)_0|^2,$$

i.e.,

$$(Kf^{[n]}, f^{[n]})_0 \le (f^{[n]}, f^{[n]})_0. \tag{1.30}$$

It follows from relations (1.29) and (1.30) that $(Af, f) \ge 0$. Therefore, in the lineal

$$\mathcal{D}_0(A) = \left\{ f \in \mathcal{D}(A) \,\middle|\, (f^{[n]}, \varphi_m)_0 = 0,\ m = 1, \dots, m_0 - 1 \right\},$$

the operator A is semibounded from below (nonnegative). Moreover, since

$$\dim \mathcal{D}(A) = m_0 \quad (\operatorname{mod} \mathcal{D}_0(A)),$$

the operator A is semibounded in the entire lineal $\mathcal{D}(A)$ (however, it is clear that this operator is not necessarily nonnegative).

■

1.4. Boundary-Value Spaces of the Operator A. Assume that A is a symmetric operator in $L_2(a, b)$ defined by relations (1.12) and (1.13). By virtue of equality (1.14), we conclude, that $\mathcal{D}(A^*) \supset \mathcal{D}$ and

$$A^* g = g^{[2n]} \quad (\forall g \in \mathcal{D}). \tag{1.31}$$

Let us show that $\mathcal{D}(A^*) = \mathcal{D}$. Indeed, suppose that $\varphi \in \mathcal{D}(A^*)$. Theorem 1.1 implies that the lineal $\mathcal{D}$ contains a vector g such that $g^{[2n]} = A^*\varphi$. Consequently, in view of (1.14) and (1.31), we obtain

$$(Af, g) = (f, g^{[2n]}) = (f, A^*\varphi) = (Af, \varphi).$$

This equality and equality (1.15) imply that $\varphi - g \in \mathfrak{N}_0$. By using the definition of the lineal $\mathfrak{N}_0$ [equality (1.10)], we conclude that $\varphi - g \in \mathcal{D}$ and, hence, $\varphi \in \mathcal{D}$.

Thus, $\mathcal{D}(A^*) = \mathcal{D}$ and the operator A^* is defined by equality (1.31).

In view of the remark to Theorem 1.1 and equality (1.31), it is easy to see that, for any $\lambda \in \mathbb{C}$, the equation $A^*g - \bar{\lambda}g = 0$ has exactly $2n$ linearly independent solutions. This means that the dimensionality of the defect subspace

$$\mathfrak{N}_\lambda = \ker(A^* - \bar{\lambda}I)$$

of the operator A is equal to $2n$.

Therefore, A is a symmetric operator with deficiency indices $(2n, 2n)$ and, hence, for this operator, we can construct a BVS in the sense of Definition I as well as BVS introduced by Definitions II and III.

First we construct a BVS (X, Γ_1, Γ_2) of the operator A in the sense of Definition I. Denote

$$f[s] = \left(f^{[0]}(s), f^{[1]}(s), f^{[2]}(s), \ldots, f^{[2n-1]}(s)\right). \tag{1.32}$$

Hence, the expression $[f, g]_s$ from equality (1.6) can be rewritten as

$$[f, g]_s = f[s]\, U g^*[s], \tag{1.33}$$

where

$$U = \begin{pmatrix} 0 & U_1 \\ -U_1 & 0 \end{pmatrix} \quad \text{and} \quad U_1 = \begin{pmatrix} 0 & 0 & \ldots & 0 & 1 \\ 0 & 0 & \ldots & 1 & 0 \\ \ldots & \ldots & \ldots & \ldots & \ldots \\ 0 & 1 & \ldots & 0 & 0 \\ 1 & 0 & \ldots & 0 & 0 \end{pmatrix}.$$

By virtue of equalities (1.8) and (1.31), we can write

$$(A^*f,g) - (f,A^*g) = [f,g]_b - [f,g]_a \quad (\{f,g\} \subset \mathcal{D}(A^*)).$$

Taking into account equality (1.33), we obtain

$$(A^*f,\ g) - (f,\ A^*g) = f[b]\,Ug^*[b] - f[a]\,Ug^*[a], \tag{1.34}$$

where the vectors $f[s]$ and $g[s]$ $(s \in \{a, b\})$ are given by equality (1.32).

Let (X, Γ_1, Γ_2) be a boundary-value space of the operator A. Then, according to the definition of BVS, we have

$$(A^*f,\ g) - (f,\ A^*g) = (\Gamma_1 f,\ \Gamma_2 g)_X - (\Gamma_2 f,\ \Gamma_1 g)_X. \tag{1.35}$$

We set $X = \mathbb{C}^{2n}$. The operators Γ_1 and Γ_2 are sought in the form

$$\Gamma_1 f = f[b]P + f[a]Q \quad \text{and} \quad \Gamma_2 f = f[b]R + f[a]S, \tag{1.36}$$

where P, Q, R, and S are fixed $2n\times 2n$ matrices.

In view of (1.35) and (1.36), we can rewrite equality (1.34) in the form

$$\begin{aligned} f[b]\,Ug^*[b] - f[a]\,Ug^*[a] &= (\Gamma_1 f,\ \Gamma_2 g)_{\mathbb{C}^{2n}} - (\Gamma_2 f,\ \Gamma_1 g)_{\mathbb{C}^{2n}} \\ &= f[b](PR^* - RP^*)\,g^*[b] + f[b](PS^* - RQ^*)\,g^*[a] \\ &\quad + f[a](QR^* - SP^*)\,g^*[b] + f[a](QS^* - SQ^*)\,g^*[a]. \end{aligned}$$

Therefore,

$$PR^* - RP^* = U, \quad PS^* - RQ^* = 0 \quad \text{and} \quad QS^* - SQ^* = -U. \tag{1.37}$$

Equations (1.37) can be rewritten in the matrix form

$$\begin{pmatrix} P & R \\ Q & S \end{pmatrix} \begin{pmatrix} 0 & E \\ -E & 0 \end{pmatrix} \begin{pmatrix} P^* & Q^* \\ R^* & S^* \end{pmatrix} = \begin{pmatrix} U & 0 \\ 0 & -U \end{pmatrix}, \tag{1.38}$$

where E is the $2n\times 2n$ identity matrix.

Note that the solutions of the matrix equation (1.38) [or, which is the same, of the system of equations (1.37)] exist. Indeed, one solution of this equation (system) has the form

$$P = \frac{1}{2}U(R^{-1})^*, \qquad Q = -P, \qquad \text{and} \qquad S = R, \tag{1.39}$$

where R is an arbitrary invertible $2n \times 2n$ matrix.

Thus, if the operators Γ_1 and Γ_2 ($\Gamma_i\colon \mathcal{D}(A^*) \to \mathbb{C}^{2n}$) are defined by equalities (1.36), where the matrices P, Q, R and S are solutions of equation (1.38), then equality (1.35) is true.

Let us show that, for all elements x and y from $\mathbb{C}^{2n}$, there exists a vector $f \in \mathcal{D}(A^*)$ such that

$$\Gamma_1 f = x \qquad \text{and} \qquad \Gamma_2 f = y.$$

In view of (1.36), for this purpose, it suffices to show that the system of equations

$$\begin{cases} f[a]Q + f[b]P = x, \\ f[a]S + f[b]R = y \end{cases} \tag{1.40}$$

is solvable with respect to $f[a]$ and $f[b]$ for all $\{x, y\} \subset \mathbb{C}^{2n}$.

Indeed, if we multiply relation (1.38) from the right by the matrix

$$\begin{pmatrix} -U & 0 \\ 0 & U \end{pmatrix}$$

and use the equality $U^2 = -E$, then we obtain

$$\begin{pmatrix} P & R \\ Q & S \end{pmatrix}\begin{pmatrix} -R^*U & S^*U \\ P^*U & -Q^*U \end{pmatrix} = \begin{pmatrix} E & 0 \\ 0 & E \end{pmatrix}.$$

This relation means that the matrix

$$\begin{pmatrix} P & R \\ Q & S \end{pmatrix}$$

is invertible and, therefore, system (1.40) is solvable for all $\{x, y\} \subset \mathbb{C}^{2n}$.

Thus, the triple $(\mathbb{C}^{2n}, \Gamma_1, \Gamma_2)$, where the operators Γ_1 and Γ_2 are defined by equalities (1.36) and the matrices $P, Q, R,$ and S are arbitrary solutions of matrix the equation (1.38), is a BVS of the symmetric operator A.

In particular, by analyzing the solution of equation (1.38) in the form (1.39), where $R = 1/\sqrt{2}\,E$, we conclude that the triple $(\mathbb{C}^{2n}, \Gamma_1, \Gamma_2)$, where the operators Γ_i $(i = 1, 2)$ are defined by the equalities

$$\Gamma_1 f = \frac{1}{\sqrt{2}}(f[b] - f[a])U \quad \text{and} \quad \Gamma_2 f = \frac{1}{\sqrt{2}}(f[b] + f[a])U \quad (f \in \mathcal{D}(A^*)),$$

is a BVS of the operator A.

If we now apply the reasoning similar to that used in the proof of Theorem 1.9 (Chapter 3) to this BVS, then we readily establish the fact that the operator $F = A^*$,

$$\mathcal{D}(F) = \left\{ f \in \mathcal{D}(A^*) \,\middle|\, \begin{array}{l} f^{[2n-k]}(a) = if^{[k-1]}(a), \\ f^{[2n-k]}(b) = -\,if^{[k-1]}(b), \end{array} \; k = 1, \ldots, n \right\}$$

is a canonical dissipative extension of the symmetric operator A. Moreover,

$$\mathcal{D}(F^*) = \left\{ g \in \mathcal{D}(A^*) \,\middle|\, \begin{array}{l} g^{[2n-k]}(a) = -\,ig^{[k-1]}(a), \\ g^{[2n-k]}(b) = ig^{[k-1]}(b), \end{array} \; k = 1, \ldots, n \right\}$$

and the quadruple $(\mathbb{C}^{2n}, \mathbb{C}^{2n}, Y_+, Y_-)$, where

$$Y_+ f = i(f[b] + f[a]) \qquad (\forall f \in \mathcal{D}(F))$$

$$Y_- g = -i(g[b] + g[a]) \qquad (\forall g \in \mathcal{D}(F^*))$$

forms the BVS of the operator A (in the sense of Definition III) associated with the canonical extension F.

By using this BVS, we can apply the results of Chapter 3 to the investigation of various classes of regular extensions of the operator A. In particular, by virtue of Corollary 2.5 (Chapter 3), we conclude that the formulas

$$A = A^* \quad \text{and} \quad \mathcal{D}(A_K) = \left\{ f + g \,\middle|\, K(f[b] + f[a]) = -(g[b] + g[a]),\ f \in \mathcal{D}(F),\ g \in \mathcal{D}(F^*) \right\}$$

establish a one-to-one correspondence between the set of dissipative extensions A_K of the operator A and the set of contracting operators K in $\mathbb{C}^{2n}$.

Remark. In this section, we use the works of Krein [3], Naimark [4, 5], and A. Kuzhel [1].

2. Boundary-Value Spaces for Model of Zero-Range Potentials with Internal Structure

In this section, we apply results obtained in Chapters 2 and 3 to the investigation of the model of zero-range potentials from the work of Pavlov [1]. In particular, we construct a boundary-value space of an Hermitian operator, define its characteristic function, and describe all dissipative, in particular, self-adjoint extensions of this operator.

2.1. Model of Zero-Range Potentials with Internal Structure. We consider a symmetric operator $-\Delta_0$ acting in the space $L_2(\mathbb{R}^3)$. This operator is a restriction of the Laplacian $-\Delta$ to the set of all functions from $W_2^2(\mathbb{R}^3)$ vanishing at the point $x = 0$, namely,

$$-\Delta_0 = -\Delta|_{\mathcal{D}(-\Delta_0)}, \qquad \mathcal{D}(-\Delta_0) = \{u(x) \in W_2^2(\mathbb{R}^3) \,|\, u(0) = 0\}. \tag{2.1}$$

Assume that A' is a self-adjoint operator acting in $\mathbb{C}^n$ and the vector $\varphi \in \mathbb{C}^n$ is not an eigenvector of the operator A'.

In the space $\mathbb{C}^n$, we consider an Hermitian operator A_0 defined as the restriction of the self-adjoint operator A' to the set $\mathbb{C}^n \ominus \langle (A' + iI) \rangle$, i.e.,

$$A_0 = A'|_{\mathcal{D}(A_0)}, \qquad \mathcal{D}(A_0) = \mathbb{C}^n \ominus \langle (A' + iI)\,\varphi \rangle. \tag{2.2}$$

In the space $\mathcal{H} = L_2(\mathbb{R}^3) \oplus \mathbb{C}^n$, we consider an Hermitian operator

$$A = -\Delta_0 \oplus A_0. \tag{2.3}$$

In what follows, the spaces $L_2(\mathbb{R}^3)$ and $\mathbb{C}^n$ are called the *external* and *internal* spaces of the operator A, respectively.

By using various self-adjoint extensions of the Hermitian operator A, we obtain various models of zero-range potentials with finite-dimensional internal structure. In applications to nonrelativistic quantum mechanics, similar models are successfully used for the investigation of the interaction between particles (in low-energy approximation) in the case where one of these particles is characterized by a complicated internal structure.[13]

[13] The detailed analysis of "physical" applications of various models of zero-range potentials with internal structure can be found in Pavlov [2].

2.2. Boundary-Value Space of the Operator A

Lemma 2.1. *If an Hermitian operator A is defined by equality (2.3), then*

$$\mathcal{D}(A) \vee \mathcal{R}(A) = \mathcal{H}.$$

Proof. Since the operator $-\Delta_0$ is densely defined in $L_2(\mathbb{R}^3)$, we have

$$\mathcal{D}(A) \vee \mathcal{R}(A) = L_2(\mathbb{R}^3) \oplus \mathcal{D}(A_0) \vee \mathcal{R}(A_0). \tag{2.4}$$

By construction, the Hermitian operator A_0 is a restriction of the self-adjoint operator A' to the subspace of finite codimensionality. Moreover, equality (2.2) coincides with equality (2.25) (Chapter 3) if, in the latter, we set

$$\mathcal{H} = \mathbb{C}^n \quad \text{and} \quad \mathcal{H}_0 = \langle (A' + iI)\varphi \rangle$$

and take into account the fact that $\mathcal{D}(A') = \mathbb{C}^n$. Hence, for the operator A_0 the statement of Corollary 2.7 (Chapter 3) is true. Therefore, since the vector φ is not an eigenvector of the operator A', we obtain

$$\mathcal{D}(A_0) \vee \mathcal{R}(A_0) = \mathbb{C}^n.$$

By virtue of (2.4), this completes the proof.

■

Lemma 2.1 shows that it is possible to construct BVS of the operator A in the sense of Definition III. To solve this problem, we first construct separate BVS for the external $-\Delta_0$ and internal A_0 parts of the operator A.

In view of (2.1), we conclude that the deficiency index of the symmetric operator $-\Delta_0$ is equal to (1, 1) and its defect subspaces $\mathfrak{N}_{-i}(-\Delta_0)$ and $\mathfrak{N}_i(-\Delta_0)$ have the form

$$\mathfrak{N}_{-i}(-\Delta_0) = \left\langle \frac{1}{|x|} e^{i\gamma|x|} \right\rangle \quad \text{and} \quad \mathfrak{N}_i(-\Delta_0) = \left\langle \frac{1}{|x|} e^{-i\bar{\gamma}|x|} \right\rangle, \tag{2.5}$$

where

$$|x| = \sqrt{x_1^2 + x_2^2 + x_3^2} \qquad \left(x = (x_1, x_2, x_3) \in \mathbb{R}^3\right)$$

and

$$\gamma = \cos\frac{\pi}{4} + i\sin\frac{\pi}{4}.$$

We set

$$e_1 = \frac{1}{4\sqrt{2}\,\pi|x|}\left(\gamma e^{-i\bar{\gamma}|x|} + \bar{\gamma}e^{i\gamma|x|}\right) \quad \text{and} \quad e_2 = \frac{i}{\sqrt{2}|x|}\left(e^{-i\bar{\gamma}|x|} - e^{i\gamma|x|}\right). \tag{2.6}$$

It is clear that any two functions $u(x)$ and $v(x)$ from $\mathcal{D}\left(-\Delta_0^*\right)$ can be represented in the form

$$u(x) = u_0(x) + \alpha_1 e_1 + \alpha_2 e_2 \quad \text{and} \quad v(x) = v_0(x) + \beta_1 e_1 + \beta_2 e_2,$$

where $\{u_0(x), v_0(x)\} \subset \mathcal{D}(-\Delta_0)$ and $\{\alpha_i, \beta_i\} \subset \mathbb{C}$. Moreover, we can easily prove the equality

$$(-\Delta_0^* u, v)_{L_2} - (u, -\Delta_0^* v)_{L_2} = \alpha_2\bar{\beta}_1 - \alpha_1\bar{\beta}_2. \tag{2.7}$$

It follows from equality (2.7) and Theorem 9.9 (Chapter 2) that the operator

$$F_{\text{ext}} = -\Delta_0^*|_{\mathcal{D}(F_{\text{ext}})}, \quad \mathcal{D}(F_{\text{ext}}) = \mathcal{D}(-\Delta_0) \dotplus \langle e_1 + ie_2\rangle \tag{2.8}$$

is a canonical dissipative extension of the operator $-\Delta_0$. The operator F_{ext}^* adjoint to the operator F_{ext} has the form

$$F_{\text{ext}}^* = -\Delta_0^*|_{\mathcal{D}(F_{\text{ext}}^*)}, \quad \mathcal{D}(F_{\text{ext}}^*) = \mathcal{D}(-\Delta_0) \dotplus \langle e_1 - ie_2\rangle. \tag{2.9}$$

By virtue of equalities (2.7)–(2.9), one can easily show that the quadruple $(\mathbb{C}, \mathbb{C}, Y_+^{\text{ext}}, Y_-^{\text{ext}})$, where the linear operators $Y_+^{\text{ext}} : \mathcal{D}(F_{\text{ext}}) \to \mathbb{C}$ and $Y_-^{\text{ext}} : \mathcal{D}(F_{\text{ext}}^*) \to \mathbb{C}$ are defined by the equalities

$$\begin{aligned} Y_+^{\text{ext}} u &= \sqrt{2}i\alpha \quad (u(x) = u_0(x) + \alpha(e_1 + ie_2) \in \mathcal{D}(F_{\text{ext}})), \\ Y_-^{\text{ext}} v &= -\sqrt{2}i\beta \quad (v(x) = v_0(x) + \beta(e_1 - ie_2) \in \mathcal{D}(F_{\text{ext}}^*)), \end{aligned} \tag{2.10}$$

is a BVS of the operator $-\Delta_0$ associated with the canonical extension F_{ext}.

Similarly, in view of (2.2), we conclude that the defect subspaces $\mathfrak{N}_{-i}(A_0)$ and $\mathfrak{N}_i(A_0)$ of the Hermitian operator A_0 have the form

$$\mathfrak{N}_{-i}(A_0) = \langle T\varphi \rangle \quad \text{and} \quad \mathfrak{N}_i(A_0) = \langle \varphi \rangle, \tag{2.11}$$

where $T = (A' + iI)(A' - iI)^{-1}$ is the Cayley transform of the operator A'.

Recall that the subspace $\mathfrak{N}_{-i}(A_0)$ of the Krein space

$$\mathcal{N}_i = \mathfrak{N}_{-i}(A_0) \dotplus \mathfrak{N}_i(A_0)$$

is maximal uniformly positive (see Section 9). Hence, the extension F_{int} of the operator A_0 defined by the decomposition

$$\mathcal{D}(F_{\text{int}}) = \mathcal{D}(A_0) \dotplus \mathfrak{N}_{-i}(A_0) \tag{2.12}$$

is a canonical dissipative extension.

On the other hand, the operator F^*_{int} adjoint to the operator F_{int} is defined by the decomposition

$$\mathcal{D}(F^*_{\text{int}}) = \mathcal{D}(A_0) \dotplus \mathfrak{N}_i(A_0).$$

We set

$$g_1 = \frac{1}{2}(T + I)\varphi \quad \text{and} \quad g_2 = \frac{1}{2i}(T - I)\varphi.$$

It is clear that

$$g_1 + ig_2 = T\varphi \quad \text{and} \quad g_1 - ig_2 = \varphi. \tag{2.13}$$

Therefore, to make this case formally similar to the case of the operator $-\Delta_0$ considered above, we represent arbitrary vectors $x \in \mathcal{D}(F_{\text{int}})$ and $y \in \mathcal{D}(F^*_{\text{int}})$ in the form

$$x = x_0 + \eta(g_1 + ig_2) \quad \text{and} \quad y = y_0 + \mu(g_1 - ig_2), \tag{2.14}$$

where $\{x_0, y_0\} \subset \mathcal{D}(A_0)$ and $\{\eta, \mu\} \subset \mathbb{C}$.

By using representation (2.14), we now introduce linear operators

$$Y^{\text{int}}_+ x = \sqrt{2}\,i\eta \quad (x \in \mathcal{D}(F_{\text{int}})) \quad \text{and} \quad Y^{\text{int}}_- y = -\sqrt{2}\,i\mu \quad (y \in \mathcal{D}(F^*_{\text{int}})), \tag{2.15}$$

acting, respectively, from $\mathcal{D}(F_{\text{int}})$ and $\mathcal{D}(F^*_{\text{int}})$ into $\mathbb{C}$.

Taking into account equalities (2.14) and (2.15), one can easily show that the quadruple $(\mathbb{C}, \mathbb{C}, Y_+^{\text{int}}, Y_-^{\text{int}})$ is a BVS of the operator A_0 associated with the canonical extension F_{int}.

Recall that the Hermitian operator A is the orthogonal sum of its external $-\Delta_0$ and internal A_0 parts. Hence, the operator

$$F = F_{\text{ext}} \oplus F_{\text{int}}, \tag{2.16}$$

where the operators F_{ext} and F_{int} are defined by equalities (2.8) and (2.12), respectively, is a canonical dissipative extension of the Hermitian operator A. Clearly, in this case,

$$F^* = F_{\text{ext}}^* \oplus F_{\text{int}}^* \tag{2.17}$$

Equalities (2.16) and (2.17) imply that the quadruple

$$\left(\mathbb{C}^2, \mathbb{C}^2, Y_+^{\text{ext}} \oplus Y_+^{\text{int}}, Y_-^{\text{ext}} \oplus Y_-^{\text{int}}\right),$$

where the operators $Y_\pm^{\text{ext}}$ and $Y_\pm^{\text{int}}$ are defined by equalities (2.10) and (2.15), respectively, is a BVS of the Hermitian operator A. This BVS is associated with the canonical dissipative extension F.

2.3. Description of Dissipative Extensions of the Operator A. By virtue of Theorem 2.3 (Chapter 3), we can describe all dissipative and, in particular, self-adjoint extensions of the operator A in terms of its already constructed BVS

$$\left(\mathbb{C}^2, \mathbb{C}^2, Y_+^{\text{ext}} \oplus Y_+^{\text{int}}, Y_-^{\text{ext}} \oplus Y_-^{\text{int}}\right).$$

However, since the operator A is not densely defined, any operator

$$K = \begin{pmatrix} K_{11} & K_{12} \\ K_{21} & K_{22} \end{pmatrix} \qquad (K_{ij} \in \mathbb{C}) \tag{2.18}$$

acting in $\mathbb{C}^2$ is not necessarily admissible with respect to this BVS.

Lemma 2.2. *An operator K defined by equality (2.18) is admissible with respect to the BVS*

$$\left(\mathbb{C}^2, \mathbb{C}^2, Y_+^{\text{ext}} \oplus Y_+^{\text{int}}, Y_-^{\text{ext}} \oplus Y_-^{\text{int}}\right)$$

if and only if at least one of the following conditions is satisfied:

(*i*) $K_{12} \neq 0$;

(*ii*) $K_{22} \neq \dfrac{(A'\varphi, \varphi) + i(\varphi, \varphi)}{(A'\varphi, \varphi) - i(\varphi, \varphi)}$.

Proof. Assume that the operator K defined by equality (2.18) is not admissible with respect to the BVS considered above. According to the definition of admissible operators (Section 2, Chapter 3), this means that one can find elements $\varphi \in \mathcal{D}(F)$ and $\psi \in \mathcal{D}(F^*)$ such that the following conditions are satisfied:

(1) $\varphi + \psi \in \mathcal{D}(A)$;

(2) $K\big(Y_+^{\text{ext}} \oplus Y_+^{\text{int}}\big)\varphi = \big(Y_-^{\text{ext}} \oplus Y_-^{\text{int}}\big)\psi$;

(3) $\big(Y_+^{\text{ext}} \oplus Y_+^{\text{int}}\big)\varphi \neq 0$.

Note that the definitions of the operators F and F^* [equalities (2.16) and (2.17)] and equalities (2.10) and (2.14) imply that the elements φ and ψ admit the representations

$$\varphi = \langle u_0(x) + \alpha(e_1 + ie_2),\ x_0 + \eta(g_1 + ig_2)\rangle,$$
$$\psi = \langle v_0(x) + \beta(e_1 - ie_2),\ y_0 + \mu(g_1 - ig_2)\rangle, \tag{2.19}$$

where the first component of an ordered pair $\langle\cdot, \cdot\rangle$ belongs to the space $L_2(\mathbb{R}^3)$ and the second component belongs to the space $\mathbb{C}$.

By using these equalities and the definitions of the operators $Y_\pm^{\text{ext}}$ and $Y_\pm^{\text{int}}$ [relations (2.10) and (2.15)], we conclude that conditions (1)–(3) can be rewritten in the form

(1′) $\alpha(e_1 + ie_2) + \beta(e_1 - ie_2) \in \mathcal{D}(-\Delta_0)$

$\eta(g_1 + ig_2) + \mu(g_1 - ig_2) \in \mathcal{D}(A_0)$,

(2′) $K\begin{pmatrix}\alpha\\ \eta\end{pmatrix} = \begin{pmatrix}-\beta\\ -\mu\end{pmatrix}$,

(3′) $\begin{pmatrix}\alpha\\ \eta\end{pmatrix} \neq 0$,

where $\binom{\cdot}{\cdot}$ denotes elements from $\mathbb{C}^2$.

Since the operator $-\Delta_0$ is densely defined in $L_2(\mathbb{R}^3)$, the first relation in (1′) is possible only for $\alpha = \beta = 0$.

Further, as follows from the definition of the operator A_0 [equality (2.2)], the second relation in (1′) is equivalent to the equality

$$(\eta(g_1 + ig_2) + \mu(g_1 - ig_2), (A' + iI)\varphi) = 0.$$

In view of (2.13), it is easy to see that this equality is true if and only if the following condition is satisfied:

$$\eta \frac{(A'\varphi, \varphi) + i(\varphi, \varphi)}{(A'\varphi, \varphi) - i(\varphi, \varphi)} = -\mu .$$

By substituting the expressions for the coefficients α, β and μ obtained above in (2′) and (3′), we conclude, that the operator K of the form (2.18) is not admissible if and only if

$$K_{12} = 0 \quad \text{and} \quad K_{22} = \frac{(A'\varphi, \varphi) + i(\varphi, \varphi)}{(A'\varphi, \varphi) - i(\varphi, \varphi)}.$$

Thus, the operator K is admissible if and only if at least one of conditions (i) or (ii) is satisfied.

■

Corollary 2.4 (Chapter 3) and Lemma 2.2 yield the following statement:

Theorem 2.3. *Any maximal dissipative (self-adjoint) extension A_K of the closed Hermitian operator A is given by the formulas*

$$\mathcal{D}(A_K) = \left\{ \varphi + \psi \;\middle|\; K\begin{pmatrix}\alpha\\ \eta\end{pmatrix} = -\begin{pmatrix}\beta\\ \mu\end{pmatrix}, \varphi \in \mathcal{D}(F), \psi \in \mathcal{D}(F^*) \right\}, \tag{2.20}$$

$$A_K(\varphi + \psi) = F\varphi + F^*\psi, \tag{2.21}$$

where the elements φ and ψ are defined by relations (2.19), the operators F and F^* are given by relations (2.16) and (2.17), respectively, and a contracting (unitary) operator K has the form (2.18) and satisfies at least one of the conditions of Lemma 2.2.

In particular, the condition of preservation of "flow" under the transition from the internal space into the external space corresponds to the self-adjoint extension A_K with

$$K = \begin{pmatrix} 0 & 1 \\ 1 & 0 \end{pmatrix}.$$

At the same time, the collection of self-adjoint extensions A_{K_α}, where

$$K_\alpha = \frac{1}{1+\alpha^2}\begin{pmatrix} 1-\alpha^2 & -2i\alpha \\ -2i\alpha & 1-\alpha^2 \end{pmatrix} \qquad (\alpha \in \mathbb{R}),$$

enables one to simulate the "divergence" of the coupling channel between the external and internal spaces by decreasing the coupling constant α $(\alpha \to 0)$ In this case, on passing to the limit, i.e., for $\alpha = 0$, in the external space, we obtain the Laplacian $-\Delta$, $\mathcal{D}(-\Delta) = W_2^2(\mathbb{R}^3)$, whereas, in the internal space, we obtain the self-adjoint operator A'.

2.4. Characteristic Function of the Operator A. We fix a nonreal number λ $(\operatorname{Im}\lambda < 0)$. It follows from (2.2) that the defect subspace $\mathfrak{N}_\lambda$ of the operator A has the form

$$\mathfrak{N}_\lambda = \mathfrak{N}_\lambda(-\Delta_0) \oplus \mathfrak{N}_\lambda(A_0),$$

where $\mathfrak{N}_\lambda(-\Delta_0)$ and $\mathfrak{N}_\lambda(A_0)$ are the defect subspaces of the operators $-\Delta_0$ and A_0 in the spaces $L_2(\mathbb{R}^3)$ and $\mathbb{C}$, respectively. This means that the canonical dissipative extension A_λ of the operator A defined by decomposition (3.1) (Chapter 3) can be represented in the form

$$A_\lambda = -\Delta_{0\lambda} \oplus A_{0\lambda}, \tag{2.22}$$

where the canonical dissipative extensions $-\Delta_{0\lambda}$ and $A_{0\lambda}$ are defined by the decompositions

$$\mathcal{D}(-\Delta_{0\lambda}) = \mathcal{D}(-\Delta_0) \dotplus \mathfrak{N}_\lambda(-\Delta_0) \quad \text{and} \quad \mathcal{D}(A_{0\lambda}) = \mathcal{D}(A_0) \dotplus \mathfrak{N}_\lambda(A_0), \tag{2.23}$$

respectively.

In view of (2.22), it is easy to see that the characteristic function $C(\lambda)$ of the Hermitian operator A in the BVS

$$\left(\mathbb{C}^2, \mathbb{C}^2, Y_+^{\text{ext}} \oplus Y_+^{\text{int}}, Y_-^{\text{ext}} \oplus Y_-^{\text{int}}\right)$$

can be represented in the form of the orthogonal sum

$$C(\lambda) = C^{\text{ext}}(\lambda) \oplus C^{\text{int}}(\lambda) \tag{2.24}$$

of the characteristic functions $C^{\text{ext}}(\lambda)$ and $C^{\text{int}}(\lambda)$ of the operators $-\Delta_0$ and A_0 in the BVS $(\mathbb{C}, \mathbb{C}, Y_+^{\text{ext}}, Y_-^{\text{ext}})$ and $(\mathbb{C}, \mathbb{C}, Y_+^{\text{int}}, Y_-^{\text{int}})$, respectively.

We now find the characteristic function $C^{\text{ext}}(\lambda)$. It is easy to see that

$$\mathfrak{N}_\lambda(-\Delta_0) = \left\langle \frac{1}{|x|} e^{i\sqrt{\bar\lambda}|x|} \right\rangle.$$

By using the representation of the elements $u(x) \in \mathcal{D}(F_{\text{ext}})$ and $v(x) \in \mathcal{D}\left(F_{\text{ext}}^*\right)$ from equality (2.10), we see that the element

$$\frac{1}{|x|} e^{i\sqrt{\bar\lambda}|x|}$$

can be represented in the form

$$\frac{1}{|x|} e^{i\sqrt{\bar\lambda}|x|} = u_0(x) + v_0(x) + \alpha(e_1 + ie_2) + \beta(e_1 - ie_2), \tag{2.25}$$

where the elements e_1 and e_2 are given by equalities (2.6), $\{u_0(x), v_0(x)\} \subset \mathcal{D}(-\Delta_0)$ and $\{\alpha, \beta)\} \subset \mathbb{C}$.

To find the coefficients α and β in (2.25), we multiply this equality by $|x|$ and pass to the limit as $|x| \to 0$. Taking into account the equalities

$$\lim_{|x|\to 0} |x|\, e_1 = \frac{1}{4\pi} \quad \text{and} \quad \lim_{|x|\to 0} |x|\, e_2 = 0$$

and the definition of the operator $-\Delta_0$ [relation (2.1)], we obtain

$$\alpha + \beta = 4\pi. \tag{2.26}$$

Similarly, if we multiply (2.25) by $|x|$, differentiate the expression obtained as a result with respect to $|x|$, and pass to the limit as $|x| \to 0$, then, in view of the equalities

$$\lim_{|x|\to 0} \frac{d(|x|\, e_1)}{d|x|} = 0 \quad \text{and} \quad \lim_{|x|\to 0} \frac{d(|x|\, e_2)}{d|x|} = 1,$$

we obtain

$$\alpha - \beta = \sqrt{\bar{\lambda}}. \tag{2.27}$$

Relations (2.26) and (2.27) imply that

$$\alpha = \frac{4\pi + \sqrt{\bar{\lambda}}}{2} \quad \text{and} \quad \beta = \frac{4\pi - \sqrt{\bar{\lambda}}}{2}.$$

Thus, in decomposition (2.25), the elements $u(x) \in \mathcal{D}(F_{\text{ext}})$ and $v(x) \in \mathcal{D}(F^*_{\text{ext}})$ have the form

$$u(x) = u_0(x) + \frac{4\pi + \sqrt{\bar{\lambda}}}{2}(e_1 + ie_2),$$
$$v(x) = v_0(x) + \frac{4\pi - \sqrt{\bar{\lambda}}}{2}(e_1 - ie_2). \tag{2.28}$$

We recall that the element

$$\frac{1}{|x|}e^{i\sqrt{\bar{\lambda}}|x|}$$

belongs to the domain of definition of the canonical dissipative extension $-\Delta_{0\lambda}$ of the operator $-\Delta_0$. By virtue of relation (3.2) (Chapter 3), this means that

$$C^{\text{ext}}(\lambda)Y_+^{\text{ext}}u = Y_-^{\text{ext}}v, \tag{2.29}$$

where the elements $u(x)$ and $v(x)$ are defined by relation (2.28).

By substituting the values of the operators $Y_\pm^{\text{ext}}$ from (2.10) in (2.29), we conclude that the action of the operator $C^{\text{ext}}(\lambda)$ in the space $\mathbb{C}$ reduces to the multiplication by the number

$$\frac{\sqrt{\bar{\lambda}} - 4\pi}{\sqrt{\bar{\lambda}} + 4\pi},$$

i.e.,

$$C^{\text{ext}}(\lambda) = \frac{\sqrt{\bar{\lambda}} - 4\pi}{\sqrt{\bar{\lambda}} + 4\pi} I \tag{2.30}$$

Let us now find the characteristic function $C^{\text{int}}(\lambda)$ of the operator A_0. In view of relation (2.2), it is easy to see that

$$\mathfrak{N}_\lambda(A_0) = \langle \gamma_\lambda \rangle,$$

where

$$\gamma_\lambda = (A' - \bar{\lambda} I)^{-1}(A' + iI)\varphi. \tag{2.31}$$

By using the representations of elements $x \in \mathcal{D}(F_{\text{int}})$ and $y \in \mathcal{D}(F^*_{\text{ext}}))$ in the form (2.14), we arrive at the following representation of the element γ_λ:

$$\gamma_\lambda = x_0 + y_0 + \eta(g_1 + ig_2) + \mu(g_1 - ig_2), \tag{2.32}$$

where $\{x_0, y_0\} \subset \mathcal{D}(A_0)$ and $\{\eta, \mu\} \subset \mathbb{C}$.

To find the unknown coefficients η and μ in decomposition (2.32), we consider the operator $A_{0\lambda}$ defined by the second decomposition in (2.23). By applying this operator to both sides of representation (2.32) and using relations (2.11) and (2.13), we obtain

$$\bar{\lambda}\gamma_\lambda = A_0(x_0 + y_0) + i\eta(g_1 + ig_2) - i\mu(g_1 - ig_2), \tag{2.33}$$

We now multiply relation (2.32) by $\bar{\lambda}$ and subtract the relation obtained as a result from (2.33). This gives

$$(A_0 - \bar{\lambda} I)(x_0 + y_0) = (\bar{\lambda} - i)\eta(g_1 + ig_2) + (\bar{\lambda} + i)\mu(g_1 - ig_2). \tag{2.34}$$

If we now apply the operator $A' - \bar{\lambda} I$ to both sides of equality (2.32), then, by virtue of (2.13), (2.31), and (2.34), we get

$$(A' + iI)\varphi = \eta(A' - iI)T\varphi + \mu(A' + iI)\varphi = (\eta + \mu)(A' + iI)\varphi.$$

Consequently,

$$\eta + \mu = 1. \tag{2.35}$$

At the same time, it follows from (2.2) and (2.32) that

$$(\gamma_\lambda - \eta(g_1 + ig_2) - \mu(g_1 - ig_2),\ (A' + iI)\varphi) = 0.$$

By virtue of (2.13), (2.31), and (2.35), this equality easily transforms into

$$(\Gamma_\lambda \varphi, \varphi) + (\mu - \eta) i (\varphi, \varphi) = 0, \tag{2.36}$$

where $\Gamma_\lambda = (I + \bar{\lambda} A')(A' - \bar{\lambda} I)^{-1}$.

Relations (2.35) and (2.36) imply that

$$\eta = \frac{i(\varphi, \varphi) + (\Gamma_\lambda \varphi, \varphi)}{2i(\varphi, \varphi)} \quad \text{and} \quad \mu = \frac{i(\varphi, \varphi) - (\Gamma_\lambda \varphi, \varphi)}{2i(\varphi, \varphi)}. \tag{2.37}$$

By using the same reasoning as in the case of the operator $-\Delta_0$, the inclusion $\gamma_\lambda \in \mathcal{D}(A_{0\lambda})$, and the definition of the characteristic function, we obtain

$$C^{\text{int}}(\lambda) Y_+^{\text{int}} (x_0 + \eta (g_1 + i g_2)) = Y_+^{\text{int}} (y_0 + \mu (g_1 - i g_2)), \tag{2.38}$$

where the coefficients η and μ are given by equality (2.37). Substituting the value of the operators $Y_\pm^{\text{int}}$ from (2.15) in (2.38), we get

$$C^{\text{int}}(\lambda) = \frac{(\Gamma_\lambda \varphi, \varphi) - i(\varphi, \varphi)}{(\Gamma_\lambda \varphi, \varphi) + i(\varphi, \varphi)} I. \tag{2.39}$$

By virtue of equalities (2.24), (2.30), and (2.39), we see that the characteristic function $C(\lambda)$ of the operator A acting in the space $\mathbb{C}^2$ has the form

$$C(\lambda) = \begin{pmatrix} \dfrac{\sqrt{\bar{\lambda}} - 4\pi}{\sqrt{\bar{\lambda}} + 4\pi} & 0 \\ 0 & \dfrac{(\Gamma_\lambda \varphi, \varphi) - i(\varphi, \varphi)}{(\Gamma_\lambda \varphi, \varphi) + i(\varphi, \varphi)} \end{pmatrix} \quad (\operatorname{Im} \lambda < 0).$$

By using this characteristic function and the results of Subsection 4.2 (Chapter 3), we obtain the description of the nonreal spectra of regular extensions of the Hermitian operator A.

In particular, Theorem 4.2 (Chapter 3) and Theorem 2.3 imply that the nonreal spectrum of an arbitrary maximal dissipative extension A_K of the operator A is a finite set of eigenvalues μ $(\operatorname{Im} \mu > 0)$ such that

$$\ker (K - C(\bar{\mu})) \neq \{0\},$$

where K is an admissible contraction which defines the extension A_K by relations (2.20) and (2.21).

Remark. The results presented in Section 2 are taken from the works by Pavlov [1] and S. Kuzhel [3].

3. Some Properties of Nonperturbed Operators

In Section 3, we define a special class $\mathfrak{M}_B$ of extensions of a symmetric operator B^2, where B is a simple maximal symmetric operator, and study some properties of these operators. We prove that any operator from $\mathfrak{M}_B$ is actually an orthogonal sum of two operators with simple structure (Theorem 3.6) and show that, for a properly chosen operator B, the classical nonperturbed Laplace operator $-\Delta$, $\mathcal{D}(-\Delta) = W_2^2(\mathbb{R}^n)$ acting in the Hilbert space $L_2(\mathbb{R}^n)$ belongs to the set $\mathfrak{M}_B$.

3.1. Properties of Simple Maximal Symmetric Operators. Recall that any symmetric operator B in a separable Hilbert space $\mathcal{H}$ is called *simple* if it does not induce a self-adjoint operator in any nonzero subspace of the space $\mathcal{H}$.

Throughout this section, we assume that the operator B is a simple maximal symmetric operator whose defect number is equal to zero in the lower half plane. The following theorem describes the structure of operators of this type:

Theorem 3.1. *Let B be a simple maximal symmetric operator with deficiency index $(m, 0)$ $(m \leq \infty)$. Then there exists a unitary operator T which maps the space $\mathcal{H}$ onto $L_2(\mathbb{R}_+, N) := L_2(\mathbb{R}_+) \otimes N$ and satisfies the relations*

$$T\mathcal{D}(B) = \overset{\circ}{W}{}_2^1(\mathbb{R}_+, N) := \overset{\circ}{W}{}_2^1(\mathbb{R}_+) \otimes N \tag{3.1}$$

$$TBu = i\frac{d}{ds}(Tu)(s) \qquad (\forall\, u \in \mathcal{D}(B)), \tag{3.2}$$

where $\mathbb{R}_+ = (0, \infty)$, N is an auxiliary m-dimensional Hilbert space, $\overset{\circ}{W}{}_2^1(\mathbb{R}_+)$ is a Sobolev space, and $\otimes$ denotes the tensor product.

The proof of Theorem 3.1 can be found in Akhiezer and Glazman [1].

Proposition 3.2. *If B is a simple maximal symmetric operator, then $A = B^2$ is a closed densely defined symmetric operator and the following statements are true:*

(i) $A^* = B^*B^*$;

(ii) $\operatorname{Ker}(A^* - z^2 I) = \operatorname{Ker}(B^* + zI) \quad (\operatorname{Im} z > 0)$;

(iii) the lineal $B\,\mathcal{D}(A)$ *is dense in* $\mathcal{H}$.

Proof. By using the "canonical" representation of the operator B from Theorem 3.1, one can easily prove this proposition. However, we prefer to present a direct proof, which does not depend on the realization of the operator B.

Since the defect number of an operator B is equal to zero in the lower half plane, for any λ ($\operatorname{Im}\lambda < 0$), there exists a bounded operator $M = (B - \lambda I)^{-1}$ defined in entire space $\mathcal{H}$. Therefore, one can easily show that the operator

$$A - \lambda^2 I = (B - \lambda I)(B + \lambda I) \tag{3.3}$$

is closed. Hence, the operator A is also closed. Moreover,

$$\mathcal{D}(A) = M\,\mathcal{D}(B), \tag{3.4}$$

whence, by using the equality

$$\operatorname{Ker} M^* = \operatorname{Ker}(B^* - \bar{\lambda} I)^{-1} = \{0\} \tag{3.5}$$

and the fact that the set $\mathcal{D}(B)$ is dense in the space $\mathcal{H}$, we conclude that the operator A is densely defined in $\mathcal{H}$.

Relation (3.3) implies that $M(A - \lambda^2 I) = B + \lambda I$, where M is a bounded operator. Therefore (see, e.g., Berezanskii, Us, and Sheftel [1]),

$$(M(A - \lambda^2 I))^* = \left(A^* - \bar{\lambda}^2 I\right)M^* = B^* + \bar{\lambda} I. \tag{3.6}$$

Since $M^* = \left(B^* - \bar{\lambda} I\right)^{-1}$, we can rewrite equality (3.6) in the form

$$A^* - \bar{\lambda}^2 I = \left(B^* + \bar{\lambda} I\right)\left(B^* - \bar{\lambda} I\right) = B^*B^* - \bar{\lambda}^2 I. \tag{3.7}$$

Therefore, statement (i) is true.

Statement (ii) immediately follows from equalities (3.5) and (3.7) with $z = \bar{\lambda}$.

Let us now prove the relation

$$\operatorname{Ker}(B^* - \mu I) = \{0\} \quad (\forall \mu \in \mathbb{R}). \tag{3.8}$$

Indeed, assume that there exists a real number μ such that

$$B^*x = \mu x \qquad (x \in \mathcal{D}(B^*)) \tag{3.9}$$

Since the operator B is simple, we have $\mu \notin \sigma_p(B)$. Therefore, $x \notin \mathcal{D}(B)$. We consider an extension B' of the operator B defined by the formulas

$$B' = B^*\big|_{\mathcal{D}(B')}, \qquad \mathcal{D}(B') = \mathcal{D}(B) \dotplus \langle x \rangle.$$

In view of relation (3.9), it is easy to see that the operator B' is a symmetric extension of the operator B. Since the operator B is maximal and symmetric, we have $B' = B$. Therefore, $x = 0$ and, hence, relation (3.8) is true. In particular, we have $\operatorname{Ker} B^* = \{0\}$ and, thus, $\overline{\mathcal{R}(B)} = \mathcal{H}$. This relation and equalities (3.4) and (3.5) imply that the set

$$B\,\mathcal{D}(A) = MB\,\mathcal{D}(B) = M\,\mathcal{R}(A)$$

is dense in $\mathcal{H}$.

■

Corollary 3.3. *If B is a simple maximal symmetric operator, then the point spectrum of any regular extension L of the symmetric operator $A = B^2$ has empty intersection with the nonnegative semiaxis.*

Proof. Let L be a regular extension of the symmetric operator $A = B^2$ and let α be a nonnegative number.

In view of Proposition 5.8 (Chapter 1) and statement (i) of Proposition 3.2, we conclude that

$$L \subset A^* = B^* B^*.$$

Hence, for any $u \in \mathcal{D}(L)$, we have

$$(L - \alpha I)u = \left(B^* + \sqrt{\alpha}\, I\right)\left(B^* - \sqrt{\alpha}\, I\right)u.$$

By using relation (3.8) (with $\mu = -\sqrt{\alpha}$ and then with $\mu = \sqrt{\alpha}$), one can easily show that

$$\operatorname{Ker}(L - \alpha I) = \{0\}.$$

■

Denote by $\mathcal{D}[B]$ ($\mathcal{D}[B^*]$) a Hilbert space formed by introducing the graph norm in the lineal $\mathcal{D}(B)$ ($\mathcal{D}(B^*)$), namely,

$$\|u\|_{\text{gr}}^2 = \|u\|^2 + \|Bu\|^2 \quad \left(\|u\|_{\text{gr}}^2 = \|u\|^2 + \left\|B^* u\right\|^2\right). \tag{3.10}$$

Proposition 3.4. *The following assertions are true:*

(i) the lineals $\mathcal{D}(B^*B)$ and $B^*\mathcal{D}(BB^*)$ are dense in $\mathcal{D}[B]$;

(ii) the lineals $\mathcal{D}(BB^*)$ and $B\mathcal{D}(B^*B)$ are dense in $\mathcal{D}[B^*]$.

Proof. Let us prove statement (i) for $B^*\mathcal{D}(BB^*)$. The remaining assertions are proved in exactly the same way.

Assume that an element u is orthogonal to the lineal $B^*\mathcal{D}(BB^*)$ in the space $\mathcal{D}[B]$. By virtue of equality (3.10), we conclude that the relation

$$0 = (u, B^*v)_{\text{gr}} = (u, B^*v) + (Bu, BB^*v) = \left(Bu, (BB^* + I)v\right). \tag{3.11}$$

holds for all $v \in \mathcal{D}(BB^*)$. In view of Corollary 6.2 (Chapter 2), the operator BB^* is self-adjoint. Clearly, this operator is nonnegative. Consequently,

$$\mathcal{R}(BB^* + I) = \mathcal{H}. \tag{3.12}$$

It follows from equalities (3.11) and (3.12) that $Bu = 0$. Since B is a simple operator, we obtain $u = 0$. Thus, the lineal $B^*\mathcal{D}(BB^*)$ is dense in the Hilbert space $\mathcal{D}[B]$.

■

3.2. Nonperturbed Operators. A self-adjoint extension L of a symmetric operator $A = B^2$ is called a *nonperturbed operator* if

$$(Lu, u) = \left\|B^* u\right\|^2. \tag{3.13}$$

for all $u \in \mathcal{D}(L)$.

It follows from the definition that the property of the operator L to be nonperturbed depends on the choice of a simple maximal symmetric operator B. In what follows, it is always clear what kind of an operator is used in every particular case.

In Subsection 3.3, we show that, for a proper choice of the operator B, the classical Laplace operator $-\Delta$, $\mathcal{D}(-\Delta) = W_2^2(\mathbb{R}^n)$ acting in the space $L_2(\mathbb{R}^n)$ is nonperturbed. Moreover, in Section 4, by using a nonperturbed operator, we deduce an abstract wave equation and show that, from the viewpoint of scattering theory (the Lax–Phillips approach), the abstract and classical wave equations have similar properties.

For this reason, nonperturbed operators can be regarded as an abstract analog of the Laplacian operator *(abstract Laplacians)*.

Denote by $\mathfrak{M}_B$ the set of all nonperturbed operators for a fixed operator B.

By using the results obtained in Section 6 (Chapter 2) and statement (i) of Proposition 3.2, we conclude that the self-adjoint operators

$$L_\mu = B^*B \quad \text{and} \quad L_M = BB^*$$

are the hard and soft extensions of the symmetric operator B^2, respectively. Clearly, the operators L_μ and L_M are nonperturbed. Let us show that the set $\mathfrak{M}_B$ is uniquely defined by operators of the same type. For this purpose, we use the canonical positive boundary-value space of the symmetric operator $A = B^2$ (see Subsection 1.1 in Chapter 3).

Clearly, the greatest lower bound $m(A)$ of the operator A is equal to zero.

We fix a self-adjoint extension L_μ of the operator A. By virtue of decomposition (1.5) (Chapter 3), we obtain

$$\mathcal{D}(A^*) = \mathcal{D}(L_\mu) \dotplus X, \tag{3.14}$$

where

$$X = \operatorname{Ker}(A^* + \alpha I) \quad (\alpha > 0). \tag{3.15}$$

By using equalities (1.6) (Chapter 3), we can construct the canonical positive BVS (X, Γ_1, Γ_2) of the operator A associated with the hard extension L_μ as

$$\Gamma_1(u+h) = P_X(L_\mu + \alpha I)u, \quad \Gamma_2(u+h) = h \quad (x = u + h \in \mathcal{D}(A^*)), \tag{3.16}$$

where $u \in \mathcal{D}(L_\mu)$, $h \in X$, and P_X is the orthoprojector onto X in $\mathcal{H}$.

It is known (see V. Gorbachuk and M. Gorbachuk [2]) that any nonnegative self-adjoint extension L of the symmetric operator A coincides with the restriction of the operator $A^* = B^*B^*$ to the lineal

$$\mathcal{D}(L) = \{x = u + h \in \mathcal{D}(A^*) \mid C\Gamma_1 x = \Gamma_2 x\}, \tag{3.17}$$

where

$$C = \left((L + \alpha I)^{-1} - (L_\mu + \alpha I)^{-1}\right)\big|_X \quad (\alpha > 0) \tag{3.18}$$

is a bounded nonnegative operator in the space X and the operators Γ_i are defined by equalities (3.15) and (3.16), respectively.

Lemma 3.5. *A self-adjoint extension L of the operator B^2 is nonperturbed if and only if the operator C defined by equality (3.18) satisfies the condition*

$$X = \operatorname{Ker} C \oplus \operatorname{Ker}\left(C - \frac{\alpha}{2} I\right). \tag{3.19}$$

In particular, the operator $C = 0$ $\left(C = \frac{\alpha}{2} I\right)$ corresponds to the hard (soft) extension of the operator B^2.

Proof. Let $y = v + \tilde{h}$, where $v \in \mathcal{D}(L_\mu)$ and $\tilde{h} \in X$. By using assertion (ii) of Proposition 3.2 (with $z = i\sqrt{\alpha}$) and equalities (3.16) and (3.17), we get

$$(Lx, y) - (B^*x, B^*y) = (G\Gamma_1 x, \Gamma_2 y), \tag{3.20}$$

where $G = C(I - 2\alpha C)$ is a bounded self-adjoint operator in X. Equalities (3.13) and (3.20) imply that the operator L is nonperturbed if and only if

$$\operatorname{Ker} G = X.$$

On the other hand, the relation

$$\operatorname{Ker} G = \operatorname{Ker} C \oplus \operatorname{Ker}\left(C - \frac{\alpha}{2} I\right)$$

follows from the fact that the operator C is self-adjoint. Hence, equality (3.19) is true.

According to the definition of the canonical positive BVS, the operator $C = 0$ corresponds to the extension L_μ.

Similarly, since

$$\mathcal{D}(L_\mu) \cap \mathcal{D}(L_M) = \mathcal{D}(B^2),$$

relation (3.19) implies that equality (3.17) with the operator $C = \frac{\alpha}{2} I$ defines the soft extension L_M.

■

Theorem 3.6. *The operator L belongs to the set $\mathfrak{M}_B$ if and only if there exist mutually orthogonal subspaces $\mathcal{H}_1$ and $\mathcal{H}_2$ of the space $\mathcal{H}$ that are invariant under the action of the operator B and the following statements are true:*

(i)
$$\mathcal{H} = \mathcal{H}_1 \oplus \mathcal{H}_2; \tag{3.21}$$

(ii) $B_i = B|_{\mathcal{D}(B)\cap\mathcal{H}_i}$ $(i = 1, 2)$ *are simple maximal symmetric operators in* $\mathcal{H}_i$;

(iii) with respect to decomposition (3.21), the operator L has the form

$$L = L_{1\mu} \oplus L_{2M}, \tag{3.22}$$

where $L_{1\mu} = B_1^*B_1$ is the hard extension of the symmetric operator B_1^2 in the Hilbert space $\mathcal{H}_1$, and $L_{2M} = B_2B_2^*$ is the soft extension of the operator B_2^2 in $\mathcal{H}_2$.

Proof. Clearly, if a self-adjoint operator L satisfies conditions (i)–(iii), then this operator is nonperturbed.

Therefore, to complete the proof, it is necessary to show that, for any nonperturbed operator L $(\in \mathfrak{M}_B)$, there exists decomposition (3.21) with properties (ii) and (iii).

In view of Theorem 3.1, it suffices to prove this assertion in the case where

$$B = i\frac{d}{ds}, \qquad \mathcal{D}(B) = \mathring{W}_2^1(\mathbb{R}_+, N)$$

is a simple maximal symmetric operator in the space $\mathcal{H} = L_2(\mathbb{R}_+, N)$. In this case,

$$B^* = i\frac{d}{ds}, \qquad \mathcal{D}(B^*) = W_2^1(\mathbb{R}_+, N),$$

and

$$X = \operatorname{Ker}(B^*B^* + I) = \{ne^{-s} \mid \forall\, n \in N\}. \tag{3.23}$$

Equalities (3.16) (with $\alpha = 1$) imply that the triple (X, Γ_1, Γ_2), where

$$\Gamma_1 x(s) = 2(x'(0) + x(0))e^{-s}, \qquad \Gamma_2 x(s) = x(0)e^{-s} \tag{3.24}$$

is a canonical positive BVS of the symmetric operator B^2 constructed by the hard extension

$$L_\mu = -\frac{d^2}{ds^2}, \quad \mathcal{D}(L_\mu) = \left\{u(s) \in W_2^2(\mathbb{R}_+, N) \,\middle|\, u(0) = 0\right\}.$$

According to Lemma 3.5, any operator $L \in \mathfrak{M}_B$ in this BVS is determined by a self-adjoint operator C that satisfies condition (3.19). By using relation (3.23), we obtain

$$\operatorname{Ker} C = \left\{n_1 e^{-s} \,\middle|\, \forall\, n_1 \in N_1\right\} \quad \text{and} \quad \operatorname{Ker}\left(C - \frac{1}{2}I\right) = \left\{n_2 e^{-s} \,\middle|\, \forall\, n_2 \in N_2\right\}, \tag{3.25}$$

where N_i are subspaces of the space N such that $N = N_1 \oplus N_2$. The last decomposition implies that

$$\mathcal{H} = \mathcal{H}_1 \oplus \mathcal{H}_2,$$

where $\mathcal{H}_i = L_2(\mathbb{R}_+, N_i)$.

Obviously, these subspaces reduce the operator B, and the operators

$$B_i = B|_{\mathcal{D}(B) \cap \mathcal{H}_i} = B|_{\mathring{W}_2^1(\mathbb{R}_+, N_i)} \quad (i = 1, 2)$$

are simple maximal symmetric operators in $\mathcal{H}_i$.

Assume that $x(s) \in \mathcal{D}(L_{1\mu})$, where $L_{1\mu} = B_1^* B_1$. Then

$$x(s) \in W_2^2(\mathbb{R}_+, N_1) \subset W_2^2(\mathbb{R}_+, N) \quad \text{and} \quad x(0) = 0.$$

Therefore, by virtue of equalities (3.24) and (3.25), we get

$$\Gamma_1 x(s) \in \operatorname{Ker} C \quad \text{and} \quad \Gamma_2 x(s) = 0.$$

Combining this with (3.17), we obtain

$$x(s) \in \mathcal{D}(L) \quad \text{and} \quad Lx = L_{1\mu} x.$$

The fact that $L_{2M} = L|_{\mathcal{D}(L_{2M})}$ can be proved by analogy.

Thus, taking into account that the operators $L_{1\mu}$ and L_{2M} are self-adjoint in the spaces $\mathcal{H}_1$ and $\mathcal{H}_2$, respectively, we conclude that decomposition (3.22) is true. Theorem 3.6 is proved.

■

The statements below are direct consequences of Theorem 3.6.

Corollary 3.7. *The spectrum $\sigma(L)$ of a nonperturbed operator L ($\in \mathfrak{M}_B$) has the following properties:*

(i)
$$\sigma(L) = [0, \infty);$$

(ii) $\sigma(L)$ is absolutely continuous with multiplicity m, where m is the nonzero defect number of the operator B.

Proof. This statement follows from Theorems 3.1 and 3.6 and the well-known assertion (see Akhiezer and Glazman [1]) that the operator

$$L_\mu(L_M) = -\frac{d^2}{ds^2}, \quad \mathcal{D}(L_\mu) = \{u(s) \in W_2^2(\mathbb{R}_+) \,|\, u(0) = 0\} \tag{3.26}$$

$\left(\mathcal{D}(L_M) = \{u(s) \in W_2^2(\mathbb{R}_+) \,|\, u'(0) = 0\}\right)$, which acts in $L_2(\mathbb{R}_+)$, has the simple and absolutely continuous spectrum, which coincides with $\mathbb{R}_+$.

■

Corollary 3.8. *If operators L and $\tilde{L}$ belong to $\mathfrak{M}_B$, then the wave operators*

$$\Omega_\pm(\tilde{L}, L) = s - \lim_{t \to \pm\infty} e^{-i\tilde{L}t} e^{iLt}$$

exist and are complete.

Proof. According to Theorem 3.1, it suffices to consider the case where the nonperturbed operators L and $\tilde{L}$ are extensions of the symmetric operator

$$B^2 = -\frac{d^2}{ds^2}, \quad \mathcal{D}(B) = \mathring{W}_2^2(\mathbb{R}_+, N),$$

which acts in the Hilbert space $\mathcal{H} = L_2(\mathbb{R}_+, N)$.

Since the operator L is nonperturbed, it follows from the proof of Theorem 3.6 that the space $L_2(\mathbb{R}_+, N)$ admits decomposition (3.21), where $\mathcal{H}_i = L_2(\mathbb{R}_+, N_i)$. Moreover, the operator L can be represented in the form (3.22).

Let $\{\psi_j\}_{j=1}^{\infty}$ be an orthonormal basis in the space N_2. Then decomposition (3.21) can be rewritten as follows:

$$L_2(\mathbb{R}_+, N) = \mathcal{H}_1 \oplus \sum_{j=1}^{\infty} \oplus L_2(\mathbb{R}_+, \langle\psi_j\rangle), \tag{3.27}$$

where

$$L_2(\mathbb{R}_+,\langle\psi_j\rangle) := L_2(\mathbb{R}_+)\otimes\langle\psi_j\rangle.$$

Assume that $\tilde{L} = L_\mu$. Then equality (3.22) implies that, with respect to (3.27), the wave operators $\Omega_\pm(L_\mu, L)$ have the following form:

$$\Omega_\pm(L_\mu, L) = I|_{\mathcal{H}_1} \oplus \sum_{j=1}^{\infty} \oplus\,\Omega_\pm(L^j_\mu, L^j), \tag{3.28}$$

where $\Omega_\pm(L^j_\mu, L^j)$ are wave operators in the space $L_2(\mathbb{R}_+,\langle\psi_j\rangle)$, and the operators L^j_μ and L^j are, respectively, the hard and soft extensions of the symmetric operator

$$B_j^2 = -\frac{d^2}{ds^2}, \quad \mathcal{D}(B_j^2) = \overset{\circ}{W}{}_2^2(\mathbb{R}_+,\langle\psi_j\rangle).$$

Since the deficiency index of the operator B_j^2 in the space $L_2(\mathbb{R}_+,\langle\psi_j\rangle)$ is $(1, 1)$, the wave operators $\Omega_\pm(L^j_\mu, L^j)$ exist and are complete in the space $L_2(\mathbb{R}_+,\langle\psi_j\rangle)$. Hence, by virtue of (3.28), we conclude that the wave operators $\Omega_\pm(L_\mu, L)$ exist and are complete in the space $L_2(\mathbb{R}_+, N)$.

Analogously, we establish a similar statement in the case where $L = L_\mu$ and $\tilde{L}$ is an arbitrary nonperturbed operator. Therefore, as follows from the "chain law" (see, e.g., Reed and Simon [2]), the wave operators $\Omega_\pm(\tilde{L}, L)$ exist and are complete.

■

Corollary 3.9. *Arbitrary operators from the set* $\mathfrak{M}_B$ *are unitary equivalent to each other.*

Proof. The proof follows directly from Corollary 3.8.

■

3.3. Examples of Nonperturbed Operators. As noted above, the property of an operator L to be nonperturbed depends on the choice of a simple maximal symmetric operator B.

In particular, if we consider the simple maximal symmetric operator

$$B = i\frac{d}{ds}, \quad \mathcal{D}(B) = \overset{\circ}{W}{}_2^1(\mathbb{R}_+),$$

that acts in the space $\mathcal{H} = L_2(\mathbb{R}_+)$, then, as follows from equality (3.13), the set $\mathfrak{M}_B$ consists of the operators L_μ and L_M defined by relations (3.26).

Let $\mathcal{H} = L_2(\mathbb{R})$. In this space,

$$B = \begin{cases} i\dfrac{d}{ds}, & s > 0, \\ -i\dfrac{d}{ds}, & s < 0, \end{cases} \tag{3.29}$$

$$\mathcal{D}(B) = \mathring{W}_2^1(\mathbb{R}_-) \oplus \mathring{W}_2^1(\mathbb{R}_+),$$

is a simple maximal symmetric operator. The soft L_μ and hard L_M extensions of the operator B^2 are the restrictions of the operator

$$A^* = B^*B^* = -\frac{d^2}{ds^2}, \quad \mathcal{D}(A^*) = \mathring{W}_2^2(\mathbb{R}_-) \oplus \mathring{W}_2^2(\mathbb{R}_+),$$

to the sets

$$\mathcal{D}(L_\mu) = \left\{u(s) \in \mathcal{D}(A^*) \,\middle|\, u(0-) = u(0+) = 0\right\}$$

and

$$\mathcal{D}(L_M) = \left\{u(s) \in \mathcal{D}(A^*) \,\middle|\, u'(0-) = u'(0+) = 0\right\},$$

respectively. Substituting an arbitrary self-adjoint extension of the operator B^2 in relation (3.13), we establish that the set $\mathfrak{M}_B$ consists of the operators L_μ, L_M, and the collection of the operators

$$L_\theta = A^*\big|_{\mathcal{D}(L_\theta)}, \quad \mathcal{D}(L_\theta) = \left\{u(s) \in \mathcal{D}(A^*) \,\middle|\, \begin{matrix} u(0-) = \overline{\theta} u(0+) \\ \theta u'(0-) = u'(0+) \end{matrix}\right\}, \tag{3.30}$$

where $\theta \in \overline{\mathbb{C}}$. In particular, for $\theta = 1$, we conclude that the operator

$$L_1 = -\frac{d^2}{ds^2}, \quad \mathcal{D}(L_1) = W_2^2(\mathbb{R}),$$

is nonperturbed.

Further, we show that the operator Δ, $\mathcal{D}(-\Delta) = W_2^2(\mathbb{R}^n)$, is nonperturbed in the space $\mathcal{H} = L_2(\mathbb{R}^n)$. We restrict our discussion to the case where n is odd. The investigation of a more complicate case where n is even can be found in S. Kuzhel [12].

To solve this problem in the case where n is odd, we use the Radon transform[14] R defined by the formula

$$(Ru)(s,\omega) = \tilde{u}(s,\omega) = (2\pi)^{-m} \int\limits_{x\cdot\omega = s} u(x)\,dS_x \quad \left(m = \frac{n-1}{2}\right), \tag{3.31}$$

where $u(x)$ belongs to the space of rapidly decreasing functions $S(\mathbb{R}^n)$, $s \in \mathbb{R}$, ω is a unit vector in $\mathbb{R}^n$, and dS_x is the Euclidean measure in the hyperplane $x \cdot \omega = s$.

One can easily verify that

$$R(\Delta u) = \partial_s^2 Ru \quad (u \in S(\mathbb{R}^n)). \tag{3.32}$$

The function $\tilde{u}(s,\omega)$ in (3.31) can be regarded as a function of s with values in the auxiliary space $N = L_2(S^{n-1})$, where S^{n-1} is the unit sphere in $\mathbb{R}^n$. For this function, the following Plancherel-type formula is true:

$$\|u(x)\|_{L_2(\mathbb{R}^n)} = \frac{1}{\sqrt{2}}\left\|\partial_s^m Ru\right\|_{L_2(\mathbb{R},N)}. \tag{3.33}$$

Following Lax and Phillips [2] and using equality (3.33), we conclude that the lineal $\partial_s^m RS(\mathbb{R}^n)$ is dense in the subspace

$$\left\{y(s,\omega) \in L_2(\mathbb{R},N) \,\middle|\, y(-s,-\omega) = (-1)^m y(s,\omega)\right\} \tag{3.34}$$

of the space $L_2(\mathbb{R},N)$. Therefore, the closure of the operator T defined by the relation

$$(Tu)(s,\omega) = \left(\partial_s^m Ru\right)(s,\omega), \quad u(x) \in S(\mathbb{R}^n), \quad s \geq 0, \tag{3.35}$$

isometrically maps the space $L_2(\mathbb{R}^n)$ into the entire space $L_2(\mathbb{R}_+, N)$.

It follows from equalities (3.32) and (3.34) that the self-adjoint operator

$$\tilde{L} = -T\Delta T^{-1} = -\frac{d^2}{ds^2}$$

---

[14] Detailed information on the Radon transformation is given in Helgason [1]

acts in the Hilbert space $L_2(\mathbb{R}_+, N)$ and has the domain of definition

$$\mathcal{D}(\tilde{L}) = \left\{ u(s,\omega) \in \overset{\circ}{W}{}_2^2(\mathbb{R}_+, N) \,\middle|\, \begin{matrix} u(0,-\omega) = (-1)^m u(0,\omega) \\ \partial_s u(0,-\omega) = (-1)^{m+1} \partial_s u(0,\omega) \end{matrix} \right\}. \tag{3.36}$$

Moreover, it is clear that

$$(-\Delta u, u)_{L_2(\mathbb{R}^n)} = \left(\tilde{L}Tu, Tu\right)_{L_2(\mathbb{R}_+, N)} = \left\| \frac{d}{ds} Tu \right\|^2. \tag{3.37}$$

Since

$$\tilde{B} = i\frac{d}{ds}, \qquad \mathcal{D}(\tilde{B}) = \overset{\circ}{W}{}_2^1(\mathbb{R}_+, N), \qquad \text{and} \qquad B = T^{-1}\tilde{B}T \tag{3.38}$$

are simple maximal symmetric operators in $L_2(\mathbb{R}_+, N)$ and $L_2(\mathbb{R}^n)$, respectively, equalities (3.36) and (3.37) imply that the operator $\tilde{L}$ belongs to the set $\mathfrak{M}_{\tilde{B}}$. Therefore, $-\Delta \in \mathfrak{M}_B$ and, hence, the operator $-\Delta$ is a nonperturbed extension of the operator B^2.

In the space $N = L_2(S^{n-1})$, we consider the orthogonal subspaces

$$N_+ = \{n_+(\omega) \in N \mid n_+(-\omega) = n_+(\omega)\}$$

and

$$N_- = \{n_-(\omega) \in N \mid n_-(-\omega) = -n_-(\omega)\}$$

of odd and even functions, respectively.

Let

$$\mathcal{H}_1 = L_2(\mathbb{R}_+, N_+) \qquad \text{and} \qquad \mathcal{H}_2 = L_2(\mathbb{R}_+, N_-).$$

Clearly, $L_2(\mathbb{R}_+, N) = \mathcal{H}_1 \oplus \mathcal{H}_2$. Taking equality (3.36) into account, we conclude that, with respect to this decomposition, the operator $\tilde{L}$ can be rewritten in the form

$$\tilde{L} = \tilde{B}_1^* \tilde{B}_1 \oplus \tilde{B}_2 \tilde{B}_2^*, \tag{3.39}$$

where

$$\tilde{B}_j = i\frac{d}{ds}, \quad \mathcal{D}(\tilde{B}_j) = \mathring{W}_2^1(\mathbb{R}_+, N_j) \quad (j = 1, 2),$$

is a simple maximal symmetric operator in the space $\mathcal{H}_j$.

Equalities (3.35) and (3.38) and decomposition (3.39) enable us to represent the operator $-\Delta$ in the form

$$-\Delta = B_1^* B_1 \oplus B_2 B_2^*, \quad B_j = T^{-1}\tilde{B}_j T, \tag{3.40}$$

with respect to the decomposition

$$L_2(\mathbb{R}^n) = L_2^{(1)}(\mathbb{R}^n) \oplus L_2^{(2)}(\mathbb{R}^n), \tag{3.41}$$

where

$$L_2^{(i)}(\mathbb{R}^n) = \left\{u(x) \in L_2(\mathbb{R}^n) \,\middle|\, u(x) = (-1)^{m+i} u(-x)\right\}.$$

Let $n = 3$. In this case, the operator $B_1^* B_1$ acts in the space $L_2^{\text{even}}(\mathbb{R}^3)$ of even functions in $\mathbb{R}^3$. This operator is the hard extension of the symmetric operator

$$B_1^2 = -\Delta|_{\mathcal{D}(B_1^2)}.$$

By using the connection between the Fourier and Radon transforms (see Theorems 4.1 and 4.5 in Stein and Weiss [1, Chapter 4]), we get

$$\mathcal{D}(B_1^2) = \left\{u(x) \in W_2^{2(\text{even})}(\mathbb{R}^3) \,\middle|\, u(0) = 0, \ \int_{\mathbb{R}^3} \frac{u(x)}{|x|^3} Y_k\left(\frac{x}{|x|}\right) dx = 0\right\},$$

where $Y_k(\omega)$ is an arbitrary spherical harmonic of order k (k is an arbitrary even number) and the integral is understood in the sense of principal value.

Similarly, the operator $B_2 B_2^*$ acts in the space $L_2^{\text{odd}}(\mathbb{R}^3)$ of odd functions in $\mathbb{R}^3$. This operator is the soft extension of the operator

$$B_2^2 = -\Delta|_{\mathcal{D}(B_2^2)}, \quad \mathcal{D}(B_2^2) = \left\{u(x) \in W_2^{2(\text{odd})}(\mathbb{R}^3) \,\middle|\, \int_{\mathbb{R}^3} \frac{u(x)}{|x|^2} Y_k\left(\frac{x}{|x|}\right) dx = 0\right\},$$

where $Y_k(\omega)$ is an arbitrary spherical harmonic of order k (k is an arbitrary odd number).

Remark. The principal results of this section were obtained by S. Kuzhel [6].

4. Abstract Wave Equation

In this section, we study the properties of an abstract wave equation $u_{tt} = -Lu$ (L is a nonperturbed operator) that are associated with the Lax–Phillips approach in scattering theory. In particular, we describe the structure of incoming and outgoing subspaces and construct a translation representation for the group of solutions of the abstract wave equation.

4.1. Definitions and Preliminary Remarks. Consider an evolving system described by the operator-differential equation

$$u_{tt} = -Lu, \tag{4.1}$$

where L is a nonperturbed operator acting in a separable Hilbert space $\mathcal{H}$. In what follows, we show that this equation and the classical wave equation in a free odd-dimensional space possess similar properties related to the Lax–Phillips approach in scattering theory.

In this connection, the operator-differential equation (4.1) with a nonperturbed operator L on the right-hand side can naturally be called an *abstract wave equation.*

Let L ($\in \mathfrak{M}_B$) be a nonperturbed operator. Since B is a simple maximal symmetric operator, by virtue of relation (3.8), we have $\operatorname{Ker} B^* = \{0\}$. Therefore, the equality

$$\|u\|_L = \sqrt{(Lu, u)} = \left\|B^* u\right\| \qquad (u \in \mathcal{D}(L))$$

defines a norm in the lineal $\mathcal{D}(L)$.

Denote by $\mathcal{H}_L$ the completion of $\mathcal{D}(L)$ in the norm $\|\cdot\|_L$. We now describe the elements of the space $\mathcal{H}_L$ in a form convenient for our discussion.

It follows from (4.2) that the sequence $\{u_n\}_{n=1}^{\infty}$ ($u_n \in \mathcal{D}(L)$) is fundamental in $\mathcal{H}_L$ if and only if the sequence $\left\{B^* u_n\right\}_{n=1}^{\infty}$ is fundamental in $\mathcal{H}$.

We say that the sequence $\{u_n\}_{n=1}^{\infty}$ converges to an element x_γ in the space $\mathcal{H}_L$ if the sequence $\left\{B^* u_n\right\}_{n=1}^{\infty}$ converges to the element γ in $\mathcal{H}$.

Assertion (iii) of Proposition 3.2 implies that the space $\mathcal{H}_L$ *can be identified with the Hilbert space*

$$\{x_\gamma \mid \forall\, \gamma \in \mathcal{H}\},$$

where

$$x_{\alpha\gamma+\beta\tilde{\gamma}} = \alpha x_\gamma + \beta x_{\tilde{\gamma}}, \quad \{\alpha, \beta\} \subset \mathbb{C}, \quad \{\gamma, \tilde{\gamma}\} \subset \mathcal{H},$$

and

$$\left(x_\gamma, x_{\tilde{\gamma}}\right)_L = (\gamma, \tilde{\gamma}). \tag{4.3}$$

It is clear that, according to this rule of identification, any element u of the lineal $\mathcal{D}(L)$ can be represented in the form x_{B^*u}. Thus,

$$\mathcal{H}_L \equiv \{x_\gamma | \forall \gamma \in \mathcal{H}\} \quad \text{and} \quad \mathcal{D}(L) \equiv \{x_{B^*u} | \forall u \in \mathcal{D}(L)\}.$$

The Hilbert space $\mathfrak{H}_L = \mathcal{H}_L \oplus \mathcal{H}$ is called the space of *initial data* or the space of *Cauchy data.* For what follows, it is convenient to write the elements of the space $\mathfrak{H}_L$ as a column vector $\begin{pmatrix} \cdot \\ \cdot \end{pmatrix}$, where the upper and lower components belong to the subspaces $\mathcal{H}_L \oplus 0$ and $0 \oplus \mathcal{H}$, respectively.

Denote $u_t = v$ and rewrite equation (4.1) in the form

$$\frac{d}{dt}\begin{pmatrix} u \\ v \end{pmatrix} = \begin{pmatrix} 0 & I \\ -L & 0 \end{pmatrix}\begin{pmatrix} u \\ v \end{pmatrix}.$$

It is easy to verify (see, e.g., Lax and Phillips [3]) that the closure $Q_L = \overline{Q}$ of the matrix operator

$$Q = \begin{pmatrix} 0 & I \\ -L & 0 \end{pmatrix}, \quad \mathcal{D}(Q) = \left\{\begin{pmatrix} u \\ v \end{pmatrix} \middle| \{u, v\} \subset \mathcal{D}(L)\right\}, \tag{4.4}$$

is a skew self-adjoint operator in the space $\mathfrak{H}_L$. Hence, Q_L is the infinitesimal generator of the unitary (in $\mathfrak{H}_L$) group

$$W_L(t) = e^{Q_L t}$$

of solutions of the Cauchy problem for equation (4.1).

In what follows, if L is the hard $L_\mu = B^*B$ or soft $L_M = BB^*$ extension of the symmetric operator B^2, we denote the generator Q_L by Q_μ or Q_M, respectively. According to Theorem 3.6, for a nonperturbed operator L, there exists decomposition (3.21) of the space $\mathcal{H}$ with respect to which the operator L has decomposition (3.22). Therefore,

$$\mathfrak{H}_L = \mathfrak{H}_{L_{1\mu}} \oplus \mathfrak{H}_{L_{2M}}, \tag{4.5}$$

where $\mathfrak{H}_{L_{1\mu}} = \mathcal{H}_{L_{1\mu}} \oplus \mathcal{H}_1$ and $\mathfrak{H}_{L_{2M}} = \mathcal{H}_{L_{2M}} \oplus \mathcal{H}_2$. With respect to decomposition (4.5), the operator Q_L can be represented in the form

$$Q_L = Q_{1\mu} \oplus Q_{2M}. \tag{4.6}$$

Decompositions (4.5) and (4.6) imply that the investigation of the properties of the operator Q_L can be reduced to a similar problem for the operators Q_μ and Q_M.

Proposition 4.1. *If $L = L_\mu$, then*

$$Q_\mu \begin{pmatrix} x_\gamma \\ p \end{pmatrix} = \begin{pmatrix} x_{Bp} \\ -B^*\gamma \end{pmatrix}, \quad \mathcal{D}(Q_\mu) = \left\{ \begin{pmatrix} x_\gamma \\ p \end{pmatrix} \middle| \gamma \in \mathcal{D}(B^*), p \in \mathcal{D}(B) \right\}.$$

On the other hand, if $L = L_M$, then

$$Q_M \begin{pmatrix} x_\gamma \\ p \end{pmatrix} = \begin{pmatrix} x_{B^*p} \\ -B\gamma \end{pmatrix}, \quad \mathcal{D}(Q_M) = \left\{ \begin{pmatrix} x_\gamma \\ p \end{pmatrix} \middle| \gamma \in \mathcal{D}(B), p \in \mathcal{D}(B^*) \right\}.$$

Proof. Let $L = L_\mu$. Since the closure of the operator Q coincides with Q_μ, for any element $\begin{pmatrix} x_\gamma \\ p \end{pmatrix} \in \mathcal{D}(Q_\mu)$, there exists a sequence

$$\left\{ \begin{pmatrix} u_n \\ v_n \end{pmatrix} \right\}_{n=1}^{\infty} \quad (\{u_n, v_n\} \subset \mathcal{D}(L_\mu))$$

with the following properties:

$$\lim_{n\to\infty} \begin{pmatrix} u_n \\ v_n \end{pmatrix} = \begin{pmatrix} x_\gamma \\ p \end{pmatrix},$$

$$\lim_{n\to\infty} Q_\mu \begin{pmatrix} u_n \\ v_n \end{pmatrix} = Q_\mu \begin{pmatrix} x_\gamma \\ p \end{pmatrix} = \begin{pmatrix} x_y \\ g \end{pmatrix} \quad (\{\gamma, p, y, g\} \subset \mathcal{H}). \tag{4.7}$$

By virtue of equalities (4.3) and (4.4) and the identification of the space $\mathcal{H}_L$ with the set of elements x_γ, we conclude that equalities (4.7) are equivalent to the following relations:

$$\lim_{n\to\infty} Bu_n = \gamma \quad \text{and} \quad \lim_{n\to\infty} B^* Bu_n = -g,$$
$$\lim_{n\to\infty} v_n = p \quad \text{and} \quad \lim_{n\to\infty} Bv_n = y. \tag{4.8}$$

Since the operator B is closed, equalities (4.8) imply that $y \in \mathcal{D}(B^*)$, $p \in \mathcal{D}(B)$, and

$$Q_\mu \begin{pmatrix} x_\gamma \\ p \end{pmatrix} = \begin{pmatrix} x_{Bp} \\ -B^*\gamma \end{pmatrix}. \tag{4.9}$$

Thus,

$$\mathcal{M} = \left\{ \begin{pmatrix} x_\gamma \\ p \end{pmatrix} \middle| \gamma \in \mathcal{D}(B^*), p \in \mathcal{D}(B) \right\} \subset \mathcal{D}(Q_\mu)$$

and the action of the operator Q_μ in the lineal $\mathcal{M}$ is defined by (4.9).

Let us show that $\mathcal{M} = \mathcal{D}(Q_\mu)$. Indeed, as follows from Proposition 3.4, for any $\gamma \in \mathcal{D}(B^*)$ and $p \in \mathcal{D}(B)$, there exist sequences $\{u_n\}_{n=1}^\infty$ and $\{v_n\}_{n=1}^\infty$ of elements from $\mathcal{D}(L_\mu)$ for which equalities (4.8) are true. Hence, $\begin{pmatrix} x_\gamma \\ p \end{pmatrix} \in \mathcal{D}(Q_\mu)$. Thus, $\mathcal{M} = \mathcal{D}(Q_\mu)$. The case $L = L_M$ can be considered analogously.

■

4.2. Outgoing and Incoming Subspaces. Recall that a subspace D_+ of a Hilbert space $\mathfrak{H}$ is called *outgoing* with respect to a unitary group $W(t)$ if the following conditions are satisfied:

(i) $$W(t)D_+ \subset D_+ \quad (t \geq 0);$$

(ii) $$\bigcap_{t\geq 0} W(t)D_+ = \{0\}.$$

A subspace D_- is called *incoming* for the group $W(t)$ if D_- is an outgoing subspace for the group $W^*(t)$.

Let an operator L ($\in \mathfrak{M}_B$) be nonperturbed. Consider the subspaces

$$D_- = \left\{ \begin{pmatrix} x_\gamma \\ -i\gamma \end{pmatrix} \middle| \gamma \in \mathcal{H} \right\} \quad \text{and} \quad D_+ = \left\{ \begin{pmatrix} x_\gamma \\ i\gamma \end{pmatrix} \middle| \gamma \in \mathcal{H} \right\} \tag{4.10}$$

in the space of initial data $\mathfrak{H}_L$. It follows from the definition of $\mathfrak{H}_L$ and equality (4.3) that these subspaces are orthogonal and

$$\mathfrak{H}_L = D_- \oplus D_+ . \tag{4.11}$$

Theorem 4.2. *The subspaces D_- and D_+ are incoming and outgoing with respect to the group of solutions $W_L(t)$ of the abstract wave equation. Moreover,*

$$\overline{\bigcup_{t\in\mathbb{R}} W_L(t)D_-} = \overline{\bigcup_{t\in\mathbb{R}} W_L(t)D_+} = \mathfrak{H}_L \tag{4.12}$$

and, for any $\gamma \in \mathcal{H}$, the following equalities are true:

$$W_L^*(t)\begin{pmatrix} x_\gamma \\ -i\gamma \end{pmatrix} = \begin{pmatrix} x_{V(t)\gamma} \\ -iV(t)\gamma \end{pmatrix} \quad \text{and} \quad W_L(t)\begin{pmatrix} x_\gamma \\ i\gamma \end{pmatrix} = \begin{pmatrix} x_{V(t)\gamma} \\ iV(t)\gamma \end{pmatrix}, \tag{4.13}$$

where $V(t) = e^{iBt}$ is an isometric semigroup in $\mathcal{H}$ with the generator iB.

Proof. It follows from Proposition 4.1 and equalities (4.6) and (4.10) that

$$\mathcal{D}_+ = \mathcal{D}(Q_L) \cap D_+ = \left\{ \begin{pmatrix} x_\gamma \\ i\gamma \end{pmatrix} \middle| \gamma \in \mathcal{D}(B) \right\}$$

and

$$Q_L\begin{pmatrix} x_\gamma \\ i\gamma \end{pmatrix} = \begin{pmatrix} x_{iB\gamma} \\ -B\gamma \end{pmatrix} \quad \left(\forall \begin{pmatrix} x_\gamma \\ i\gamma \end{pmatrix} \in \mathcal{D}_+ \right).$$

By using these relations, one can easily verify that the operator $Q_+ = Q_L|_{\mathcal{D}_+}$ is a maximal skew-symmetric operator in the Hilbert space D_+. Hence, this operator is the infinitesimal generator of the isometric semigroup $U(t) = e^{Q_+ t}$.

By using Cook's recipe, we obtain

$$\left(W^*(t)U(t) - I\right)d_+$$

$$= \int_0^t (W^*(s)U(s)d_+)' ds = \int_0^t W^*(s)(Q_+ - Q)U(s)d_+ ds = 0 \tag{4.14}$$

for any $d_+ \in \mathcal{D}_+ = \mathcal{D}(Q_+)$. This proves the equality $U(t)d_+ = W(t)d_+$ for a dense subset $\mathcal{D}_+$ in D_+. Therefore,

$$U(t) = W(t)|_{D_+}. \tag{4.15}$$

This means that relation (i) in the definition of outgoing subspaces is true for the subspace D_+.

Denote by $V(t) = e^{iBt}$ an isometric semigroup of operators in $\mathcal{H}$ with the infinitesimal generator iB. In the subspace D_+, we consider the semigroup $\tilde{U}(t)$ defined by the relation

$$\tilde{U}(t)\begin{pmatrix} x_\gamma \\ i\gamma \end{pmatrix} = \begin{pmatrix} x_{V(t)\gamma} \\ iV(t)\gamma \end{pmatrix}. \tag{4.16}$$

It is easy to see that the semigroup $\tilde{U}(t)$ has the infinitesimal generator Q_+. Therefore, $\tilde{U}(t) = U(t)$ and by virtue of relations (4.15) and (4.16), we conclude that the second relation in (4.13) is true.

Assume that

$$d = \begin{pmatrix} x_\gamma \\ i\gamma \end{pmatrix} \in \bigcap_{t \geq 0} W(t)D_+.$$

Then it follows from (4.10) and the second relation in (4.13) that

$$\gamma \in \bigcap_{t \geq 0} V(t)\mathcal{H}. \tag{4.17}$$

Since the operator B is simple, the semigroup $V(t)$ is completely nonunitary. But then relation (4.17) yields $\gamma = 0$ and, consequently, relation (ii) holds for the subspace D_+.

Similarly, we establish that the action of the semigroup $W_L^*(t)$ in D_- is determined by the first relation in (4.13). Hence, the subspace D_- satisfies conditions (i) and (ii).

Assume that

$$g \perp \bigcup_{t \in \mathbb{R}} W_L(t)D_+.$$

Then, by virtue of equality (4.11), we get $g \in W_L^*(t)D_-$ $(t \in \mathbb{R})$. By using condition (ii) for the group $W_L^*(t)$, we conclude that $g = 0$. Thus, equality (4.12) holds for the subspace D_+.

The case of the subspace D_- can be considered by analogy.

∎

4.3. Translation Representation. Let L be a nonperturbed operator and let the subspaces D_- and D_+ be defined by equalities (4.10). Below, for the group of solutions $W_L(t)$ of the abstract wave equation (4.1), we construct a translation representation associated with these subspaces.

Let T be a unitary mapping of $\mathcal{H}$ onto $L_2(\mathbb{R}_+, N)$ for which relations (3.1) and (3.2) hold.

First, we consider the case $L = L_\mu$. For any element $d = \begin{pmatrix} x_\gamma \\ p \end{pmatrix} \in \mathfrak{H}_L$, we introduce

$$T_\mu d = \begin{cases} \dfrac{1}{\sqrt{2}}(T(i\gamma + p))(s) & (s > 0), \\ \dfrac{1}{\sqrt{2}}(T(i\gamma - p))(-s) & (s < 0). \end{cases} \tag{4.18}$$

In particular, for $d_+ = \begin{pmatrix} x_\gamma \\ i\gamma \end{pmatrix}$ and $d_- = \begin{pmatrix} x_\gamma \\ -i\gamma \end{pmatrix}$, we obtain

$$T_\mu d_+ = \begin{cases} i\sqrt{2}\,(T\gamma)(s) & (s > 0), \\ 0 & (s < 0), \end{cases}$$

and

$$T_\mu d_- = \begin{cases} 0 & (s > 0), \\ i\sqrt{2}(T\gamma)(-s) & (s < 0), \end{cases}$$

respectively. Therefore,

$$\left\| T_\mu d_\pm \right\|^2_{L_2(\mathbb{R}, N)} = 2\int_0^\infty \|(T\gamma)(s)\|_N^2 ds = 2\|\gamma\|^2 = \left\| d_\pm \right\|^2_{\mathfrak{H}_{L_\mu}}. \tag{4.19}$$

Combining equalities (4.19) and (4.11), we establish that the operator T_μ is a unitary mapping of $\mathfrak{H}_{L_\mu}$ onto $L_2(\mathbb{R}, N)$ and, moreover,

$$T_\mu D_+ = L_2(\mathbb{R}_+, N) \quad \text{and} \quad T_\mu D_- = L_2(\mathbb{R}_-, N).$$

Assume that $d \in \mathcal{D}(Q_\mu)$. By virtue of Proposition 4.1 and equalities (3.1), (3.2), and (4.18), we get

$$T_\mu \mathcal{D}(Q_\mu) = W_2^1(\mathbb{R}, N) \quad \text{and} \quad T_\mu Q_\mu d = -\frac{d}{ds}(T_\mu d)(s). \tag{4.20}$$

It is clear that the operator

$$-\frac{d}{ds}, \quad \mathcal{D}\left(-\frac{d}{ds}\right) = W_2^1(\mathbb{R}, N),$$

is a generator of the unitary group $\mathcal{F}(t)$ in $L_2(\mathbb{R}, N)$ that acts as the translation to the right by t units. Thus, equalities (4.20) imply that

$$T_\mu W_{L_\mu}(t) = \mathcal{F}(t)T_\mu .$$

This means that the operator T_μ determines a translation representation of the unitary group $W_{L_\mu}(t)$ which is both incoming and outgoing.

Similarly, we establish that, in the case $L = L_M$, the operator

$$T_M\begin{pmatrix} x_\gamma \\ p \end{pmatrix} = \begin{cases} \frac{1}{\sqrt{2}}(T(i\gamma + p))(s) & (s > 0), \\ \frac{1}{\sqrt{2}}(T(-i\gamma + p))(-s) & (s < 0) \end{cases} \tag{4.21}$$

defines a translation representation of the group $W_{L_M}(t)$.

Let L be an arbitrary nonperturbed operator. According to Theorem 3.6, there exists decomposition (3.21) of the space $\mathcal{H}$ with respect to which the operator L has decomposition (3.22). Therefore, by virtue of equalities (4.5) and (4.6), the operator

$$T_L = T_{1\mu} \oplus T_{2M} \tag{4.22}$$

defines a translation representation of the group $W_L(t)$.

Note that, in equality (4.22), the operators $T_{1\mu}$ and T_{2M} are defined by formulas (4.18) and (4.21) and determine translation representations for the groups $W_{L_{1\mu}}(t)$ and $W_{L_{2M}}(t)$, respectively.

It follows from equality (4.22) that

$$T_L Q_L d = -\frac{d}{ds}(T_L d)(s) \quad (d \in \mathcal{D}(Q_L)) \tag{4.23}$$

and

$$T_L W_L(t) = \mathcal{F}(t)T_L . \tag{4.24}$$

4.4. Spaces of Test and Generalized Functions. Translation representations are a convenient tool for the investigation of the properties of the group $W_L(t)$. To completely solve this problem, it is necessary to use spaces of test functions $\mathcal{D}(\mathbb{R}, N)$ and generalized functions (distributions) $\mathcal{D}'(\mathbb{R}, N)$ of a translation representation. In this subsection, we present necessary definitions and investigate the structure of these spaces.

Denote by $C_0^\infty(\mathbb{R}, N)$ the set of infinitely differentiable functions with compact support from $L_2(\mathbb{R}, N)$.

We say that the sequence of functions $\{y_n(s)\}_{n=1}^\infty$ from $C_0^\infty(\mathbb{R}, N)$ converges to a function $y(s)$ $(\in C_0^\infty(\mathbb{R}, N))$ if the following conditions are satisfied:

(i) there exists $\rho > 0$ such that

$$\operatorname{supp} y_n \subset [-\rho, \rho] \qquad \forall n \in \mathbb{N}, \tag{4.25}$$

where $\operatorname{supp} y = \overline{\{ s \in \mathbb{R} \mid \|y(s)\|_N \neq 0 \}}$ is the support of the function $y(s)$;

(ii) the equality

$$\lim_{n \to \infty} \left\| \frac{d^m}{ds^m}(y_n(s) - y(s)) \right\|_{L_2(\mathbb{R}, N)} = 0 \tag{4.26}$$

holds for any $m \in \mathbb{N} \cup \{0\}$.

The linear space $C_0^\infty(\mathbb{R}, N)$ with convergence thus defined is called the *space of test functions* $\mathcal{D}(\mathbb{R}, N)$. Denote by $\mathcal{D}'(\mathbb{R}, N)$ the space of distributions over the space $\mathcal{D}(\mathbb{R}, N)$.

Let L $(\in \mathfrak{M}_B)$ be a nonperturbed operator and let $V(t) = e^{iBt}$. Denote by $C_0^\infty(L)$ the linear set of elements

$$u \in \bigcap_{m=1}^{\infty} \mathcal{D}(L^m)$$

for which there exists $\rho_u > 0$ such that $V^*(\rho_u)u = 0$.

We say that a sequence $\{u_n\}_{n=1}^\infty$ $(u_n \in C_0^\infty(L))$ converges to an element u $(\in C_0^\infty(L))$ if

(i) there exists $\rho > 0$ such that

$$V^*(\rho)u_n = 0 \qquad \forall n \in \mathbb{N};$$

(ii) the equality

$$\lim_{n \to \infty} \left\| L^m(u_n - u) \right\| = 0 \tag{4.27}$$

holds for any $m \in \mathbb{N} \cup \{0\}$.

The linear space $C_0^\infty(L)$ with the convergence defined above is denoted by $\mathcal{D}_L$.

Clearly, the elements of the space $\mathcal{D}_L$ belong to $\mathcal{H}$. In order to define a similar structure in $\mathcal{H}_L$, we use the identification $\mathcal{H}_L \equiv \{x_\gamma \mid \forall \gamma \in \mathcal{H}\}$ (see Subsection 4.1).

Denote

$$X_{\mathcal{D}_L} := \{x_u \mid \forall u \in \mathcal{D}_L\}. \tag{4.28}$$

We say that a sequence $\{x_{u_n}\}_{n=1}^\infty$ converges to an element x_u in the space $X_{\mathcal{D}_L}$ if and only if the sequence $\{u_n\}_{n=1}^\infty$ converges to the element u in the space $\mathcal{D}_L$.

Denote by $\mathcal{D}_L'$ and $X_{\mathcal{D}_L'}$ the spaces of distributions over the spaces $\mathcal{D}_L$ and $X_{\mathcal{D}_L}$, respectively. Taking equality (4.3) into account, we get

$$X_{\mathcal{D}_L'} = \{x_\tau \mid \forall \tau \in \mathcal{D}_L'\}, \tag{4.29}$$

where the distribution x_τ is defined by the relation

$$\langle x_\tau, x_u \rangle := \langle \tau, u \rangle \quad (u \in \mathcal{D}_L). \tag{4.30}$$

Here, the ordered pairs $\langle x_\tau, x_u \rangle$ and $\langle \tau, u \rangle$ indicate the action of the linear continuous functionals x_τ and τ on the elements x_u and u, respectively.

Let $\tilde{L}$ be a nonperturbed operator such that L and $\tilde{L}$ are mutually simple (i.e., $\mathcal{D}(L) \cap \mathcal{D}(\tilde{L}) = \mathcal{D}(B^2)$). By virtue of Theorem 3.6, there exists decomposition (3.21) of the space $\mathcal{H}$ with respect to which the operator L can be represented in form (3.22). Clearly, with respect to decomposition (3.21), the operator $\tilde{L}$ can be represented in the following form:

$$\tilde{L} = L_{1M} \oplus L_{2\mu}. \tag{4.31}$$

Consider the linear space

$$\mathcal{M}_{(L,\tilde{L})} := X_{\mathcal{D}_{\tilde{L}}} \oplus \mathcal{D}_L. \tag{4.32}$$

Obviously, the space of distributions $\mathcal{M}'(L, \tilde{L})$ has the form

$$\mathcal{M}'_{(L,\tilde{L})} = \left\{ f = \begin{pmatrix} x_\tau \\ \psi \end{pmatrix} \middle| \forall \tau \in \mathcal{D}'_{\tilde{L}}, \ \forall \psi \in \mathcal{D}'_L \right\} \tag{4.33}$$

and the action of the distribution f on the element $d = \begin{pmatrix} x_u \\ v \end{pmatrix} \in \mathcal{M}_{(L,\tilde{L})}$ is determined by the relation

$$\langle f, d \rangle := \langle \tau, u \rangle + \langle \psi, v \rangle \quad (u \in \mathcal{D}_{\tilde{L}}, \ v \in \mathcal{D}_L). \tag{4.34}$$

The statement below establishes the relationship between $\mathcal{M}_{(L,\tilde{L})}$ and the space of test function $\mathcal{D}(\mathbb{R}, N)$.

Theorem 4.3. *The operator* T_L *defined by (4.22) maps the set* $\mathcal{M}_{(L,\tilde{L})}$ *onto the set* $\mathcal{D}(\mathbb{R}, N)$. *Moreover, the convergence of a sequence* $\{d_n\}_{n=1}^{\infty}$ *to an element* d *in the space* $\mathcal{M}_{(L,\tilde{L})}$ *is equivalent to the convergence of the sequence* $\{T_L d_n\}_{n=1}^{\infty}$ *to the element* $T_L d$ *in the space* $\mathcal{D}(\mathbb{R}, N)$.

Proof. It follows from relations (3.22), (4.31), and (4.32) that, with respect to decomposition (4.5), the space $\mathcal{M}_{(L,\tilde{L})}$ can be represented in the form

$$\mathcal{M}_{(L,\tilde{L})} = \mathcal{M}_{(L_{1\mu}, L_{1M})} \oplus \mathcal{M}_{(L_{2M}, L_{2\mu})}.$$

Combining this relation with equality (4.22), we conclude that, to complete the proof, it remains to consider the following cases:

1. The operator $L = L_{1\mu}$ is the hard extension of the operator B_1^2 in the space $\mathcal{H}_1$.

2. The operator $L = L_{2M}$ is the soft extension of the operator B_2^2 in the space $\mathcal{H}_2$.

Consider case 1 (below, we omit the subscript 1).
It follows from Proposition 4.1 that

$$\mathcal{M}_{(L_\mu, L_M)} \subset C^\infty(Q_\mu) = \bigcap_{m=1}^{\infty} \mathcal{D}(Q_\mu).$$

Taking into account equalities (4.20), we conclude that the operator T_μ maps the set $\mathcal{M}_{(L_\mu, L_M)}$ into the set $C^\infty(\mathbb{R}, N)$.

According to the definitions of the spaces $\mathcal{D}_L$ and $X_{\mathcal{D}_L}$, for all

$$d = \begin{pmatrix} x_\gamma \\ p \end{pmatrix} \in \mathcal{M}_{(L_\mu, L_M)},$$

there exists $\rho > 0$ such that

$$V^*(\rho)\gamma = V^*(\rho)p = 0. \tag{4.35}$$

Theorem 3.1 implies that the operator $T V(t) T^{-1}$ acts as a translation to the right by t in the space $L_2(\mathbb{R}_+, N)$. Therefore, relation (4.35) means that the supports of the functions $(T\gamma)(s)$ and $(Tp)(s)$ lie in $[0, \rho]$ and, hence, in view of equality (4.18), we can write

$$\operatorname{supp}(T_\mu d)(s) \subset [-\rho, \rho]. \tag{4.36}$$

Thus, $T_\mu \mathcal{M}_{(L_\mu, L_M)} \subset C_0^\infty(\mathbb{R}, N)$.

Conversely, assume that $(T_\mu d)(s) \in \mathcal{D}(\mathbb{R}, N)$ and relation (4.36) is true. In this case, equality (4.18) implies that

$$(T\gamma)(s) = (Tp)(s) = 0, \quad s > \rho,$$

and we arrive at equality (4.35).

Since $(T_\mu d)(s) \in C^\infty(-d/ds)$, by using relation (4.20) and Proposition 4.1, we get $d \in \mathcal{M}_{(L_\mu, L_M)}$. Thus, we have proved the equality

$$T_\mu \mathcal{M}_{(L_\mu, L_M)} = C_0^\infty(\mathbb{R}, N).$$

In view of the definitions of the spaces $\mathcal{D}_L$ and $X_{\mathcal{D}_L}$ and equality (4.20), we conclude that the convergence of a sequence $\{d_n\}_{n=1}^\infty$ in the space $\mathcal{M}_{(L_\mu, L_M)}$ to the element d is equivalent to the convergence of the sequence $\{T_\mu d_n\}_{n=1}^\infty$ in the space $\mathcal{D}(\mathbb{R}, N)$ to the element $T_\mu d$. Therefore, Theorem 4.3 is proved for $L = L_\mu$.

For $L = L_M$, the proof is similar.

■

Theorem 4.3 means that the operator T_L can be extended to the space of distributions $\mathcal{M}'_{(L, \tilde{L})}$ by the formula

$$\langle T_L f, y(s)\rangle := \left\langle f, T_L^{-1}(y(s))\right\rangle \qquad \left(f \in \mathcal{M}'_{(L,\tilde{L})}, y(s) \in \mathcal{D}(\mathbb{R}, N)\right) \tag{4.37}$$

Moreover,

$$T_L \mathcal{M}'_{(L,\tilde{L})} = \mathcal{D}'(\mathbb{R}, N). \tag{4.36}$$

Similarly, it follows from Theorem 4.3 and equalities (4.23) and (4.24) that

$$W_L(t)\mathcal{M}_{(L,\tilde{L})} = \mathcal{M}_{(L,\tilde{L})} \quad \text{and} \quad \mathcal{M}_{(L,\tilde{L})} \subset \mathcal{D}(Q_L).$$

Therefore, by using the formulas

$$\langle W_L(t) f, d\rangle := \left\langle f, W_L^*(t) d\right\rangle \quad \text{and} \quad \langle Q_L f, d\rangle := \langle f, -Q_L d\rangle,$$
$$f \in \mathcal{M}'_{(L,\tilde{L})}, \quad d \in \mathcal{M}_{(L,\tilde{L})} \tag{4.39}$$

we can correctly extend the group $W_L(t)$ and its generator Q_L to the space $\mathcal{M}'_{(L,\tilde{L})}$. For the extended operators, we preserve the same notation $W_L(t)$ and Q_L.

By using equalities (4.23), (4.37), and (4.39), we obtain

$$\langle T_L Q_L f, y(s)\rangle = \left\langle f, -Q_L T_L^{-1}(y(s))\right\rangle = \left\langle f, T^{-1}\left(\frac{d}{ds}y(s)\right)\right\rangle = \left\langle -\frac{d}{ds}Tf, y(s)\right\rangle,$$

where $f \in \mathcal{M}'_{(L,\tilde{L})}$ and $y(s) \in \mathcal{D}(\mathbb{R}, N)$. Consequently,

$$T_L Q_L f = -\frac{d}{ds}T_L f \qquad (\forall f \in \mathcal{M}'_{(L,\tilde{L})})$$

and, hence, the extended operator T_L determines a "translation representation" of the group $W_L(t)$ in the space of distributions $\mathcal{M}'_{(L,\tilde{L})}$.

4.5. Eventually Outgoing and Initially Incoming Elements. We say that an element $f \in \mathcal{M}'_{(L,\tilde{L})}$ is *eventually outgoing (initially incoming)* if there exists $\rho \geq 0$ such that the support of the distribution $T_L W_L(\rho) f$ $(T_L W_L(-\rho) f)$ lies on the positive (negative) semiaxis.

Proposition 4.4. *An element* $f = \begin{pmatrix} x_\tau \\ \psi \end{pmatrix} \in \mathcal{M}'_{(L,\tilde{L})}$ *is eventually outgoing (initially incoming) if and only if there exists* $\rho \geq 0$ *such that*

$$\langle \tau - i\psi, V(\rho)u \rangle = 0 \qquad (\langle \tau + i\psi, V(\rho)u \rangle = 0),$$

where $V(t) = e^{iBt}$, $\tau \in \mathcal{D}'_{\tilde{L}}$, *and* $\psi \in \mathcal{D}'_L$ *holds for all* $u \in \mathcal{D}_L \cap \mathcal{D}_{\tilde{L}} = \mathcal{D}_{B^2}$.

Proof. Assume that an element f is eventually outgoing. Then there exists $\rho \geq 0$ such that the equality

$$\langle T_L W_L(\rho) f, y(s) \rangle = 0$$

holds for all $y(s) \in \mathcal{L} = \mathcal{D}(\mathbb{R}, N) \cap L_2(\mathbb{R}_-, N)$. According to Theorem 4.3, this means that

$$\langle W_L(\rho) f, d \rangle = 0, \tag{4.40}$$

for any $d \in T_L^{-1}\mathcal{L} = \mathcal{M}_{(L, \tilde{L})} \cap D_-$.

It follows from equalities (4.10) and (4.32) that

$$\mathcal{M}_{(L, \tilde{L})} \cap D_- = \left\{ \begin{pmatrix} x_u \\ -iu \end{pmatrix} \middle| \, u \in \mathcal{D}_L \cap \mathcal{D}'_L = \mathcal{D}_{B^2} \right\}.$$

By using the first relations in (4.12) and (4.39) and equality (4.34), we conclude that equality (4.40) is equivalent to the following equality:

$$\langle \tau - i\psi, V(\rho)u \rangle = 0 \qquad \left(\forall\, u \in \mathcal{D}_{B^2} \right).$$

The case of an initially incoming element f is similar.

■

We say that $f \in \mathcal{M}'_{(L, \tilde{L})}$ is an element with *compact support* if the support of the distribution $(T_L f)(s)$ is compact in $\mathbb{R}$.

Theorem 4.5. *Let* $g \in \mathcal{M}'_{(L, \tilde{L})}$ *be an arbitrary element with compact support and let* μ *be a complex number. Then the equation*

$$(Q_L - \mu I) f = g \tag{4.41}$$

possesses a unique eventually outgoing (initially incoming) solution $f \in \mathcal{M}'_{(L, \tilde{L})}$.

Proof. Our proof is based on the main ideas of the proof of a similar statement for the classical wave equation (see Theorem 4.1 in Lax and Phillips [1, Chapter 4]). To make our presentation more complete, we now outline the principal ideas of the proof.

In view of the definition of eventually outgoing elements, relations (4.23), (4.24), and (4.38) imply that equation (4.41) possesses an eventually outgoing solution if and only if the equation

$$\left(\frac{d}{ds}+\mu I\right)k(s) = -(T_L g)(s) \tag{4.42}$$

possesses a solution $k(s) \in \mathcal{D}'(\mathbb{R}, N)$ such that $\operatorname{supp} k(s) \subset (-\rho, \infty)$.

Consider a linear continuous functional $k_{\text{ev}}(s) \in \mathcal{D}'(\mathbb{R}, N)$ defined as

$$\langle k_{\text{ev}}(s), y(s)\rangle := \left\langle (T_L g)(s), -\eta(s)e^{\mu s}\int_s^{\infty} y(\delta)e^{-\mu\delta}d\delta\right\rangle \quad (y(s) \in \mathcal{D}(\mathbb{R}, N)), \tag{4.43}$$

where $\eta(s) \in C_0^{\infty}(\mathbb{R})$ is a function such that $\eta(s) = 1$ for $s \in (-\rho, \rho)$ and ρ is a positive number chosen from the condition

$$\operatorname{supp}(T_L g)(s) \subset (-\rho, \rho).$$

It is easy to show that the distribution $k_{\text{ev}}(s)$ is a solution of equation (4.41) and $\operatorname{supp} k_{\text{ev}}(s) \subset (-\rho, \infty)$. Therefore, equation (4.42) possesses the eventually outgoing solution $f_{\text{ev}} = T_L^{-1} k_{\text{ev}}(s)$.

To prove its uniqueness, it suffices to show that if f is eventually outgoing and $Q_L f = \mu f$, then $f = 0$.

Indeed, let $x(t) = \langle W_L(t)f, d\rangle$, where $d \in \mathcal{M}_{(L, \bar{L})}$. The equality $Q_L f = \mu f$ and relations (4.39) imply that $x'(t) = \mu x(t)$ and, hence,

$$x(t) = \langle f, d\rangle e^{\mu t}. \tag{4.44}$$

Further, since the element f is eventually outgoing, there exists $\rho > 0$ such that

$$\operatorname{supp}(T_L W_L(\rho) f) \subset \mathbb{R}_+. \tag{4.45}$$

On the other hand, the relation

$$W_L(\rho - t)d \in D_- \tag{4.46}$$

holds for sufficiently large t. By using relations (4.39), (4.45), and (4.46), we conclude that

$$x(t) = \langle W_L(\rho)f, W_L(\rho - t)d\rangle = \langle T_L W_L(\rho)f, T_L W_L(\rho - t)d\rangle = 0.$$

By combining this equality with (4.44), we obtain $f = 0$. Thus, we have proved the statement of Theorem 4.5 for the case of eventually outgoing solutions.

Similarly, we find that the element $f_{in} = T_L^{-1} k_{in}(s)$, where the distribution $k_{in}(s) \in \mathcal{D}'(\mathbb{R}, N)$ is defined by the formula

$$\langle k_{in}(s), y(s)\rangle := \left\langle (T_L g)(s), \eta(s) e^{\mu s} \int_{-\infty}^{s} y(\delta) e^{-\mu\delta} d\delta \right\rangle,$$

is a unique initially incoming solution of equation (4.41).

■

Corollary 4.6. *For* $s \in \mathbb{R} \setminus [-\rho, \rho]$, *the translation representations* $k_{ev}(s)$ *and* $k_{in}(s)$ *of the eventually outgoing and initially incoming solutions of equation (4.41) have the form*

$$k_{ev}(s) = \begin{cases} e^{-\mu s} n, & s > \rho, \\ 0, & s < -\rho, \end{cases} \quad \text{and} \quad k_{in}(s) = \begin{cases} 0, & s > \rho, \\ e^{-\mu s} n, & s < -\rho, \end{cases}$$

where $n \in \mathbb{N}$ *and* $\rho > 0$ *is such that* $\operatorname{supp}(T_L g)(s) \subset [-\rho, \rho]$.

Proof. Equality (4.43) implies that $k_{ev}(s) = 0$ for $s < -\rho$.

Let $y(s) \in \mathcal{D}(\mathbb{R}, N)$ and let $\operatorname{supp} y(s) \subset (\rho, \infty)$. Then the right-hand side of equality (4.43) can be represented in the form

$$\left\langle e^{\mu s} (T_L g)(s), -\eta \int_{-\infty}^{\infty} y(\delta) e^{-\mu\delta} d\delta \right\rangle. \tag{4.47}$$

Note that

$$\int_{-\infty}^{\infty} y(\delta) e^{-\mu\delta} d\delta \in N \quad \text{and} \quad \eta(s) = 1 \quad (\forall s \in \operatorname{supp}(T_L g)(s)).$$

Therefore, relation (4.47) determines a distribution over the Hilbert space N. Then

$$\left\langle e^{\mu s} (T_L g)(s), -\eta \int_{-\infty}^{\infty} y(\delta) e^{-\mu\delta} d\delta \right\rangle = \left(n, \int_{-\infty}^{\infty} y(\delta) e^{-\mu\delta} d\delta \right)_N,$$

where n is a uniquely determined element of the space N.

Thus, by using (4.43), we obtain

$$\langle k_{ev}, y \rangle = \int_{-\infty}^{\infty} (n, y(\delta))_N e^{-\mu\delta} d\delta = \left\langle n e^{-\mu\delta}, y(\delta) \right\rangle.$$

Hence, $k_{ev}(s) = e^{-\mu s} n$ for $s > \rho$.

The analysis of the case of an initially incoming solution is similar.

■

4.6. Wave Equation in $\mathbb{R}^n$. In the space $\mathcal{H} = L_2(\mathbb{R}^n)$, where n is odd, the Laplace operator

$$L = -\Delta, \quad \mathcal{D}(L) = W_2^2(\mathbb{R}^n)$$

is nonperturbed (see Subsection 3.3). Substituting this operator in the right-hand side of equation (4.1), we obtain the classical wave equation in the free space

$$u_{tt} = \Delta u. \tag{4.48}$$

Let us show that the subspaces D_- and D_+ defined by equalities (4.10) coincide with the well-known outgoing and incoming subspaces of the classical wave equation (4.48).

Let $u \in \mathcal{D}(L)$. Then the element

$$d_+ = \begin{pmatrix} x_{B^* u} \\ iB^* u \end{pmatrix},$$

where the operator B is defined by relations (3.39), belongs to the subspace D_+.

Since the elements $x_{B^* u}$ and u are identified in $\mathcal{H}_L$ (see Subsection 4.1), we conclude that the element d_+ can be written in the form

$$d_+ = \begin{pmatrix} u \\ iB^* u \end{pmatrix}. \tag{4.49}$$

It is known[15] that the solution $u(x, t)$ of the Cauchy problem for the wave equation (4.48) can be represented in the form

[15] See, e.g., Theorem 2.2 and Corollary 2.1 in Lax and Phillips [1, Chapter 4].

$$u(x,t) = \frac{1}{2}\frac{1}{(2\pi)^m}\int_{S^{n-1}} h(x\cdot\omega - t, \omega)\,d\omega, \qquad \omega \in S^{n-1}, \qquad m = \frac{n-1}{2}, \tag{4.50}$$

where S^{n-1} is the unit sphere in $\mathbb{R}^n$. Here, the function $h(s, \omega)$ depends on the choice of initial data d from $\mathfrak{H}_L$. In particular, if an element $d = d_+$ is defined by relation (4.49), then the function $h(s, \omega)$ admits the following representation:

$$h(s,\omega) = \left(\partial_s^{2m} Ru\right)(s,\omega) - i\left(\partial_s^{2m-1} RB^* u\right)(s,\omega), \tag{4.51}$$

where R denotes the Radon transform defined by (3.31).

It is clear that

$$\begin{aligned} \left(\partial_s^{2m} Ru\right)(s,\omega) &= \left(\partial_s^{2m} Ru\right)(-s,-\omega), \\ \left(\partial_s^{2m-1} RB^* u\right)(s,\omega) &= -\left(\partial_s^{2m-1} RB^* u\right)(-s,-\omega). \end{aligned} \tag{4.52}$$

By virtue of relations (4.51) and (4.52), we conclude that, for $s < 0$, the function $h(s, \omega)$ has the form

$$h(s,\omega) = (-1)^m\left(\left(\partial_s^m Tu\right)(-s,-\omega) + i\left(\partial_s^{m-1} TB^* u\right)(-s,-\omega)\right), \tag{4.53}$$

where the operator T is defined by relation (3.35).

By using relations (3.38), we get

$$TB^* u(-s,-\omega) = -i\partial_s (Tu)(-s,-\omega).$$

Therefore, it follows from equality (4.53) that

$$h(s,\omega) = 0 \quad \text{for} \quad s < 0.$$

This fact and representation (4.50) imply that the solution $u(x, t)$ of the wave equation (4.48) with the Cauchy data d_+ vanishes in the forward cone $|x| < t$.

Similarly, we show that any solution $u(x, t)$ of the wave equation with the Cauchy data $d_- = \begin{pmatrix} u \\ -iB^* u \end{pmatrix}$ vanishes in the backward cone $|x| < -t$.

Taking into account statement (iii) in Proposition 3.2 and equalities (4.3) and (4.10), we conclude that the set of the initial data d_+ (d_-) introduced above is dense in the subspace D_+ (D_-). Therefore, in view of decomposition (4.11), the subspace D_+ (D_-)

coincides with the set of initial data in $\mathfrak{H}_L$ such that the corresponding solutions $u(x, t)$ of the wave equation vanish in the forward (backward) cone $|x| < t$ $(|x| < -t)$.

Further, it is easy to see that the operator $T_{-\Delta}$ defined by equality (4.22) determines the well-known translation representation of the group of solutions of the wave equation in the free space (see Lax and Phillips [1]). In this case, the operator T is determined by relation (3.35) and the operators $Q_{1\mu}$ and Q_{2M} are constructed for given decompositions (3.40) and (3.41).

Recall that $-\Delta \in \mathfrak{M}_B$, where B is the simple maximal symmetric operator defined by the second equality in (3.38). The semigroup $V(t) = e^{iBt}$ generated by the operator B is characterized by a quite complicated structure.

By using equalities (3.38), we can now write the semigroup $V(t)$ in the form

$$V(t) = T^{-1}\mathcal{F}_+(t)T,$$

where $\mathcal{F}_+(t)$ acts as an operator of translation to the right by t in $L_2(\mathbb{R}_+, N)$.

In what follows, without loss of generality, we assume that $n = 3$. By using the Schwartz theorem and the "support theorem" (see Helgason [1]), we arrive at the following assertion:

Proposition 4.7. *Let* $\rho > 0$ *be fixed. Then the subspace* $\operatorname{Ker} V^*(\rho)$ *consists of all functions* $u(x) = L_2(\mathbb{R}^3)$ *such that, for* $|x| > \rho$,

$$\int_{S^2} u(x) Y_k\left(\frac{x}{|x|}\right) d\omega = \begin{cases} 0, & k = 0, \\ \sum_{i=1}^{\frac{k+1}{2}} \frac{\alpha_{ik}}{|x|^{2i}}, & k \text{ odd}, \\ \sum_{i=1}^{\frac{k}{2}} \frac{\beta_{ik}}{|x|^{2i+1}}, & k \text{ even}, \end{cases} \tag{4.53}$$

where S^2 *is the unit sphere in* $\mathbb{R}^3$, $\omega = \dfrac{x}{|x|} \in S^2$, $\{\alpha_{ik}, \beta_{ik}\} \subset \mathbb{C}$, *and* Y_k *is an arbitrary spherical harmonic of order* k.

It follows from Proposition 4.7 that the set of elements $\mathcal{D}_{-\Delta}$ coincides with the set of infinitely differentiable functions $u(x)$ such that

(i) the functions $u(x)$ and all their derivatives belong to the space $L_2(\mathbb{R}^3)$;

(ii) for any function $u(x)$, there exists $\rho > 0$ such that $u(x)$ satisfies relations (4.53).

By using these properties and the definition of convergence in the space $\mathcal{D}_L$ (see Subsection 4.4), we obtain

$$\mathcal{D}(\mathbb{R}^3) \subset \mathcal{D}_{-\Delta}, \quad \mathcal{D}'_{-\Delta} \subset \mathcal{D}'(\mathbb{R}^3),$$

where $\mathcal{D}(\mathbb{R}^3)$ is the classical Schwartz space of test functions.

Remarks. The results presented in this section were obtained by S. Kuzhel in [9]. Our aim was to show that the properties of the classical wave equation important for the Lax–Phillips scattering theory (see Lax and Phillips [1, Chapter 4]) can be generalized (in a natural way) to the case of an abstract wave equation. The other results established in this direction (μ-outgoing and μ-incoming solutions of the reduced abstract wave equation, the analog of the Sommerfeld radiation conditions, etc.) can be found in the work by S. Kuzhel [4].

5. Elements of the Lax–Phillips Scattering Theory for ρ-Perturbed Abstract Wave Equations

5.1. Preliminary Remarks. The possibility of the application of the Lax–Phillips scattering theory to the investigation of the dynamics of systems described by operator-differential equations of the form

$$u_{tt} = -\tilde{L}u, \tag{5.1}$$

where $\tilde{L}$ is a self-adjoint operator in a Hilbert space $\mathcal{H}$, depends on the existence of outgoing and incoming subspaces for the group of solutions of equation (5.1) in the space of initial data. In the case of the classical wave equation, the application of the Lax–Phillips method makes it possible to obtain excellent results concerning the relationship between the properties of the scattering matrix and the nature of scatterers (i.e., scattering by obstacles, potentials with compact support, and inhomogeneities) (see, e.g., the surveys by Reed and Simon [2], Petkov [1], and Baumgartel and Wollenberg [1]). These investigations are based on the examination, in each special case, of the operator $\tilde{L}$ in equation (5.1) as a perturbation of the nonperturbed Laplacian and construction of the required incoming and outgoing subspaces as certain modifications of the well-known incoming and outgoing subspaces for the wave equation in the free space.

In this connection, it is of interest to construct the Lax–Phillips theory for equation (5.1), where the operator $\tilde{L}$ is a *perturbation of the nonperturbed operator* L.

By using this abstract approach, one can significantly extend the applicability of the Lax–Phillips method and consider various new scattering problems for the classical wave equation.

In this section, we present a brief sketch of a new approach to these problems based on the concept of ρ-perturbations of a nonperturbed operator. These perturbations can be regarded as an abstract analog of perturbations of the Laplace operator in bounded domains.

For a more detailed presentation and some other results established in this direction, see S. Kuzhel [5, 7, 10, 11].

5.2. ***ρ–Perturbed Abstract Wave Equation.*** Let B be a simple maximal symmetric operator acting in a separable Hilbert space $\mathcal{H}$. We consider a family of subspaces $\mathcal{H}_\rho$ of the space $\mathcal{H}$ defined as

$$\mathcal{H}_0 \equiv \mathcal{H}, \quad \mathcal{H}_\rho = V(\rho)\mathcal{H} \quad (\rho > 0), \tag{5.2}$$

where $V(t) = e^{iBt}$ is an isometric semigroup in $\mathcal{H}$ and iB is its generator.

It is clear that

$$\mathcal{H}_{\rho_1} \supset \mathcal{H}_{\rho_2} \quad \text{as} \quad \rho_1 < \rho_2 .$$

In the space $\mathcal{H}_\rho$, we define a simple maximal symmetric operator B_ρ as

$$B_0 \equiv B, \quad B_\rho = B\Big|_{\mathcal{D}(B_\rho)}, \quad \mathcal{D}(B_\rho) = V(\rho)\mathcal{D}(B) \quad (\rho > 0). \tag{5.3}$$

Assume that $\tilde{\mathcal{H}}$ is a subspace of the space $\mathcal{H}$. A self-adjoint operator $\tilde{L}$ with finitely many eigenvalues on the negative semiaxis acting in the subspace $\tilde{\mathcal{H}}$ is called *ρ-perturbed* if $0 \notin \sigma_p(\tilde{L})$ and there exists $\rho \geq 0$ such that

$$\tilde{\mathcal{H}} \supset \mathcal{H}_\rho \quad \text{and} \quad \tilde{L} \supset B_\rho^2 . \tag{5.4}$$

The operator-differential equation (5.1) with a ρ-perturbed operator $\tilde{L}$ on the right-hand side is called a *ρ-perturbed abstract wave equation.*

Assume that the operator $\tilde{L}$ is ρ-perturbed. Then, for any $f \in \mathcal{D}(\tilde{L})$ and $u \in \mathcal{D}(B_\rho^2)$, we get

$$\left(P_\rho \tilde{L} f, u\right) = \left(f, B_\rho^2 u\right) = \left(P_\rho f, B_\rho^2 u\right),$$

where P_ρ is the orthoprojector onto $\mathcal{H}_\rho$ in $\mathcal{H}$.

Therefore,

$$P_\rho \mathcal{D}(\tilde{L}) \subset \mathcal{D}\left(B_\rho^{2^*}\right) = \mathcal{D}(B_\rho^* B_\rho^*) \tag{5.5}$$

and

$$P_\rho \tilde{L} f = (B_\rho^*)^2 P_\rho f \quad (\forall f \in \mathcal{D}(\tilde{L})). \tag{5.6}$$

In what follows, for the sake of simplicity, we assume that the operator $\tilde{L}$ is *positive.* The general case was studied by S. Kuzhel in [10].

Denote by $\mathcal{H}_{\tilde{L}}$ the completion of the lineal $\mathcal{D}(\tilde{L})$ in the norm

$$\|f\|_{\tilde{L}}^2 = (\tilde{L} f, f) \quad (f \in \mathcal{D}(\tilde{L})). \tag{5.7}$$

Since the operator $\tilde{L}$ is ρ-perturbed, the equality

$$\|u\|_{\tilde{L}} = \left\|B_\rho u\right\| \tag{5.8}$$

holds for all $u \in \mathcal{D}(B_\rho^2)$.

Denote by $\mathcal{H}_{B_\rho^2}$ the completion of the lineal $\mathcal{D}(B_\rho^2)$ in norm (5.8). Acting as in Subsection 4.1 (with the operator B_ρ and the space $\mathcal{H}_\rho$ instead of B and $\mathcal{H}$, respectively), we conclude that the subspace $\mathcal{H}_{B_\rho^2}$ of the Hilbert space $\mathcal{H}_{\tilde{L}}$ *can be identified* with the Hilbert space

$$\{x_\gamma \mid \forall \gamma \in \mathcal{H}_\rho\},$$

where the scalar product $(\cdot,\cdot)_{\tilde{L}}$ is defined by equality (4.3).

According to this rule of identification, any element $u \in \mathcal{D}(B_\rho^2)$ can be represented in the form $x_{B_\rho u}$. Thus,

$$\mathcal{H}_{B_\rho^2} \equiv \{x_\gamma \mid \forall \gamma \in \mathcal{H}_\rho\} \quad \text{and} \quad \mathcal{D}(B_\rho^2) \equiv \{x_{B_\rho u} \mid \forall u \in \mathcal{D}(B_\rho^2)\}.$$

Further, by virtue of the definition of an element $x_\gamma \in \mathcal{H}_{B_\rho^2}$, there exists a sequence $\{u_n\}_{n=1}^\infty$ $(u_n \in \mathcal{D}(B_\rho^2))$ such that

$$\lim_{n\to\infty} B_\rho u_n = \gamma.$$

Therefore, by using equalities (5.5)–(5.7), we conclude that

$$(f, x_\gamma)_{\tilde L} = \lim_{n\to\infty}(f, u_n)_{\tilde L} = \lim_{n\to\infty}(f, B_\rho^2 u_n)$$

$$= \lim_{n\to\infty}(B_\rho^* P_\rho f, B_\rho u) = (B_\rho^* P_\rho f, \gamma) \tag{5.9}$$

for all $f \in \mathcal{D}(\tilde L)$ and $x_\gamma \in \mathcal{H}_{B_\rho^2}$.

In the space of initial data $\mathfrak{H}_{\tilde L} = \mathcal{H}_{\tilde L} \oplus \tilde{\mathcal{H}}$, the operator

$$Q = \begin{pmatrix} 0 & I \\ -\tilde L & 0 \end{pmatrix}, \quad \mathcal{D}(Q) = \left\{ \begin{pmatrix} u \\ v \end{pmatrix} \,\middle|\, \{u, v\} \subset \mathcal{D}(\tilde L) \right\}, \tag{5.10}$$

is essentially skew-self-adjoint. Therefore, its closure $Q_{\tilde L} = \overline{Q}$ is the generator of the unitary (in $\mathfrak{H}_{\tilde L}$) group $W_{\tilde L}(t) = e^{Q_{\tilde L} t}$ of solutions of the Cauchy problem for the ρ-perturbed abstract wave equation (5.1).

Let L be a nonperturbed operator acting in the space $\mathcal{H}$. We set

$$D_-^\rho = W_L(-\rho) D_- \quad \text{and} \quad D_+^\rho = W_L(\rho) D_+, \tag{5.11}$$

where the subspaces $D_\pm$ are defined by equalities (4.10) and $W_L(t)$ is the group of solutions of the abstract wave equation (4.1).

It follows from Theorem 4.2 and equality (5.2) that

$$D_-^\rho = \left\{ \begin{pmatrix} x_\gamma \\ -i\gamma \end{pmatrix} \,\middle|\, \gamma \in \mathcal{H}_\rho \right\} \quad \text{and} \quad D_+^\rho = \left\{ \begin{pmatrix} x_\gamma \\ i\gamma \end{pmatrix} \,\middle|\, \gamma \in \mathcal{H}_\rho \right\}. \tag{5.12}$$

The definitions of the ρ-perturbed operator $\tilde L$ and the space of initial data $\mathfrak{H}_{\tilde L}$ imply that the mutually orthogonal subspaces $D_\pm^\rho$ are contained in $\mathfrak{H}_{\tilde L}$.

Note that, unlike the case of an abstract wave equation (Subsection 4.2), the orthogonal sum

$$D_-^\rho \oplus D_+^\rho = \mathcal{H}_{B_\rho^2} \oplus \mathcal{H}_\rho$$

does not coincide with the space of initial data $\mathfrak{H}_{\tilde L}$.

Theorem 5.1. *The subspaces* D_-^ρ *and* D_+^ρ *are, respectively, incoming and outgoing subspaces with respect to the group of solutions* $W_{\tilde{L}}(t)$. *Moreover, for any* $\gamma \in \mathcal{H}_\rho$ *and* $t \geq 0$,

$$W_{\tilde{L}}^*(t)\begin{pmatrix} x_\gamma \\ -i\gamma \end{pmatrix} = \begin{pmatrix} x_{V(t)\gamma} \\ -iV(t)\gamma \end{pmatrix} \quad \textit{and} \quad W_{\tilde{L}}(t)\begin{pmatrix} x_\gamma \\ i\gamma \end{pmatrix} = \begin{pmatrix} x_{V(t)\gamma} \\ iV(t)\gamma \end{pmatrix}, \tag{5.13}$$

where $V(t) = e^{iBt}$.

Proof. As follows from equalities (5.11), the subspaces D_-^ρ and D_+^ρ are, respectively, incoming and outgoing subspaces for the group of solutions $W_L(t)$ of the abstract wave equation (4.1) with a nonperturbed operator L on the right-hand side.

Let

$$d_+ = \begin{pmatrix} x_{B_\rho u} \\ iB_\rho u \end{pmatrix}, \tag{5.14}$$

where $u \in \mathcal{D}(B_\rho^3)$. By virtue of Theorem 4.2 and relation (5.3), we obtain

$$W_L(s)\,d_+ = \begin{pmatrix} x_{V(s)B_\rho u} \\ iV(s)B_\rho u \end{pmatrix} = \begin{pmatrix} x_{B_\rho V(s)u} \\ iB_\rho V(s)u \end{pmatrix} \quad (s > 0), \tag{5.15}$$

where the element $x_{B_\rho V(s)u} = V(s)\,u$ belongs to $\mathcal{D}(B_\rho^3)$. It follows from the definition of the ρ-perturbed operator $\tilde{L}$ and equalities (4.4) and (5.10) that

$$W_L(s)\,d_+ \in \mathcal{D}(Q_{\tilde{L}}) \cap \mathcal{D}(Q_L) \quad \text{and} \quad (Q_L - Q_{\tilde{L}})\,W_L(s)d_+ = 0.$$

Therefore,

$$\begin{aligned}\left(W_{\tilde{L}}(-t)W_L(t) - I\right)d_+ &= \int_0^t \left(W_{\tilde{L}}(-s)W_L(s)d_+\right)' ds \\ &= \int_0^t W_{\tilde{L}}(-s)\left(Q_L - Q_{\tilde{L}}\right)W_L(s)d_+ ds = 0.\end{aligned}$$

Thus,

$$W_{\tilde{L}}(t)d_+ = W_L(t)d_+ \quad (t \geq 0). \tag{5.16}$$

By virtue of equalities (3.1), (3.2), and (4.3), it easy to see that the set of elements d_+ in the form (5.14) is dense in D_+^ρ. Therefore, equalities (5.15) and (5.16) imply the second equality in (5.13).

Thus, condition (i) in the definition of outgoing subspaces (see Subsection 4.2) is satisfied for the subspace D_+^ρ.

Acting as in the proof of Theorem 4.2, we conclude that condition (ii) is also true for the subspace D_+^ρ. Therefore, the subspace D_+^ρ is outgoing.

Similarly, for the set of elements

$$d_- = \begin{pmatrix} x_{B_\rho u} \\ -iB_\rho u \end{pmatrix} \qquad (u \in \mathcal{D}(B_\rho^3)),$$

by using the same reasoning as above, we conclude that the subspace D_-^ρ is incoming for the group $W_{\tilde{L}}(t)$ and the first equality in (5.13) is true.

■

Proposition 5.2. *Assume that the operators* L *and* $\tilde{L}$ *are, respectively, nonperturbed and* ρ*-perturbed. Then the wave operators*

$$\Omega_\pm(Q_{\tilde{L}}, Q_L) = s-\lim_{t\to\pm\infty} W_{\tilde{L}}(-t)W_L(t)$$

exist and

$$\Omega_\pm(Q_{\tilde{L}}, Q_L)\mathfrak{H}_L = M_\pm, \quad \textit{where} \quad M_\pm = \overline{\bigcup_{t\in\mathbb{R}} W_{\tilde{L}}(t)D_\pm^\rho}. \tag{5.17}$$

Proof. In view of (5.16), we obtain

$$\Omega_+(Q_{\tilde{L}}, Q_L)d_+ = d_+ \qquad (\forall\, d_+ \in D_+^\rho).$$

Therefore, for all $s \in \mathbb{R}$, we have

$$\Omega_+(Q_{\tilde{L}}, Q_L)W_L(s)d_+ = W_{\tilde{L}}(s)\Omega_+(Q_{\tilde{L}}, Q_L)d_+ = W_{\tilde{L}}(s)d_+$$

Taking into account equality (4.12), we conclude that the wave operator $\Omega_+(Q_{\tilde{L}}, Q_L)$ exists and

$$\Omega_+(Q_{\tilde{L}}, Q_L)\mathfrak{H}_L = \overline{\bigcup_{t\in\mathbb{R}} W_{\tilde{L}}(t)D_+^\rho}.$$

The case of the wave operator $\Omega_-(Q_{\tilde{L}}, Q_L)$ is similar.

■

Recall that the operator T_L defined by equality (4.22) determines a translation representation of the group $W_L(t)$ of solutions of the abstract wave equation (4.1).

We set

$$\tilde{T}_- = \mathcal{F}(\rho)T_L\Omega_-^{-1}(Q_{\tilde{L}}, Q_L) \quad \text{and} \quad \tilde{T}_+ = \mathcal{F}(-\rho)T_L\Omega_+^{-1}(Q_{\tilde{L}}, Q_L), \tag{5.18}$$

where the group $\mathcal{F}(t)$ acts as a translation to the right by t in the space $L_2(\mathbb{R}, N)$.

By virtue of equalities (4.23) and (4.24) and Proposition 5.2, one can easily show that the operator $\tilde{T}_-$ ($\tilde{T}_+$) determines an incoming (outgoing) translation representation of the restriction of the group $W_{\tilde{L}}(t)$ to the subspace M_- (M_+). Moreover,

$$\tilde{T}_- D_-^\rho = L_2(\mathbb{R}_-, N) \quad \text{and} \quad \tilde{T}_+ D_+^\rho = L_2(\mathbb{R}_+, N).$$

5.3. Definitions of Scattering Operator and Scattering Matrix. Assume that $\tilde{L}$ is a positive ρ-perturbed operator. For any $f \in M_-$, we set

$$g = P_{M_+} f,$$

where P_{M_+} is the orthoprojector onto M_+ in $\mathfrak{H}_{\tilde{L}}$. According to Subsection 5.2, the elements f and g possess the incoming $(\tilde{T}_- f)(s)$ and outgoing $(\tilde{T}_+ g)(s)$ translation representations.

The operator

$$S : (\tilde{T}_- f)(s) \to (\tilde{T}_+ g)(s) \tag{5.19}$$

acting in the space $L_2(\mathbb{R}, N)$ is called the *scattering operator.*

It is easy to see that the scattering operator S is a contracting in $L_2(\mathbb{R}, N)$ and commutes with the operators of right shift in this space.

We note that the definition of the scattering operator in the Lax–Phillips approach [equality (5.19)] is not directly related to nonperturbed $W_L(t)$ and perturbed $W_{\tilde{L}}(t)$ groups of unitary operators.

In applications, it is often convenient to define the scattering operator via the wave operators, namely,

$$\hat{S} = \Omega_+^*(Q_{\tilde{L}}, Q_L)\Omega_-(Q_{\tilde{L}}, Q_L).$$

To establish the relationship between the operators S and $\hat{S}$, we note that, by virtue of equalities (5.18) and (5.19),

$$S\,\mathcal{F}(\rho)T_L\Omega_-^{-1}(Q_{\bar{L}}, Q_L)f = \mathcal{F}(-\rho)T_L\Omega_+^{-1}(Q_{\bar{L}}, Q_L)g.$$

Since the operator S commutes with $\mathcal{F}(t)$, we have

$$S = \mathcal{F}(-2\rho)T_L\Omega_+^{-1}(Q_{\bar{L}}, Q_L)P_{M_+}\Omega_-(Q_{\bar{L}}, Q_L)T_L^{-1}.$$

In view of Proposition 5.2, we can write

$$\Omega_+^*(Q_{\bar{L}}, Q_L) = \Omega_+^{-1}(Q_{\bar{L}}, Q_L)P_{M_+}.$$

Thus,

$$S = \mathcal{F}(-2\rho)T_L\Omega_+^*(Q_{\bar{L}}, Q_L)\Omega_-(Q_{\bar{L}}, Q_L)T_L^{-1} = \mathcal{F}(-2\rho)T_L\hat{S}T_L^{-1}.$$

Let F be the Fourier transformation in the space $L_2(\mathbb{R}, N)$. It is known (see Lax and Phillips [1]) that the operator

$$\mathfrak{S} = FSF^{-1}$$

can be realized as an operator of multiplication by a function $\mathfrak{S} \equiv \mathfrak{S}(\delta)$ $(\delta \in \mathbb{R})$ whose values are contracting operators in the space N for almost all $\delta \in \mathbb{R}$.

The operator-valued function $\mathfrak{S}(\delta)$ is called the *scattering matrix*. Note that the scattering matrix $\mathfrak{S}(\delta)$ is the restriction to the real axis of a contracting operator-valued analytic function $\mathfrak{S}(z)$ holomorphic in the lower half plane (for more details, see Adamjan an Arov [2] and Kuzhel and Tretyakov [1]).

The function $\mathfrak{S}(z)$ $(\operatorname{Im} z < 0)$ is called the *Heisenberg scattering matrix*.

5.4. Description of Singularities of the Scattering Matrix. It is clear that the special features of the scattering matrix $\mathfrak{S}(\delta)$ are determined by the properties of the ρ-perturbed unitary group $W_{\bar{L}}(t)$.

We say that the scattering matrix *has a singularity*

(i) at a point z in the lower half plane $(\operatorname{Im} z < 0)$ if the operator $\mathfrak{S}(z)$ is not completely invertible in the space N (i.e., $0 \notin \rho(\mathfrak{S}(z))$);

(ii) at a real point δ if the operator-valued function $\mathfrak{S}(\delta)$ is not analytic[16] at this point.

Theorem 5.3. *Let* $\tilde{\mathcal{H}} = \mathcal{H}$ *and let an operator* $\tilde{L}$ *be nonperturbed. Then the corresponding scattering matrix* $\mathfrak{S}(\delta)$ *for the group* $W_{\tilde{L}}(t)$ *has no singularities.*

Proof. In view of the definition of a ρ-perturbed operator (Subsection 5.2), we conclude that, under the conditions of the theorem, the operator $\tilde{L}$ is 0-perturbed. By virtue of equalities (4.12) and (5.18), we establish that the operators

$$\tilde{T}_- = T_L \Omega_-^{-1}(Q_{\tilde{L}}, Q_L) \quad \text{and} \quad \tilde{T}_+ = T_L \Omega_+^{-1}(Q_{\tilde{L}}, Q_L)$$

determine, respectively, the incoming and outgoing translation representations for the group of solutions $W_{\tilde{L}}(t)$ acting in the space of initial data $\mathfrak{H}_{\tilde{L}}$. Moreover, equality (5.16) implies

$$\Omega_-(Q_{\tilde{L}}, Q_L)d_- = d_- \quad \text{and} \quad \Omega_+(Q_{\tilde{L}}, Q_L)d_+ = d_+ \quad \left(d_\pm \in D_\pm^0 = D_\pm\right).$$

Therefore,

$$\tilde{T}_- d_- = T_L d_- \quad \text{and} \quad \tilde{T}_+ d_+ = T_L d_+. \tag{5.20}$$

Since the operator $\tilde{L}$ is nonperturbed, we conclude that the operator $T_{\tilde{L}}$ defined by equality (4.22) also determines a translation representation for the group $W_{\tilde{L}}(t)$.

Let us clarify the relationship between the operators $T_{\tilde{L}}$ and $\tilde{T}_\pm$.

Taking equalities (4.18), (4.21) and (4.22) into account, one can directly check that

$$\left(T_{\tilde{L}} d_-\right)(s) = J_{\tilde{L}}\left(T_{L_\mu} d_-\right)(s) \quad \text{and} \quad \left(T_{\tilde{L}} d_+\right)(s) = \left(T_{L_\mu} d_+\right)(s) \tag{5.21}$$

$$(\forall\, d_+ \in D_+,\ d_- \in D_-),$$

where the self-adjoint unitary operator $J_{\tilde{L}}$ acts in N and is defined by the equalities

$$J_{\tilde{L}}\big|_{N_1} = I \quad \text{and} \quad J_{\tilde{L}}\big|_{N_2} = -I,$$

and the mutually orthogonal subspaces N_1 and N_2 of the space N are determined by the conditions

[16] We say that the scattering matrix is analytic at the point δ if the function $\mathfrak{S}(\delta)$ can be analytically extended outside the real axis in the neighborhood of this point.

$$T\mathcal{H}_1 = L_2(\mathbb{R}_+, N_1) \quad \text{and} \quad T\mathcal{H}_2 = L_2(\mathbb{R}_+, N_2).$$

Here, T is a unitary mapping of $\mathcal{H}$ into $L_2(\mathbb{R}_+, N)$ for which relations (3.1) and (3.2) are satisfied, and the subspaces $\mathcal{H}_1$ and $\mathcal{H}_2$ form decomposition (3.22) for the nonperturbed operator $\tilde{L}$.

It follows from equality (5.21) and an analogous equality for the nonperturbed operator L that

$$(T_L d_-)(s) = E(T_{\tilde{L}} d_-)(s) \quad \text{and} \quad (T_L d_+)(s) = (T_{\tilde{L}} d_+)(s), \tag{5.22}$$

where the operator $E = J_L J_{\tilde{L}}$ is unitary in the space N.

Equalities (5.20)–(5.22) imply that

$$\tilde{T}_- W_{\tilde{L}}(t) d_- = \mathcal{F}(t)\tilde{T}_- d_- = \mathcal{F}(t) T_L d_- = E\mathcal{F}(t) T_{\tilde{L}} d_- = E T_{\tilde{L}} W_{\tilde{L}}(t) d_- \quad (t \in \mathbb{R})$$

and

$$\tilde{T}_+ W_{\tilde{L}}(t) d_+ = T_{\tilde{L}} W_{\tilde{L}}(t) d_+.$$

These equalities and relations (4.12) (with $\tilde{L} = L$) yield

$$\tilde{T}_- = E T_{\tilde{L}} \quad \text{and} \quad \tilde{T}_+ = T_{\tilde{L}}. \tag{5.23}$$

According to equalities (5.19) and (5.23), we conclude that the scattering operator S acts in the space $L_2(\mathbb{R}, N)$ as the operator of multiplication by the unitary (in the space N) operator E^{-1}. Consequently, the scattering matrix $\mathfrak{S}(\delta)$ can be analytically extended to the entire complex plane according to the formula

$$\mathfrak{S}(z) = E^{-1} \quad (\forall\, z \in \mathbb{C})$$

Thus, the scattering matrix for the nonperturbed group $W_{\tilde{L}}(t)$ has no singularities.

■

Let $\tilde{L}$ be an arbitrary ρ-perturbed operator. We set

$$\mathcal{K}_\rho = \mathfrak{H}_{\tilde{L}} \ominus \left(D_-^\rho \oplus D_+^\rho\right),$$

where the incoming D_-^ρ and outgoing D_+^ρ subspaces for the group $W_{\tilde{L}}(t)$ are defined by equalities (5.12). It is well known (see, e.g., Lax and Phillips [1]) that the operators

$$Z(t) = P_{\mathcal{K}_\rho} W_{\tilde{L}}(t)\Big|_{\mathcal{K}_\rho}, \tag{5.24}$$

where $P_{\mathcal{K}_\rho}$ is the orthoprojector onto $\mathcal{K}_\rho$ in $\mathfrak{H}_{\tilde{L}}$, form a semigroup of contractions that act in the Hilbert space $\mathcal{K}_\rho$.

Denote by A the generator of $Z(t)$. In the Lax–Phillips approach, there is a close connection between the spectrum of the generator A and the set of singularities of the scattering matrix $\mathfrak{S}(\delta)$ of the ρ-perturbed group $W_{\tilde{L}}(t)$. It is convenient to investigate this relationship by using the representation of the scattering matrix as the characteristic function of the generator A (for more details, see Adamjan and Arov [1, 2]).

In particular, by using the results presented in S. Kuzhel [8], we can show that the following statement is true:

Theorem 5.4. *Let a positive operator $\tilde{L}$ be ρ-perturbed and let $\mathfrak{S}(\delta)$ be a scattering matrix for the group of solutions $W_{\tilde{L}}(t)$. Then*

(i) $0 \in \sigma_\alpha(\mathfrak{S}(z))$ *if and only if* $-iz \in \sigma_\alpha(A^*)$ $(\alpha \in \{p, r, c\},\ \operatorname{Im} z < 0)$;

(ii) the scattering matrix $\mathfrak{S}(\delta)$ is analytic and unitary on the set $\mathbb{R} \setminus \{i\sigma(A_s^*)\}$, where A_s^* is the completely nonskew-self-adjoint part[17] *of the operator A^*.*

Corollary 5.5. *The set of the points of singularity of the scattering matrix $\mathfrak{S}(\delta)$ coincides with the set $i\sigma(A_s^*)$.*

Note that, by construction, the generator A does not depend on the choice of a non-perturbed operator L. Therefore, it follows from Corollary 5.5 that the set of the points of singularity of the scattering matrix $\mathfrak{S}(\delta)$ *does not depend* on the choice of a non-perturbed group $W_L(t)$.

5.5. ***Case of 0-Perturbations.*** In this case, we can directly describe the singularities of a scattering matrix in terms of positive BVS. It should be noted that, for obtaining these results, it is not necessary to construct translation representations for the group of solutions.

Let $\tilde{L}$ be a 0-perturbed operator. By virtue of equalities (5.2) and (5.4), this means that the operator $\tilde{L}$ (acting in the space $\tilde{\mathcal{H}} = \mathcal{H}$) is a self-adjoint extension with finitely many eigenvalues on the negative semiaxis of the operator B^2 and $0 \notin \sigma_p(\tilde{L})$.

[17] By analogy with the canonical decomposition of a contraction into unitary and completely nonunitary parts, the generator A can be uniquely decomposed into skew-self-adjoint and completely nonskew-self-adjoint parts.

As above, we assume that the operator $\tilde{L}$ is positive. At the end of this subsection, we give a sketch of the proof in the general case.

In view of Theorem 5.1, the subspaces $D_-^0 \equiv D_-$ and $D_+^0 \equiv D_+$ defined by (5.12) are incoming and outgoing subspaces with respect to the group of solutions $W_{\tilde{L}}(t)$.

Proposition 5.5. *In the case of a* 0*-perturbed operator* $\tilde{L}$, *the incoming* D_- *and outgoing* D_+ *subspaces satisfy the following relation:*

$$\overline{\bigcup_{t \in \mathbb{R}} W_{\tilde{L}}(t) D_-} = \overline{\bigcup_{t \in \mathbb{R}} W_{\tilde{L}}(t) D_+} = \mathfrak{H}_{\tilde{L}}. \tag{5.25}$$

Proof. To prove (5.25) in the space

$$\mathcal{K}_0 = \mathfrak{H}_{\tilde{L}} \ominus (D_- \oplus D_+), \tag{5.26}$$

we consider the contracting operator

$$Y = P_{\mathcal{K}_0} U \big|_{\mathcal{K}_0}, \tag{5.27}$$

where $P_{\mathcal{K}_0}$ is the orthoprojector onto $\mathcal{K}_0$ in $\mathfrak{H}_{\tilde{L}}$, and the operator

$$U = \left(Q_{\tilde{L}} + \xi I\right)\left(Q_{\tilde{L}} - \xi I\right)^{-1} \quad (\xi > 0) \tag{5.28}$$

is the cogenerator of the group $W_{\tilde{L}}(t)$. For the Lax–Phillips scheme, it is known (see Lax and Phillips [1]) that the properties of the operator Y uniquely determine all characteristics of the scattering process. In particular, relations (5.25) are equivalent to the requirement that

$$Y^n \hat{y} \to 0 \quad \text{and} \quad (Y^*)^n \hat{y} \to 0 \quad \text{as} \quad n \to \infty \tag{5.29}$$

for all $\hat{y} \in \mathcal{K}_0$.

Let us establish some properties of the operator Y. It follows from equalities (5.10) and (5.28) that, for all

$$\hat{u} = \begin{pmatrix} u \\ v \end{pmatrix} \quad (u \in \mathcal{D}(\tilde{L}),\ v \in \mathcal{H}),$$

the operator U admits the representation

$$U\hat{u} = \begin{pmatrix} (\tilde{L}-\xi^2 I)R & -2\xi R \\ 2\xi\tilde{L}R & (\tilde{L}-\xi^2 I)R \end{pmatrix}\hat{u}, \tag{5.30}$$

where $R = (\tilde{L}+\xi^2 I)^{-1}$ is the resolvent of the operator $\tilde{L}$.

Equalities (5.12) with $\rho = 0$ yield

$$\begin{pmatrix} 0 \\ \mathcal{H} \end{pmatrix} \subset D_- \oplus D_+, \tag{5.31}$$

where

$$\begin{pmatrix} 0 \\ \mathcal{H} \end{pmatrix} = \left\{ \begin{pmatrix} 0 \\ \gamma \end{pmatrix} \middle| \gamma \subset \mathcal{H} \right\}.$$

Taking relation (5.30) into account, we conclude that

$$\begin{pmatrix} \mathcal{D}(\tilde{L}) \\ 0 \end{pmatrix} \subset \bigvee_{n\in\mathbb{Z}} U^n(D_- \oplus D_+). \tag{5.32}$$

Equalities (5.31) and (5.32) imply that

$$\bigvee_{n\in\mathbb{Z}} U^n(D_- \oplus D_+) = \mathfrak{H}_{\tilde{L}}. \tag{5.33}$$

It follows from (5.33) that Y is a completely nonunitary contraction (see Szökefalvi-Nagy and Foias [1]). Moreover, the operator Y is self-adjoint. Indeed, equality (5.30) implies that

$$P_{\mathcal{H}_{\tilde{L}}} U P_{\mathcal{H}_{\tilde{L}}} = P_{\mathcal{H}_{\tilde{L}}} U^* P_{\mathcal{H}_{\tilde{L}}},$$

where $P_{\mathcal{H}_{\tilde{L}}}$ is the orthoprojector onto $\mathcal{H}_{\tilde{L}}$ in $\mathfrak{H}_{\tilde{L}}$. Further, by virtue of the definition of the space $\mathcal{K}_0$, we can conclude that $\mathcal{K}_0 \subset \mathcal{H}_{\tilde{L}}$. Therefore,

$$Yy = P_{\mathcal{K}_0} U y = P_{\mathcal{K}_0} P_{\mathcal{H}_{\tilde{L}}} U y = P_{\mathcal{K}_0} P_{\mathcal{H}_{\tilde{L}}} U^* y = P_{\mathcal{K}_0} U^* y = Y^* y$$

for all $y \in \mathcal{K}_0$.

The fact that the operator Y is self-adjoint means that the intersection of its spectrum with the unit circle is a set of measure zero and, therefore, relation (5.29) is true (see

Proposition 6.7 in Szökefalvi-Nagy and Foias [1, Chapter 2]). But then relations (5.25) are also true.

■

In view of (5.28), the cogenerator U of the group $W_{\tilde{L}}(t)$ can be written in the form

$$U = I + 2\xi(Q_{\tilde{L}} - \xi I)^{-1}.$$

By using the Laplace transform representation of $(Q_{\tilde{L}} - \xi I)^{-1}$, we get

$$U\hat{y} = \hat{y} - 2\xi\int_0^\infty e^{-\xi t} W_{\tilde{L}}(t)\hat{y}\,dt$$

for all $\hat{y} \in \mathfrak{H}_{\tilde{L}}$. In particular, for $\hat{y} \in \mathcal{K}_0$, taking equalities (5.24) and (5.27) into account, we obtain

$$Y\hat{y} = \hat{y} - 2\xi\int_0^\infty e^{-\xi t} Z(t)\hat{y}\,dt = \left(I + 2\xi(A - \xi I)^{-1}\right)\hat{y} = (A + \xi I)(A - \zeta I)^{-1}\hat{y}.$$

Thus, the operator Y is the cogenerator of the group $Z(t)$. By virtue of Theorem 5.4, this means that the set of singularities of the scattering matrix $\mathfrak{S}(\delta)$ for the group $W_{\tilde{L}}(t)$ is determined by the spectrum of the cogenerator Y.

Consider the structure of the space $\mathcal{K}_0$ and the properties of the operator Y in more detail.

Recall (see Subsection 5.2) that, in the space $\mathcal{H}_{\tilde{L}}$, the lineal $\mathcal{D}(B^2)$ can be identified with the lineal $\{x_{Bu} \mid \forall\, u \in \mathcal{D}(B^2)\}$, and the subspace $\mathcal{H}_{B^2}$ can be identified with the Hilbert space $\{x_\gamma \mid \forall\, \gamma \in \mathcal{H}\}$, where

$$\left(x_{\gamma_1}, x_{\gamma_2}\right)_{\tilde{L}} = (\gamma_1, \gamma_2). \tag{5.34}$$

Further, according to (5.9), we have

$$(f, x_\gamma)_{\tilde{L}} = (B^* f, \gamma) \tag{5.35}$$

for any $f \in \mathcal{D}(\tilde{L})$ and $\gamma \in \mathcal{H}$.

By using equalities (5.12), (5.26), (5.34) and (5.35), one can easily show that the closure of the lineal

$$\left\{ \hat{y}_f = \begin{pmatrix} y_f \\ 0 \end{pmatrix} \,\middle|\, y_f = f - x_{B^* f} \quad \forall f \in \mathcal{D}(\tilde{L}) \right\} \tag{5.36}$$

in $\mathfrak{H}_{\tilde{L}}$ coincides with $\mathcal{K}_0$ and, moreover,

$$\left\| y_f \right\|_{\tilde{L}}^2 = (\tilde{L}f, f) - \left\| B^* f \right\|^2. \tag{5.37}$$

In the space $\mathcal{H}$, we consider the canonical positive BVS (X, Γ_1, Γ_2) of the operator B^2 defined by equalities (3.15) and (3.16) (with $\alpha = \xi^2$). Since the 0-perturbed positive operator $\tilde{L}$ is a self-adjoint extension of the operator B^2, by virtue of equality (3.17), we conclude that

$$\tilde{L} = B^* B^* \big|_{\mathcal{D}(\tilde{L})}, \qquad \mathcal{D}(\tilde{L}) = \left\{ f = u + h \,\middle|\, \tilde{C}\Gamma_1 f = \Gamma_2 f \right\}, \tag{5.38}$$

where $\tilde{C}$ is a bounded nonnegative operator in X.

It follows from equalities (3.20) and (5.37) that

$$\left(y_f, y_g \right)_{\tilde{L}} = (G\Gamma_1 f, \Gamma_1 g), \qquad \{f, g\} \subset \mathcal{D}(\tilde{L}), \tag{5.39}$$

where $G = \tilde{C}(I - 2\xi^2 \tilde{C})$ is a bounded self-adjoint operator in X.

Let us clarify the action of the operator Y on the elements $\hat{y}_f$ of the space $\mathcal{K}_0$.

Lemma 5.6. *If* $\{f, g\} \subset \mathcal{D}(L)$, *then*

$$\left(Y\hat{y}_f, \hat{y}_g \right)_{\mathfrak{H}_{\tilde{L}}} = \left((I - 4\xi^2 \tilde{C}) G\Gamma_1 f, \Gamma_1 g \right).$$

Proof. Assume that the sequence $\{u_n\}_{n=1}^{\infty}$ $(u_n \in \mathcal{D}(B^2))$ converges in $\mathcal{H}_{\tilde{L}}$ to the element $x_{B^* f}$. Then, by using relations (5.27), (5.30), and (5.36), we obtain

$$\begin{aligned} \left(Y\hat{y}_f, \hat{y}_g \right)_{\mathfrak{H}_{\tilde{L}}} &= \lim_{n \to \infty} \left((\tilde{L} - \xi^2 I) R(f - u_n), y_g \right)_{\tilde{L}} \\ &= \left(y_f, y_g \right)_{\tilde{L}} - 2\xi^2 \lim_{n \to \infty} \left(R(f - u_n), y_g \right)_{\tilde{L}}. \end{aligned} \tag{5.40}$$

We set $R(f - u_n) = t_n$. By virtue of equalities (3.16) and (3.17), we conclude that

$$\tilde{C}\Gamma_1 t_n = \Gamma_2 t_n = \tilde{C}P_X(f - u_n).$$

Therefore, taking into account the decomposition $\mathfrak{H}_{\tilde{L}} = D_- \oplus \mathcal{K}_0 \oplus D_+$ and equality (5.39), we can write

$$(t_n, y_g)_{\tilde{L}} = (y_{t_n}, y_g)_{\tilde{L}} = (f - u_n, G\Gamma_1 g) \tag{5.41}$$

In this case, by virtue of assertion (ii) in Proposition 3.2 (with $z = i\xi$), we get

$$(u_n, G\Gamma_1 g) = -\frac{i}{\xi}(Bu_n, G\Gamma_1 g)$$

and, hence, equality (5.41) implies that

$$\lim_{n\to\infty}(t_n, y_g)_{\tilde{L}} = \frac{i}{\xi}\big((B^* - i\xi I)f, G\Gamma_1 g\big). \tag{5.42}$$

We represent the element f as $f = u + h$, where $u \in \mathcal{D}(L_\mu)$ and $h \in X$. By using equalities (3.16) and (3.17), we can now rewrite the right-hand side of equality (5.42) in the form

$$\frac{i}{\xi}((B - i\xi I)u - 2i\xi h, G\Gamma_1 g) = 2(\Gamma_2 f, G\Gamma_1 g) = (2\tilde{C}\Gamma_1 f, G\Gamma_1 g).$$

Thus,

$$\lim_{n\to\infty}(t_n, y_g)_{\tilde{L}} = (2\tilde{C}\Gamma_1 f, G\Gamma_1 g).$$

In view of (5.39) and (5.40), this completes the proof of Lemma 5.6.

■

By using Lemma 5.6, we can easily establish the relationship between the operators Y and $\tilde{C}$. Indeed, since the operator $\tilde{L}$ is positive, equality (5.39) implies that the operator G is nonnegative in X. Further, in view of equality (5.39), we can also show that the closure of the operator Ω defined on the elements $\hat{y}_f$ by the formula

$$\Omega\hat{y}_f = \sqrt{G}\,\Gamma_1 f \qquad (f \in \mathcal{D}(\tilde{L}))$$

isometrically maps the space $\mathcal{K}_0$ onto $\overline{\mathcal{R}(G)}$. Moreover, Lemma 5.6 implies that

$$\Omega Y \hat{y}_f = \left(I - 4\xi^2 \tilde{C}\right)\Omega \hat{y}_f$$

and, hence,

$$Y\hat{y} = \Omega^{-1}\left(I - 4\xi^2 \tilde{C}\right)\Omega \hat{y} \tag{5.43}$$

for any $\hat{y} \in \mathcal{K}_0$.

The operator $\tilde{C}$ that determines the operator $\tilde{L}$ in the canonical positive BVS (X, Γ_1, Γ_2) characterizes, in a certain sense, the degree of perturbation of this operator as compared with the nonperturbed operator $L_\mu = B^*B$. Therefore, it is natural to expect that the spectral properties of the operator $\tilde{C}$ determine the singularities of the scattering matrix $\mathfrak{S}(\delta)$ for the group $W_{\tilde{L}}(t)$.

Theorem 5.7. *Let a positive operator* $\tilde{L}$ *be* 0*-perturbed. Then the scattering matrix* $\mathfrak{S}(\delta)$ *of the group* $W_{\tilde{L}}(t)$ *is analytic and unitary for* $\delta \in \mathbb{R} \setminus \{0\}$. *Furthermore,* $\mathfrak{S}(\delta)$ *is the boundary value of an operator-valued function* $\mathfrak{S}(z)$ *analytic in the lower half plane and, moreover,*

(i) $\mathfrak{S}(\delta)$ *is analytic for* $\delta = 0$ *if and only if*

$$\frac{1}{2\xi^2} \in \rho\left(\tilde{C}\big|_{X_1}\right),$$

where $X_1 = \overline{\mathcal{R}\left(\tilde{C} - \frac{1}{2\xi^2} I\right)}$;

(ii) *in the lower half plane, the set of singular points of the function* $\mathfrak{S}(z)$ *is located on the imaginary axis and consists of all points* $z = -i|z|$ *at which either the operator* $\mathfrak{S}(z)$ *is not invertible or its inverse operator is unbounded and*

$$0 \in \sigma_\alpha(\mathfrak{S}(z)) \Leftrightarrow \frac{1}{2\xi(\xi + |z|)} \in \sigma_\alpha(\tilde{C}), \qquad \alpha \in \{p, c\}.$$

Proof. The proof of Theorem 5.7 is based on the use of the well-known results from the Lax–Phillips scattering theory and the Szökefalvi-Nagy–Foias theory of characteristic functions. For this reason, we only outline its basic idea.

Since $\tilde{L}$ is a 0-perturbed operator, it follows from Subsection 5.2 and Proposition 5.5 that the group $W_{\tilde{L}}(t)$ admits the incoming and outgoing translation representations constructed in terms of the subspaces D_- and D_+.

As indicated in Subsection 5.3, the relevant scattering matrix $\mathfrak{S}(\delta)$ is the boundary value of an operator-valued function $\mathfrak{S}(z)$ analytic in the lower half plane (the Heisenberg scattering matrix). The function $\mathfrak{S}(z)$ is, in fact, the characteristic function of the generator A of the associated semigroup $Z(t)$ or (to within a bilinear transformation) the characteristic function of the contraction Y. Thus, to complete the proof, it remains to use Theorem 5.4 and equalities (5.33) and (5.43).

■

Remark. By construction, the subspaces $D_{\pm}$ that determine the scattering matrix $\mathfrak{S}(\delta)$ in the Lax–Phillips scattering scheme do not depend on the choice of a nonperturbed operator L. Therefore, the choice of the canonical positive BVS (X, Γ_1, Γ_2) that corresponds to the hard extension $L_\mu = B^*B$ for the description of the properties of the Heisenberg scattering matrix $\mathfrak{S}(z)$ is explained only by a simpler proof and more convenient formulas in assertions (i) and (ii) of Theorem 5.7. At the same time, one can easily establish results similar to Theorem 5.7 by using any nonperturbed operator L $(\in \mathfrak{M}_B)$ for the construction of the canonical positive BVS.

Indeed, let L be a nonperturbed operator. In the canonical positive BVS (X, Γ_1, Γ_2), the operator L is determined by the self-adjoint operator

$$C = \left((L+\xi^2 I)^{-1} - (L_\mu + \xi^2 I)^{-1}\right)\Big|_X \tag{5.44}$$

(see (3.17) and (3.18)). In what follows, to explicitly indicate the relationship between L and C, we denote the operator L by L_C.

In decomposition (3.14), we replace the operator L_μ by L_C and determine the canonical positive BVS $(X, \Gamma_1', \Gamma_2')$ corresponding to the nonperturbed operator L_C by using equalities (3.15) and (3.16). As a result, we get

$$X = \operatorname{Ker}(B^*B^* + \xi^2 I)$$

and

$$\Gamma_1'(u+h) = P_X(L_C + \xi^2 I) \quad \text{and} \quad \Gamma_2'(u+h) = h \quad (f = u + h \in \mathcal{D}(B^*B^*)),$$

where $u \in \mathcal{D}(L_C)$, $h \in X$, and P_X is the orthoprojector onto X in $\mathcal{H}$. In this BVS, the 0-perturbed operator $\tilde{L}$ is defined by the formulas

$$\tilde{L} = B^*B^*\big|_{\mathcal{D}(\tilde{L})}, \quad \mathcal{D}(\tilde{L}) = \{f = u + h \mid C'\Gamma_1' f = \Gamma_2' f\},$$

where the self-adjoint operator

$$C' = \left(\left(\tilde{L}+\xi^2 I\right)^{-1} - \left(L_C+\xi^2 I\right)^{-1}\right)\Big|_X \tag{5.45}$$

can be regarded as the degree of perturbation of the operator $\tilde{L}$ *as compared with the nonperturbed operator* L_C.

By using (5.44) and (5.45), we establish that the operators $\tilde{C}$ and C' (which determine the operator $\tilde{L}$ in the BVS (X, Γ_1, Γ_2) and $(X, \Gamma_1', \Gamma_2')$, respectively) satisfy the equality

$$\tilde{C} = \left(\left(\tilde{L}+\xi^2 I\right)^{-1} - \left(L_\mu+\xi^2 I\right)^{-1}\right)\Big|_X = C' + C.$$

Hence, if the operator L is given in the BVS $(X, \Gamma_1', \Gamma_2')$, then the operator $\tilde{C}$ in assertions (i) and (ii) of Theorem 5.7 should be replaced by $C' + C$.

Assume that the spectrum of the 0-perturbed operator $\tilde{L}$ on the negative semiaxis consists of finitely many normal eigenvalues $-\eta_1^2, \ldots, -\eta_n^2$. Let us briefly discuss what modifications should be made for establishing results similar to Theorem 5.7.

First, we note that equality (5.7) defines the indefinite metric

$$[f, f]_{\tilde{L}} = \left(\tilde{L}f, f\right) \quad (f \in \mathcal{D}(\tilde{L})) \tag{5.46}$$

in the lineal $\mathcal{D}(\tilde{L})$. The completion of the lineal $\mathcal{D}(\tilde{L})$ in the norm

$$\|f\|_{|\tilde{L}|} = \sqrt{\left(|\tilde{L}|f, f\right)},$$

where $|\tilde{L}|$ is the absolute value of the operator $\tilde{L}$, is denoted by $\mathcal{H}_{\tilde{L}}$. The space $\mathcal{H}_{\tilde{L}}$ is a Pontryagin space[18] with the indefinite metric (5.46) whose rank of indefiniteness is equal to the multiplicity of the spectrum of the operator $\tilde{L}$ on the negative semiaxis.

Similarly, the space of initial data $\mathfrak{H}_{\tilde{L}} = \mathcal{H}_{\tilde{L}} \oplus \mathcal{H}$ is also a Pontryagin space with the indefinite metric $[\cdot,\cdot]_{\mathfrak{H}_{\tilde{L}}}$ defined for the elements

$$\begin{pmatrix} u \\ v \end{pmatrix} \in \mathfrak{H}_{\tilde{L}}, \quad u \in \mathcal{D}(\tilde{L}), \quad \text{and} \quad v \in \mathcal{H}$$

as follows:

---

[18] For the definition of Pontryagin spaces and conventional "indefinite" terminology, see A. Kuzhel [7].

$$\left[\begin{pmatrix} u \\ v \end{pmatrix}, \begin{pmatrix} u \\ v \end{pmatrix}\right]_{\mathfrak{H}_{\tilde{L}}} = [u, u]_{\tilde{L}} + \|v\|^2.$$

In the Pontryagin space $\mathfrak{H}_{\tilde{L}}$, the operator $Q_{\tilde{L}}$ is the generator of the π-unitary group $W_{\tilde{L}}(t)$ and its spectrum consists of some points of the continuous spectrum on the imaginary axis and $2n$ real eigenvalues $\pm\eta_1, \ldots, \pm\eta_n$ $(\eta_i > 0)$.

Let P_0 denote the π-orthogonal projector onto the nondegenerate subspace

$$\mathcal{P} = \sum_{i=1}^{n} \dot{+} \left(\operatorname{Ker}(Q_{\tilde{L}} - \eta_i I) \dot{+} \operatorname{Ker}(Q_{\tilde{L}} + \eta_i I)\right)$$

in $\mathfrak{H}_{\tilde{L}}$.

Proposition 5.5′. *The subspaces* D_- *and* D_+ *remain incoming and outgoing for the group* $W_{\tilde{L}}(t)$. *In this case, relation (5.25) is modified as follows:*

$$(I - P_0)\overline{\bigcup_{t \in \mathbb{R}} W_{\tilde{L}}(t) D_-} = (I - P_0)\overline{\bigcup_{t \in \mathbb{R}} W_{\tilde{L}}(t) D_+} = (I - P_0)\mathfrak{H}_{\tilde{L}}. \qquad (5.25')$$

Proof. We choose ξ such that $\xi \notin \{\eta_1, \ldots, \eta_n\}$. By analogy with the proof of Theorem 5.1, we can show that the subspaces D_- and D_+ are incoming and outgoing, respectively.

In the Pontryagin space

$$\mathcal{K}_0 = \mathfrak{H}_{\tilde{L}} [-] (D_- \dot{+} D_+),$$

we consider the π–nonexpanding operator

$$Y = P_{\mathcal{K}_0} U \big|_{\mathcal{K}_0},$$

where $P_{\mathcal{K}_0}$ is the π-orthoprojector onto $\mathcal{K}_0$ in $\mathfrak{H}_{\tilde{L}}$. The operator Y is also π-self-adjoint, and its spectrum outside the unit circle consists of n normal eigenvalues

$$\gamma_i = -\frac{\xi + \eta_i}{\xi - \eta_i} \quad (i = 1, \ldots, n).$$

The subspace

$$\mathcal{K}_0^- = \sum_{i=1}^{n} \dotplus \operatorname{Ker}(Y - \gamma_i I)$$

is the maximal negative subspace of the space $\mathcal{K}_0$ and, therefore, its π-orthogonal complement

$$\mathcal{K}_0^+ = \mathcal{K}_0 [-] \mathcal{K}_0^-$$

is the maximal positive subspace.

Taking into account equality (5.33) and the results presented in S. Kuzhel [8], we conclude that (5.25′) is equivalent to the relation

$$Y^n \hat{y} \to 0 \qquad \left(\forall \hat{y} \in \mathcal{K}_0^+\right),$$

which can be proved in exactly the same way as relation (5.29) in Proposition 5.5. This completes the proof of Proposition 5.5′.

■

By using assertion (ii) of Proposition 3.2, one can easily show (see S. Kuzhel [10]) that the indefinite metric (5.46) in $\mathcal{D}(B^2)$ is equivalent to the scalar product $(\cdot, \cdot)_{|\tilde{L}|}$. Thus, the description of the elements of the spaces $\mathcal{H}_{B^2}$ and $\mathcal{K}_0$ and Lemma 5.6 remain valid. By using this lemma, one can easily establish the relationship between the operators Y and $\tilde{C}$. This and Theorem 20 in S. Kuzhel [8] enable us to prove the following statement:

Theorem 5.7′. *The scattering matrix $\mathfrak{S}(\delta)$ of the group $W_{\tilde{L}}(t)$ is analytic and unitary for $\delta \in \mathbb{R} \setminus \{0\}$, and assertion (i) of Theorem 5.7 is true. The scattering matrix $\mathfrak{S}(\delta)$ is the boundary value of an operator-valued function $\mathfrak{S}(z)$ meromorphic in the lower half plane and such that the set of its singular points is located on the imaginary axis and consists of the points $\{-i\eta_j\}_{j=1}^{n}$, where $\mathfrak{S}(z)$ has first-order poles, and all points $z = -i|z|$ described in assertion (ii) of Theorem 5.7.*

5.6. Examples

1. *Wave equation on a semiaxis.* In the space $\mathcal{H} = L_2(\mathbb{R}_+)$, we consider a simple maximal symmetric operator of the following form:

$$B = i\frac{d}{ds}, \qquad \mathcal{D}(B) = \overset{\circ}{W}{}^1_2(\mathbb{R}_+). \tag{5.47}$$

In Subsection 3.3, we have shown that, in this case, the class $\mathfrak{M}_B$ of nonperturbed operators consists of only two elements, namely, the hard extension

$$L_\mu = -\frac{d^2}{ds^2}\bigg|_{\mathcal{D}(L_\mu)}, \qquad \mathcal{D}(L_\mu) = \left\{u(s) \in W_2^2(\mathbb{R}_+) \,|\, u(0) = 0\right\},$$

and the soft extension

$$L_M = -\frac{d^2}{ds^2}\bigg|_{\mathcal{D}(L_M)}, \qquad \mathcal{D}(L_M) = \left\{u(s) \in W_2^2(\mathbb{R}_+) \,|\, u'(0) = 0\right\}.$$

Equation (5.1) with the operator L_μ (or L_M) on the right-hand side defines the free dynamics and is, in fact, the wave equation on the positive semiaxis with the boundary condition $u(0, t) = 0$ (or $\partial_s u(0, t) = 0$).

By virtue of equalities (5.2) and (5.4), the self-adjoint operator

$$L_\theta = -\frac{d^2}{ds^2}\bigg|_{\mathcal{D}(L_\theta)}, \qquad \mathcal{D}(L_\theta) = \left\{u(s) \in W_2^2(\mathbb{R}_+) \,|\, u'(0) + \theta u(0) = 0\right\}, \quad \theta \in \overline{\mathbb{R}},$$

is 0-perturbed. Therefore, the wave equation on the positive semiaxis with the boundary condition

$$\partial_s u(0, t) + \theta u(0, t) = 0$$

determines a 0-perturbed dynamics.

As follows from (5.47), the semigroup $V(t) = e^{iBt}$ is the semigroup of the right shift by t, i.e.,

$$V(t)u(s) = \begin{cases} u(s-t), & s \geq t, \\ 0, & 0 \leq s < t. \end{cases}$$

Therefore, a self-adjoint operator $\tilde{L}$ with finitely many eigenvalues on the negative semiaxis that acts in a space $\tilde{\mathcal{H}} \subseteq L_2(\mathbb{R}_+)$ is ρ-perturbed if and only if $0 \notin \sigma_p(\tilde{L})$ and the following conditions are satisfied:

$$\mathcal{D}(\tilde{L}) \supset \overset{\circ}{W}{}_2^2(\rho, \infty) \quad \text{and} \quad \tilde{L}\big|_{\overset{\circ}{W}{}_2^2(\rho,\infty)} = -\frac{d^2}{ds^2}.$$

In particular, this means that the equation

$$\partial_t^2 u(s,t) = \partial_s^2 u(s,t) + g(s)u(s,t), \qquad \partial_s u(0,t) + \theta u(0,t) = 0,$$

where $g(s) \geq 0$ and $\operatorname{supp} g(s) \subset (0, \rho)$, is ρ-perturbed.

By using equalities (3.15) and (3.16), we conclude that the triple (X, Γ_1, Γ_2), where $X = \langle e^{-\xi s} \rangle$ and $\Gamma_i \colon W_2^2(\mathbb{R}_+) \to X$ are such that

$$\Gamma_1 u = 2\xi(u'(0) + \xi u(0))e^{-\xi s} \quad \text{and} \quad \Gamma_2 u = u(0)e^{-\xi s}, \tag{5.48}$$

is the canonical positive BVS for the symmetric operator B^2 that corresponds to the hard extension L_μ.

It is known that, for $\theta > 0$, the operator L_θ possesses the normal eigenvalue $-\theta^2$ on the negative semiaxis and, for $\theta \leq 0$, this operator is positive. Thus, for all $\theta \in \overline{\mathbb{R}} \setminus \{\xi\}$, the operator L_θ is defined in the BVS (X, Γ_1, Γ_2) by relation (5.38), and it can readily be shown that

$$\tilde{C} = \frac{1}{2\xi(\xi - \theta)} I.$$

Let $\mathfrak{S}_\theta(\delta)$ be the scattering matrix for the group $W_{L_\theta}(t)$ determined by the operator L_θ in equation (5.1). By virtue of Theorems 5.7 and 5.7′ and the fact that ξ is arbitrary, the scattering matrix $\mathfrak{S}_\theta(\delta)$ is analytic and unitary on the real axis and has a unique singular point in the lower half plane, whose location depends on θ.

Thus, for $\theta < 0$, the operator-valued function $\mathfrak{S}_\theta(z)$ is not invertible at the point $i\theta$, and for $\theta > 0$, it has a first-order pole at the point $-i\theta$.

For $\theta \in \{0, \infty\}$, the function $\mathfrak{S}_\theta(z)$ has no singularities and corresponds to free dynamics.

Note that, even in this simple example, the application of the results obtained in Chapter 4 enables us to avoid the quite cumbersome (see, e.g., Lax and Phillips [1]) procedure of explicit description of incoming and outgoing subspaces and construction of translation representations.

2. *Wave equation in* $\mathbb{R}^3$. In Subsection 3.3, we have shown that the operator

$$-\Delta, \qquad \mathcal{D}(-\Delta) = W_2^2(\mathbb{R}^3)$$

is nonperturbed in the space $\mathcal{H} = L_2(\mathbb{R}^3)$ and, hence, determines free dynamics in equation (5.1). Moreover, the operator $-\Delta$ is a self-adjoint extension of the operator B^2, where the operator B is defined by the second equality in (3.38),

$$B^2 = -\Delta|_{\mathcal{D}(B^2)}, \quad \mathcal{D}(B^2) = \left\{ u(x) \in W_2^2(\mathbb{R}^3) \,|\, u(0) = 0, \int\limits_{\mathbb{R}^3} \frac{u(x)}{|x|^{\alpha_k}} Y_k\left(\frac{x}{|x|}\right) dx = 0 \right\},$$

$k \in \mathbb{N}$, $\alpha_k = 2$ for odd k, and $\alpha_k = 3$ for even k (for even k, the integral is understood in the sense of principal value).

Thus, if the right-hand side of equation (5.1) is an arbitrary self-adjoint extension $-\tilde{\Delta}$ of the operator B^2 (with finitely many eigenvalues on the negative semiaxis), the results presented in Subsection 5.5 enable us to describe singularities of the corresponding scattering matrix for 0-perturbed dynamics.

By virtue of (3.38), the operator

$$\tilde{L} = -T\tilde{\Delta}T^{-1},$$

where the isometric operator T is defined by (3.35), is a self-adjoint extension of the operator

$$\tilde{B}^2 = -\frac{d^2}{ds^2}, \quad \mathcal{D}(\tilde{B}^2) = \mathring{W}_2^2(\mathbb{R}_+, N).$$

Therefore, the scattering matrices for the groups of solutions of equation (5.1) with the operators $-\tilde{\Delta}$ and $\tilde{L}$ coincide for a proper choice of translation representations. Thus, the investigation of the singularities of the scattering matrix for the 0-perturbed dynamics determined by the self-adjoint extension $-\tilde{\Delta}$ of the operator B^2 in (5.1) can be reduced to a similar problem for the self-adjoint extension $\tilde{L}$ of the operator $\tilde{B}^2$, which can be represented in a more convenient form in terms of canonical positive BVS.

We set

$$X = \operatorname{Ker}(\tilde{B}^*\tilde{B}^* + \xi^2 I) = \left\{ n(\omega)e^{-\xi s} \,\middle|\, n(\omega) \in N = L_2(S^2) \right\}. \tag{5.49}$$

It is clear that the operator

$$L = -T\Delta T^{-1}$$

is nonperturbed ($\in \mathfrak{M}_{\tilde{B}}$) in the space $L_2(\mathbb{R}_+, N)$. In Subsection 3.3, we have shown that

$$\mathcal{D}(L) = \left\{ u(s,\omega) \in W_2^2(\mathbb{R}_+, N) \;\middle|\; \begin{array}{l} u(0,-\omega) = -u(0,\omega) \\ \partial_s u(0,-\omega) = \partial_s u(0,\omega) \end{array} \right\}. \tag{5.50}$$

Relations (3.16), (5.49), and (5.50) immediately imply that the triple $(X, \Gamma_1', \Gamma_2')$, where the operators $\Gamma_i' : \mathcal{D}(\tilde{B}^* \tilde{B}^*) = W_2^2(\mathbb{R}_+, N) \to X$ are defined by the relations

$$\Gamma_1' f(s) = 2\xi(f'(0) + \xi f(0))e^{-\xi s} \quad \left(f(s) \in W_2^2(\mathbb{R}_+, N)\right),$$

$$\Gamma_2' f(s) = \left(P_{\text{ev}} f(0) - \frac{1}{\xi} P_{\text{od}} f'(0)\right) e^{-\xi s},$$

and P_{ev} (P_{od}) is the orthoprojector onto the set of even (odd) functions in $N = L_2(S^2)$, is the canonical positive BVS of the operator $\tilde{B}^2$ associated with the nonperturbed operator L.

Note that the operator L is not the hard extension of the operator $\tilde{B}^2$. In view of (5.50), we conclude that, in the canonical positive BVS (X, Γ_1, Γ_2) corresponding to the hard extension $L_\mu = \tilde{B}^* \tilde{B}$ (it can be constructed as in the case of the wave equation on a semiaxis; see equalities (5.48)), the operator L is determined by the operator C defined as follows:

$$C\left(n(\omega)e^{-\xi s}\right) = \frac{1}{2\xi^2}(P_{\text{od}}\, n(\omega))e^{-\xi s}. \tag{5.51}$$

Let $\mathfrak{S}(\sigma)$ be the scattering matrix for the group $W_{-\tilde{\Delta}}(t)$ determined by the operator $-\tilde{\Delta}$ in equation (5.1). Without loss of generality, we can choose ξ such that $-\xi^2 \in \rho\left(-\tilde{\Delta}\right)$.

Then the self-adjoint operator $\tilde{L} = -T\tilde{\Delta}T^{-1}$ is determined in the canonical positive BVS $(X, \Gamma_1', \Gamma_2')$ by a bounded operator C'. By using Theorems 5.7 and 5.7′ and the remark to Theorem 5.7, we establish the relationship between the singularities of $\mathfrak{S}(\sigma)$ and the spectral properties of the operator $\tilde{C} = C' + C$.

For example, consider the particular case where the operator $-\tilde{\Delta}$ is a self-adjoint extension of the symmetric Laplace operator

$$-\Delta_0 = -\Delta\big|_{\mathcal{D}(\Delta_0)}, \quad \mathcal{D}(\Delta_0) = \left\{u(x) \in W_2^2(\mathbb{R}^3) \mid u(0) = 0\right\}.$$

Note that $B^2 \subset -\Delta_0$ and, hence, any self-adjoint extension $-\tilde{\Delta}$ of the operator $-\Delta_0$ is also a self-adjoint extension of the operator B^2.

As is known (see, e.g., Koshmanenko [1]), any self-adjoint extension $-\tilde{\Delta}$ of the operator $-\Delta_0$ can be represented in the form

$$\mathcal{D}(-\tilde{\Delta}) = \left\{ f(x) = u(x) + u(0)\beta\frac{e^{-\xi|x|}}{|x|} \;\middle|\; \beta \in \mathbb{R},\ u(x) \in W_2^2(\mathbb{R}^3) \right\},$$

$$-\tilde{\Delta}f = -\Delta_0 u - \xi^2 u(0)\beta\frac{e^{-\xi|x|}}{|x|}.$$

In the canonical positive BVS $(X, \Gamma_1', \Gamma_2')$, the operator $\tilde{L} = -T\tilde{\Delta}T^{-1}$ is determined by the operator

$$C' = \frac{\beta}{2\xi}P_1,$$

where P_1 is the orthoprojector onto $\left\langle e^{-\xi s}\right\rangle$ in X. By virtue of (5.51), the spectrum of the operator $\tilde{C} = C' + C$ consists of the eigenvalues

$$0, \quad \frac{1}{2\xi^2}, \quad \text{and} \quad \frac{\beta}{2\xi}$$

and, hence, by virtue of Theorems 5.7 and 5.7′, the scattering matrix $\mathfrak{S}(\delta)$ has no singularities for $\beta \in \{0, 1/\xi\}$ (the case of nonperturbed operators).

For $\beta \in (0, 1/\xi)$, the function $\mathfrak{S}(z)$ is not invertible at the point

$$z = -i\left(\frac{1}{\beta} - \xi\right)$$

and, for $\beta \in \mathbb{R}\setminus[0, 1/\xi]$, the function $\mathfrak{S}(z)$ has a first-order pole at the point

$$z = -i\left(\xi - \frac{1}{\beta}\right).$$

Remark. The results presented in Section 5 are obtained by S. Kuzhel in [5, 7, 10].

References

Adamjan, V. M. and Arov, D. Z.

[1]. *About one class of scattering operators and characteristic operator functions of contractions,* Dokl. Akad. Nauk SSSR **160** (1965), No. 1, 9–12 *(in Russian)*

[2] *Unitary couplings of semi-unitary operators,* Mat. Issled. **1** (1966), No. 2, 3–64 *(in Russian).*

Adamjan, V. M. and Pavlov, B. S.

[1] *Zero-radius potentials and Krein formula of generalized resolvents,* Zap. LOMI **149** (1986), 7–23 *(in Russian).*

Akhiezer, N. I. and Glazman, I. M.

[1] Theory of Linear Operators in Hilbert Space, Pitman (APP), 1981.

Albeverio, S., Gesztesy, F., Høegh-Krohn, R., and Holden, H.

[1] Solvable Models in Quantum Mechanics, Springer, Berlin, 1988.

Ando, T. and Nishio, K.

[1] *Positive self-adjoint extensions of positive symmetric operators,* Tohoku Math. J. **22** (1970), No. 1, 65–75.

Arlinskii, Y. M.

[1] *Positive spaces of boundary values and sectorial extensions of nonnegative operators,* Ukr. Mat. Zh. **40** (1988), No. 1, 8–14 *(in Russian).*

[2] *On a class of extensions of a* $C(\alpha)$*-suboperator,* Dokl. Akad. Nauk Ukr. (1992), No. 8, 12–16 *(in Russian).*

[3] *Extremal extensions of sectorial linear relations,* Mat. Studii **7** (1997), No. 1, 81–96.

Arlinskii, Y. M. and Tsekanovskii, E. R.

[1] *Maximal sectorial extensions of positive Hermitian operators and their resolvents,* Dokl. Akad. Nauk Arm. SSR **79** (1984), No. 5, 12–16 *(in Russian).*

Azizov, T. Ya. and Iokhvidov, I. S.

[1] Linear Operators in Spaces with an Indefinite Metric, Wiley, 1989.

Balakrishnan, A. V.

[1] Applied Functional Analysis, Springer, Berlin, 1976.

Baumgartel, H. and Wollenberg, M.

[1] Mathematical Scattering Theory, Akademie-Verlag, Berlin, 1983

Berezanskii, Yu. M., Us, G. F. and Sheftel, Z. G.

[1] Functional Analysis, Vyshcha Shkola, Kiev, 1990 *(in Russian).*

Bojnar, J.

[1] Indefinite Inner Product Spaces, Springer-Verlag, Berlin, 1974.

Bruk, V. M.

[1] *On a class of problems with a spectral parameter in a boundary condition,* Mat. Sb. **100** (1976), No. 2, 210–216 *(in Russian).*

[2] *On extensions of symmetric relations,* Mat. Zametki **22** (1977), No. 6, 825–834 *(in Russian).*

Calkin, J. W.

[1] *Abstract symmetric boundary conditions,* Trans. Am. Math. Soc. **45** (1939), 369–442.

Coddington, E. A.

[1] *Self-adjoint subspace extensions of nondensely defined symmetric operators,* Adv. Math. **14** (1974), No. 3, 309–332.

Coddington, E. A. and de Snoo, H. S.

[1] *Positive self-adjoint extensions of positive symmetric subspaces,* Mat. Z. **159** (1978), 203–214

Coddington, E. A. and Dijksma A.

[1] *Self-adjoint subspaces and eigenfunction expansions of ordinary differential subspaces,* J. Diff. Equat. **20** (1976), No. 2, 473–526.

Derkach, V. A.

[1] *On extensions of an Hermitian operator in a Krein space,* Dokl. Akad. Nauk Ukr. SSR, Ser. A, (1988), No. 5, 5–9 *(in Russian).*

[2] *On extensions of a nondensely defined Hermitian operator in a Krein space,* Dokl. Akad. Nauk Ukr. SSR, Ser. A, (1990), No. 10, 14–18 *(in Russian).*

Derkach, V. A. and Malamud, M. M.

[1] *Weyl function of Hermitian operators and its relation to characteristic function,* Preprint No. 85-9 (104), Physicotechnical Institute, Donetsk, 1985 *(in Russian).*

[2] *Generalized resolvents and the boundary-value problems for Hermitian operators with gap,* J. Funct. Anal. **95** (1991), No. 1, 1–95.

[3] *Characteristic functions of linear operators,* Russian Acad. Sci. Dokl. Math. **45** (1992), No. 2, 417–424

Dunford, N. and Schwartz, J. T.

[1] Linear Operators, vol. II: Spectral Theory. Self-Adjoint Operators in Hilbert Spaces, Interscience, New York–London, 1963

Evans, W. D. and Knowles, J.

[1] *On the extensions problem for accretive differential operators,* J. Funct. Anal. **63** (1985), No.3, 276–298.

Freudenthal, H.

[1] *Über die Friedrichsche halbbeschrankter Hermitescher operatoren,* Proc. Acad. Amsterdam **39** (1936), No. 7, 832–833.

Friedrichs, K.

[1] Spectraltheorie halbbeschraanter operatoren und anwendungauf die spektralzerlegung von differential operatoren, Match. Ann. **109** (1934), No. 4, 465–487.

Gorbachuk, M. L. and Gorbachuk, V. I.

[1] Boundary-Value Problems for Operator-Differential Equations, Kluwer, Dordrecht, 1991.

Gorbachuk, M. L., Gorbachuk, V. I., and Kochubei, A.N.

[1] *The theory of extensions of symmetric operators and boundary-value problems for differential equations,* Ukr. Mat. Zh. **41** (1989), No. 10, 1299–1313 *(in Russian).*

Grandall, M. G.

[1] *Norm preserving extensions of linear transformations on a Hilbert spaces,* Proc. Am. Math. Soc. (1969), **21:2**, 335–340.

Helgason, S.

[1] The Radon Transform, Birkhaüser, Boston, Basel, 1980.

Kato, T.

[1] Perturbation Theory for Linear Operators, Springer-Verlag, Berlin, 1966.

Kochubei, A. N.

[1] *On extensions of symmetric operators and symmetric binary relations,* Mat. Zametki **17** (1975), No. 1, 41–48 *(in Russian).*

[2] *On extensions of nondensely defined symmetric operators,* Sib. Mat. Zh. **18** (1977), No. 2, 314–320 *(in Russian).*

[3] *About symmetric operators commuting with a family of unitary operators,* Funk. Anal. Prilozh. **13** (1979), No. 4, 77–78 *(in Russian)*

[4] *Symmetric operators and nonclassical spectral problems,* Mat. Zametki **25** (1979), No. 3, 425–434 *(in Russian)*

[5] *On extensions and characteristic functions of symmetric operators,* Izv. Akad. Nauk Arm. SSR **15** (1980), No. 3, 219–232 *(in Russian)*

[6] *Elliptic operators with boundary conditions on a subset of measure zero,* Funk. Anal. Prilozh. **16** (1982), No. 2, 74–75 *(in Russian).*

[7] *On extensions of positive definite symmetric operators,* Trans. Am. Math. Soc. **124** (1984), 139–142

[8] *Self-adjoint extensions of Schrödinger operator with singular potential,* Sib. Mat. Zh. **187** (1991), No. 3, 60–70 *(in Russian).*

Koshmanenko, V. D.

[1] Singular Bilinear Forms in Perturbation Theory for Self-Adjoint Operators, Naukova Dumka, Kiev, 1993 *(in Russian).*

Krasnosel'skii, M. A.

[1] *On self-adjoint extensions of Hermitian operators,* Ukr. Mat. Zh. **1** (1949), No. 1, 21–28 *(in Russian).*

Krein, M. G.

[1] *Resolvents of Hermitian operators with deficiency index* (m, m), Dokl. Akad. Nauk SSSR **52** (1946), No. 8, 657–660 *(in Russian).*

[2] *Theory of self-adjoint extensions of semibounded operators and its applications I,* Mat. Sb. **20** (1947), No. 3, 431–498 *(in Russian).*

[3] *Theory of self-adjoint extensions of semibounded operators and its applications II,* Mat. Sb. **21** (1947), No. 3, 365–404 *(in Russian).*

Krein, M. G. and Shmulyan, Yu. L.

[1] *On linear fractional transformations with operator coefficients,* Mat. Issled. **2** (1967), No. 3, 64–96 *(in Russian).*

Kuzhel, A. V.

[1] *On the spectrum of a regular quasidifferential operator,* Dokl. Akad. Nauk SSSR **156** (1964), No. 4, 731–733 *(in Russian).*

[2] *Regular extensions of Hermitian operators,* Dokl. Akad. Nauk SSSR **251** (1980), No. 1, 30–33 *(in Russian).*

[3] *An analog of Krein formula for resolvents of nonself-adjoint extensions of an Hermitian operator,* Teor. Funk. Funk. Anal. Prilozh. (1981), No. 36, 49–55 *(in Russian).*

[4] *Regular extensions of Hermitian operators in a space with indefinite metric,* Dokl. Akad. Nauk SSSR **265** (1982), No. 5, 1059–1061 *(in Russian).*

[5] *Classification of the spectrum in algebras,* Dokl. Akad. Nauk Ukr. SSR, Ser. A (1983), No. 10, 11–14 *(in Russian).*

[6] Extensions of Hermitian Operators, Vyshcha Shkola, Kiev, 1989 *(in Russian).*

[7] Characteristic Functions and Models of Nonself-Adjoint Operators, Kluwer, Dordrecht, 1996.

[8] *On regular U-invariant extensions of Hermitian operators,* Mat. Studii **7** (1997), No. 2, 193–200 *(in Ukrainian).*

Kuzhel, A. V. and Rotkevich, E. S.

[1] *Normal extensions of Hermitian operators with preserving of the bound,* in: Funct. Anal. Linear Operators (1983), No. 21, 100–107 *(in Russian).*

[2] *On normal extensions of Hermitian contractions with preserving of the norm,* in: Funct. Anal. Spectral Theory (1987), No. 27, 104–107 *(in Russian).*

Kuzhel, A. V. and Rudenko, L. I.

[1] *Regular extensions of Hermitian and isometric operators,* Ukr. Mat. Zh. **33** (1981), No. 6, 810–814 *(in Russian).*

[2] *Description of regular extensions of Hermitian operators,* Funk. Anal. Prilozh. **16** (1982), No. 1, 74–75 *(in Russian).*

Kuzhel, A. V. and Tretyakov, V. D.

[1] *On a generalization of the Lax–Phillips scheme in scattering theory,* Dokl. Akad. Nauk Ukr. SSR, Ser. A. (1982), No.2, 19–21 *(in Russian).*

Kuzhel, S. A.

[1] *On boundary-value spaces and regular extensions of Hermitian operators,* Ukr. Mat. Zh. **42** (1990), No. 6, 860–865 *(in Russian).*

[2] *On regular extensions of Hermitian operators,* in: Funct. Anal. Spectral Theory (1990), No. 31, 91–100 *(in Russian).*

[3] *On boundary-value spaces for the model of zero-range potential with internal structure,* in: Boundary-Value Problems for Operator-Differential Equations, Institute of Mathematics, Kiev (1991), 49–56 *(in Russian).*

[4] *Abstract Wave Equation; Definition and Properties of Solutions,* Preprint No. 14.96, Institute of Mathematics, Ukrainian Academy of Sciences, Kiev 1996 *(in Russian).*

[5] *Abstract Lax–Phillips scattering scheme for second-order operator-differential equations,* Ukr. Mat. Zh. **48** (1996), No. 4, 452–464 *(in Russian).*

[6] *On some properties of unperturbed operators*, Meth. Funct. Anal. Topology, **2** (1996), No. 3–4, 78–85.

[7] *On abstract Lax–Phillips scattering scheme for second-order operator-differential equations*, Funk. Anal. Prilozh. **30** (1996), No. 1, 70–73 *(in Russian).*

[8] *Abstract Lax–Phillips scattering scheme in Pontryagin spaces*, Mat. Sb. **187** (1996), No. 10, 87–108 *(in Russian).*

[9] *On some properties of abstract wave equation*, Meth. Funct. Anal. Topology, **3** (1997), No. 1, 82–87.

[10] *On the elements of Lax–Phillips scattering scheme for the* ρ*-perturbation of an abstract wave equation*, Ukr. Mat. Zh. (to appear).

[11] *On the form of scattering matrix for the* ρ*–perturbations of an abstract wave equation*, Ukr. Mat. Zh. (to appear).

[12] *On the form of incoming and outgoing subspaces for the acoustic equation in* $\mathbb{R}^n$, Ukr. Mat. Zh. (to appear).

Lax, P. D. and Phillips, R. S.

[1] Scattering Theory, Academic Press, New York, 1969.

[2] *Scattering theory for the acoustic equation in an even number of space dimensions*, Indiana Univ. Math. J., **22** (1972), No. 2, 101–134

[3] Scattering Theory for Automorphic Functions, Princeton University Press, Princeton, 1974

Livsic, M. S.

[1] *On a spectral decomposition of linear self-adjoint operators*, Mat. Sb. **34** (1954), 144–199 *(in Russian).*

Malamud, M. M.

[1] *On extensions of Hermitian sectorial operators and dual pairs of contractions*, Sov. Math. Dokl. **39** (1989), No. 2.

[2] *On new approach to the theory of extensions of nondensely defined Hermitian operators*, Dokl. Akad. Nauk Ukr. SSR (1990), No. 3, 20–25 *(in Russian).*

Mikhailets, V. A.

[1] *Point interactions on the line*, Rep. Math. Phys., **33** (1993), No. 1–2, 131–135.

Naimark, M. A.

[1] *On self-adjoint extensions of the second kind of symmetric operator,* Izv. Akad. Nauk SSSR **4** (1940), No. 1, 53–104 *(in Russian).*

[2] *On spectral functions of a symmetric operator,* Izv. Akad. Nauk SSSR **4** (1940), No. 3, 277–318 *(in Russian).*

[3] *On spectral functions of a symmetric operator,* Izv. Akad. Nauk SSSR **7** (1943), No. 6, 285–296 *(in Russian).*

[4] Linear Differential Operators, Part 1, Ungar, New York, 1957.

[5] Linear Differential Operators, Part 2, Ungar, New York, 1968.

Neumann, J. von

[1] *Allgemeine eigenwer theorie Hermitesher Funktional operatoren,* Math. Ann. **102** (1929), 49–131 (German)

Pavlov, B. S.

[1] *Model of zero-range potential with internal structure,* Teor. Mat. Fiz. **59** (1984), No. 3, 345–354 *(in Russian).*

[2] *Theory of extensions and explicitly solvable models,* Uspekhi Mat. Nauk **42** (1987), No. 6, 99–131 *(in Russian).*

Pavlov, B. S. and Shushkov, A. A.

[1] *Theory of extensions and zero-radius potentials with internal structure,* Mat. Sb. (1989), No. 2, 147–183 *(in Russian).*

Petkov, V.

[1] Scattering Theory for Hyperbolic Operators, North-Holland, Amsterdam, 1989.

Phillips, R. S.

[1] *Dissipative operators and hyperbolic systems of partial differential equations,* Trans. Am. Math. Soc. **90** (1959), 195–254

[2] *The extension of dual subspaces invariant under an algebra,* Proceedings of the International Symposium (Jerusalem, 1960), Pergamon Press, 1961.

Prokaj, V. and Sebestyen, Z.

[1] *On Friedrichs extensions of operators,* Acta Sci. Math. **62** (1996), 243–247.

Reed, M. and Simon, B.

[1] Methods of Modern Mathematical Physics, II; Fourier Analysis, Self-Adjointness, Academic Press, New York, 1975.

[2] Methods of Modern Mathematical Physics, III; Scattering Theory, Academic Press, New York, 1979.

Richtmyer, R. D.

[1] Principles of Advanced Mathematical Physics, Berlin, Springer, 1978.

Riesz, F. and Szökefalvi-Nagy, B.

[1] Functional Analysis, New York, Ungar, 1955.

Sebestyen, Z. and Stochel, J.

[1] *Restrictions of positive self-adjoint operators,* Acta Sci. Math. **55** (1991), No. 1–2, 149–155.

Shtraus, A. V.

[1] *On the theory of Hermitian operators,* Dokl. Akad. Nauk SSSR **67** (1949), No. 4, 611–614 *(in Russian).*

[2] *On extensions and characteristic functions of symmetric operators,* Izv. Akad. Nauk SSSR, Ser. Mat. **32** (1968), No. 1, 186–207 *(in Russian).*

[3] *On extensions of semibounded operators,* Dokl. Akad. Nauk SSSR **211** (1973), No. 3, 543–546 *(in Russian).*

[4] *On extensions of bounded symmetric operators with preserving of the bound,* Funct. Anal. Linear Operators (1976), No. 6, 165–173 *(in Russian).*

[5] *To the theory of extensions of semibounded operators,* Funct. Anal. Linear Operators (1977), No. 9, 167–173 *(in Russian).*

Stein, E. and Weiss, G.

[1] Introduction to Fourier Analysis on Euclidean Spaces, Princeton University Press, Princeton, 1971

Stone, M.

[1] Linear Transformation in Hilbert Space, Am. Math. Soc. Colloq. Publ., **15**, Am. Math. Soc., New York, 1932

Storozh, O. G.

[1] *Extensions of symmetric operators with different defect numbers,* Mat. Zametki **36** (1984), No. 5, 791–795 *(in Russian).*

[2] *Extremal extensions of nonnegative operator and accretive boundary-value problems,* Ukr. Mat. Zh. **42** (1990), No. 6, 857–860 *(in Russian).*

Szökefalvi-Nagy, B. and Foias, C.

[1] Harmonic Analysis of Operators in Hilbert Spaces, North Holland, Amsterdam, 1970.

Taljush, M. O.

[1] *The typical structure of dissipative operators,* Dokl. Akad. Nauk Ukr. SSR, Ser. A (1973), No. 11, 993–996 *(in Ukrainian).*

Tsekanovskii, E. R. and Shmulyan, Yu. L.

[1] *The theory of biextensions of operators in rigged Hilbert spaces. Unbounded operator couplings and characteristic functions,* Uspekhi Mat. Nauk **32** (1977), No. 5, 69–124 *(in Russian).*

Vishik, M. I.

[1] *On general boundary-value problems for elliptic differential equations,* Tr. Mosk. Mat. Obshch. **1** (1952), 187–246 *(in Russian).*

Subject index

Abstract wave equation, 215
  ρ-perturbed, 235

Characteristic function, 159
Cogenerator, 59

Decomposition
  canonical, 120
  fundamental, 119

Defect numbers, 16
Defect subspace, 11
Deficiency index, 16
Dilation, 52

Element
  eventually outgoing, 227
  initially incoming, 227
Extension
  canonical dissipative, 126
  defined by decomposition, 124
  Friedrichs, 66
  hard, 88
  quasi-Hermitian, 31
  regular, 28
  soft, 88

Generator, 59
Greatest lower bound, 62

Involution, 17

J-space, 119

Kernel, 1

Lineal
  B_--dense, 79
  maximal uniformly positive, 120
  positive, 119

Linear span, 2

N-operator, 2

Operator
  accretive, 53
  accumulative, 53
  admissible, 147
  angular, 120
  dissipative, 45
  Hermitian, 10
  Hermitian domain of, 26
  Hermitian part of, 26
  isometric, 22
  lower (upper) bound of, 61
  maximal accretive, 56
  maximal dissipative, 49
  maximal symmetric, 23
  nonperturbed, 204
  positively closable, 99
  ρ-perturbed, 235
  scattering, 241
  sectorial, 56
  self-adjoint, 10
  semibounded from below, 61
  simple Hermitian, 163
  simple symmetric, 201
  symmetric, 10
  transition, 120
  $\mathcal{U}$-invariant, 106
  unitary, 22

Point
  of regular type, 4
  regular, 4

Quasiderivative, 177

Resolvent, 5
Restriction, 3

Set
    regularity, 4
    resolvent, 4
Scattering matrix, 241
    Heisenberg, 241
    singularity of, 541
Space
    boundary-value, 129, 134, 137
        canonical positive, 132
    external, 189
    internal, 189
    Krein, 119
    of Cauchy data, 216
    of initial data, 216
    of test functions, 223
Spectrum
    continuous, 7
    point, 6
    residual, 7
    stable, 6
    unstable, 6
Strong Hermitian contraction, 69
Subspace
    incoming, 218
    outgoing, 218

Translation representation, 221

Notation

$\mathcal{B}[X]$, 1
$\mathcal{D}(T)$, 1
$\mathcal{R}(T)$, 1
$\mathcal{L} \vee \mathcal{M}$, 2
$\rho(T)$, 4
$\pi(T)$, 4
$\sigma(T)$, 4
R_λ, 5
M^*, 6
$\mathfrak{N}_\lambda$, 11
$\mathcal{D}$, 11
G_A, 26
$\mathcal{P}(H)$, 28
$T_{\mu\lambda}$, 32
$\mathfrak{M}_\lambda$, 38
Γ_λ, 38
$m(A)$, 62
$\mathcal{D}_0$, 65
S_-, 77
S_+, 77
B_-, 77
B_+, 77
$\mathcal{P}_0(S)$, 77
$\mathcal{P}(S; K)$, 81
$R \cup T$, 81
A_μ, 88
A_M, 88
$[\dot{+}]$, 119
$\mathcal{N}_\lambda$, 123
(X, Γ_1, Γ_2), 129, 134
(X_+, X_-, Y_+, Y_-), 137
$C(\lambda)$, 159
$\Phi_T(z)$, 160
A_K, 167
$f^{[k]}$, 177
$\mathfrak{M}_B$, 205
L_μ, 205
L_M, 205
$\mathcal{H}_L$, 215
x_γ, 215
$\mathfrak{H}_L$, 216
Q_L, 216
Q_μ, 216
Q_M, 216
D_-, 218
D_+, 218
T_μ, 221
T_M, 222
T_L, 222
$\mathcal{F}(t)$, 222
$\mathcal{D}(\mathbb{R}, N)$, 223
$\mathcal{D}'(\mathbb{R}, N)$, 223
$\mathcal{D}_L$, 224
$\mathcal{D}'_L$, 224
$X_{\mathcal{D}_L}$, 224
$X_{\mathcal{D}'_L}$, 224
$\mathcal{M}(L, \tilde{L})$, 224
$\mathcal{M}'(L, \tilde{L})$, 225
$\mathcal{H}_\rho$, 235
B_ρ, 235
P_ρ, 236
D_-^ρ, 237
D_+^ρ, 237
$\mathfrak{S}(\delta)$, 241
$\mathcal{K}_\rho$, 244

Bei Fragen zur Produktsicherheit wenden Sie sich bitte an:
If you have any questions regarding product safety,
please contact:

Walter de Gruyter GmbH
Genthiner Straße 13
10785 Berlin
productsafety@degruyterbrill.com